Hydrologic Measurements with Flexible Liners and Other Applications

This book provides hydrologists the information needed for the characterization of contaminated subsurface hydrologic sites. It explains how to seal boreholes, map contaminant distribution in a formation, map the flow zones, and measure the hydraulic head distribution using a single flexible liner. Results of the measurement methods provided demonstrate the reality and reliability of the unique FLUTe techniques. These measurements help to predict contaminant migration and aid in the design of a groundwater remedy. The limitations of several methods are provided to allow an intelligent choice of methods and a well-informed selection of devices among the alternative methods. The mechanics of flexible liner systems are explained with examples of applications beyond the hydrologic measurements such as relining of piping.

Features include:

- The first book on a modern technology that is replacing traditional technology globally
- Written by the inventor of the FLUTe technology with 25 years' experience with successful applications
- Describes FLUTe technology in detail, including the theory behind the tools, how to use the tools, and the mathematics used to interpret the data generated by the tools
- Provides step-by-step explanations of how to conduct fieldwork and how to analyze the data gathered
- Minimizes reliance on mathematical explanations and uses illustrations and examples that allow readers to understand the technology

This book is of interest to environmental professionals, mine operators, petroleum engineers, geophysicists who use these methods or are considering using these methods for remediation of groundwater contamination, academics, students, and regulators.

Hydrologic Measurements with Flexible Liners and Other Applications

Carl Keller

CRC Press
Taylor & Francis Group
Boca Raton London New York

CRC Press is an imprint of the
Taylor & Francis Group, an **informa** business

First edition published 2023
by CRC Press
6000 Broken Sound Parkway NW, Suite 300, Boca Raton, FL 33487-2742

and by CRC Press
4 Park Square, Milton Park, Abingdon, Oxon, OX14 4RN

CRC Press is an imprint of Taylor & Francis Group, LLC

ISBN: 978-1-032-21262-3 (hbk)
ISBN: 978-1-032-21427-6 (pbk)
ISBN: 978-1-003-26837-6 (ebk)

DOI: 10.1201/9781003268376

Typeset in Times
by KnowledgeWorks Global Ltd.

*Dedicated to my wife Lisa V. and our children,
Craig, Julia, Matt, Keith, and Buster who tolerated
my distraction of this effort over many years.*

Contents

Foreword (by Joe Rossabi)..xvii
Preface...xix
Acknowledgments... xxiii
Author xxv
List of Abbreviations..xxvii

Chapter 1   Introduction/Purpose 1

Chapter 2   Brief History of Flexible Liner Underground Technologies
(FLUTe) Methods...3

Chapter 3   The Mechanics of Flexible Liners.....................................7
            3.1   The Flexible Liner Characteristics 7
            3.2   The Eversion of a Flexible Liner 8
                   3.2.1   The Towing Force..8
                   3.2.2   Drag (Friction Effects) 9
                   3.2.3   Eversion into a Borehole............................ 10
                   3.2.4   Other Factors Influential on the Liner
                           Propagation.. 12
                           3.2.4.1   Hole/Liner Diameter.................... 12
                           3.2.4.2   Wet Film Adhesion 12
                           3.2.4.3   The Minimum Tension 13
                           3.2.4.4   The Difference between the Eversion
                                      and the Inversion of the Liner.................... 13
                           3.2.4.5   The Air Balloon Drag and the Air Vent 15
                           3.2.4.6   Effect of Breakouts on Liner Eversion 17
                           3.2.4.7   The Impermeable Borehole Installations....... 17
                   3.2.5   Stretch of the Liner.................................... 19
            3.3   The Liner Removal Methods................................20
                   3.3.1   The Normal Inversion from a Permeable Borehole 20
                   3.3.2   The Pump and Drag Removal 21
                   3.3.3   The Impermeable Borehole Removal........................22
            3.4   The Liner Seal 23
                   3.4.1   Interior View of the Sealing Liner 23
                   3.4.2   The Highest Head Measurement Method 24
                   3.4.3   Artesian Conditions....................................26
                   3.4.4   Liner Seal Comparison with Packers 29
            3.5   Liner Installation Devices 30
                   3.5.1   Air Pressure Canisters.................................30

3.5.2   Hose Canisters..34
3.5.3   Gravity-Driven Installations........................35
3.5.4   Magic Gland...35
3.5.5   The Drop-in-Place Liner Installation..........36
3.5.6   The Bulbous Wellhead for Artesian Installations......37
3.5.7   Mud-Filled Liners..37
         3.5.7.1   Purpose of Mud Fill....................37
         3.5.7.2   An Example of the Mud Pressure
                    Calculation.................................40
         3.5.7.3   In Summary, How the Heavy Mud
                    Is Used.......................................41

Chapter 4   Chemistry of the Liners.......................................43
4.1   Arsenic...43
4.2   Toluene..43
4.3   1,4-Dioxane...44
4.4   Polyfluoronated Alkyl Substances (PFAS)............44
4.5   N-Nitrosodimethylamine (NDMA)........................45

Chapter 5   Kinds of Blank Liners...47
5.1   Different Diameters..47
5.2   Fabrics..48
       5.2.1   Nylon Liners...48
       5.2.2   Polyester Liners...48
       5.2.3   Silicon Rubber Liners.................................48
       5.2.4   Transparent Liners and Geophysical Logging...........48
       5.2.5   Different Fabric Weight Liners.....................52
       5.2.6   Tubular Plastic Film Liners.........................53
5.3   Carrier Liners for Coverings.................................53
5.4   Lay Flat Hose Liners..54

Chapter 6   Novel Applications of Blank Liners......................55
6.1   Surface Extensions..55
6.2   Eversions on or under Water................................55
6.3   Vertical Upward Unsupported Extensions.............55
6.4   Eversions through Crooked Piping Systems...........56
6.5   Lining Boreholes to Prevent Grout Loss or Grout
       Shrinkage Outside of a Casing..............................57

Chapter 7   General Advantages of Flexible Blank Liners.......61

Chapter 8   Hazards to the Liner and Precautions 63

Chapter 9   Special Devices Designed for Use with Liners 65
    9.1   Green Machine 65
    9.2   Linear Capstan 67
        9.2.1   Background 67
        9.2.2   The Linear Capstan Design 69
    9.3   T Profiler 70
    9.4   Braking Devices of Several Kinds 71
    9.5   The Air-Coupled Water-Level Meter Systems 71
        9.5.1   The ACT (Air-Coupled Transducer) 71
            9.5.1.1   ACT Purpose 71
            9.5.1.2   Background/Comparisons 71
            9.5.1.3   The ACT Design and Theory 73
            9.5.1.4   The Range of Pressure Changes for the
                         ACT Transducers 76
            9.5.1.5   The Temperature Effect 76
            9.5.1.6   First Result of the ACT Measurement 77
            9.5.1.7   The Field Measurements 78
            9.5.1.8   Input Data and Apparatus 79
            9.5.1.9   Usual Applications of the ACT System 81
            9.5.1.10 Resolution of the ACT Method 82
            9.5.1.11 Barometric Corrections 83
            9.5.1.12 How Is the Raw Data Used? 83
            9.5.1.13 Advantages and Limitations of the
                         Method 84
        9.5.2   The Vacuum Water-Level Meter (VWLM) 85
        9.5.3   The Air-Coupled Water-Level Meter (ACWLM) 86
    9.6   Eversion/Inversion AIDS 87

Chapter 10 Theory and Application of FLUTe Liner Methods 89
    10.1   Blank Sealing Liners 89
        10.1.1 Installation of a Blank Liner 90
        10.1.2 Transparent Blank Liners 91
        10.1.3 Measurements by Others Using FLUTe Flexible
               Liners 91
    10.2   FLUTe Blank Liners with Special Coverings 91
        10.2.1 The NAPL FLUTe 91
            10.2.1.1 History of NAPL FLUTe Development 91
            10.2.1.2 How the NAPL FLUTe Is Installed in
                    Direct Push Rods 92

              10.2.1.3 NAPL FLUTe Installations in an Open
                        Stable Borehole 96
              10.2.1.4 NAPL FLUTe Covers over Core 97
              10.2.1.5 NAPL FLUTe Sand Bags 98
              10.2.1.6 Examples of NAPL FLUTe Stains 99
        10.2.2 The FACT Application 104
              10.2.2.1 History and Experience 104
              10.2.2.2 The FACT Method 105
              10.2.2.3 Assessment of the FACT Method 110
              10.2.2.4 Quantitative FACT Assessment at the
                        NAWC Site 113
              10.2.2.5 Comparisons of the FACT with Other
                        Methods 128
              10.2.2.6 The daFACT 128
              10.2.2.7 Advantages and Limitations of the
                        FACT Measurement 134
        10.2.3 Absorbers of Other Kinds on Blank Liners 134
              10.2.3.1 Pore Water Collection in the Vadose
                        Zone 134
              10.2.3.2 Radioactive Contamination Absorbers 135
  10.3 The Transmissivity Measurement Method 135
        10.3.1 History of the Transmissivity Profile Method..........135
        10.3.2 The Transmissivity Measurement Method..............136
              10.3.2.1 The Liner Behavior...................................136
              10.3.2.2 The Calculational Model 140
              10.3.2.3 When to Terminate the T Profile 142
        10.3.3 The T Profile Results..142
        10.3.4 Examples of Other T Profiles 145
        10.3.5 Calculation of the Effective Fracture Aperture
                Using the T Profile Results....................................147
        10.3.6 Corrections to the Simple T Profile Calculational
                Model..148
              10.3.6.1 Transient Correction 148
              10.3.6.2 The Borehole Diameter Correction 151
              10.3.6.3 The Vertical Head Correction 151
        10.3.7 The Transmissivity Profiling Equipment 152
              10.3.7.1 Maintaining a Constant Tension on the
                        Liner..153
              10.3.7.2 Maintaining the Constant Driving Head .. 154
        10.3.8 Effect of Well Development on the T Profile 155
        10.3.9 A Special Design for T Profiles of Boreholes
                with Very High Artesian Heads 156
        10.3.10 T Profile Comparison with Straddle Packer Results 157
        10.3.11 Advantages and Limitations of the T Profile 161
              10.3.11.1 Advantages 161
              10.3.11.2 Limitations 161

10.4   RHP (Reverse Head Profile) Measurement of a
       Head Profile 161
       10.4.1   The History of the RHP Method 161
       10.4.2   The Purpose of the Formation Head
                Measurement 162
       10.4.3   The RHP Calculation 163
                10.4.3.1   The Times to Equilibration for Each
                           Step of the RHP 165
                10.4.3.2   The Use of the RHP to Refine the
                           Transmissivity Profile 167
                10.4.3.3   Selection of the RHP Intervals to Be
                           Measured 167
       10.4.4   A Result of the RHP Method 168
       10.4.5   Calculation of the Synthetic Flow Log 170
       10.4.6   RHP Profile Summary 171
       10.4.7   Advantages and Limitations of the RHP 171
10.5   FLUTE MLS (Multilevel Sampling) Systems 172
       10.5.1   Water FLUTe 172
                10.5.1.1   History of Water FLUTes 172
                10.5.1.2   The Geometry of the Water FLUTe
                           Design 174
                10.5.1.3   Function of the Water FLUTe 179
                10.5.1.4   Transducer Options for Monitoring
                           Head History 184
                10.5.1.5   The Tracer Monitoring Capability of
                           the Water FLUTe Design 184
                10.5.1.6   Materials in the Water FLUTe
                           Construction 185
                10.5.1.7   Installation and Removal Procedure for
                           Water FLUTes 187
                10.5.1.8   Advantages and Limitations of Water
                           FLUTe System 187
       10.5.2   The SWF (Shallow Water FLUTe) 188
                10.5.2.1   The Design and Function 188
                10.5.2.2   Other Advantages and Limitations of
                           the SWF 189
       10.5.3   CHS (Cased Hole Sampler) Systems 191
                10.5.3.1   Background and History 191
                10.5.3.2   Geometry of the CHS 193
                10.5.3.3   Installation Procedure for CHS 193
                10.5.3.4   Purging and Sampling 196
                10.5.3.5   The Removal Procedure 196
                10.5.3.6   Special CHS Design for Potassium
                           Permanganate 196
       10.5.4   The pdCHS (Positive Displacement CHS) 197
                10.5.4.1   The Design of the pdCHS System 197

10.5.4.2 Simultaneous Purging and Sampling
of the pdCHS 199
10.5.4.3 Installation and Removal of the
pdCHS...200
10.5.4.4 Installation of CHS and pdCHS with
Mud or Grout-Filled Liner 200
10.5.4.5 Installation of CHS Systems in
Uncased Holes....................................202
10.5.5   Use of ACT Systems with the CHS Systems 203
10.5.6   Depth Limitations for CHS and pdCHS Systems ..204
10.5.6.1 Depth Limits for CHS Systems 204
10.5.6.2 Depth Limits for pdCHS Systems 205
10.5.7   Relative Cost of the CHS Based Systems...............206
10.5.8   Advantages and Limitations of Both CHS Systems .. 206
10.5.9   Use of FLUTe MLS Systems in General 207
10.5.9.1 Water FLUTe (In Use Since 1996) 207
10.5.9.2 Shallow Water FLUTe (SWF) (In
Use Since 2014)....................................207
10.5.9.3 CHS Systems (In Use Since 2018).........207
10.5.9.4 Mapping Cross-Hole Connection
with FLUTe MLS Systems 208
10.5.10 Comparison of FLUTe MLS Systems with
Other MLS Systems 213
10.5.11 The DEIL 214
10.5.11.1 The Purpose and Design of the DEIL
(Discrete Extraction and Injection
Liner)... 214
10.5.11.2 The Geometry of the DEIL Liner.......... 215
10.5.11.3 The DEIL Design Advantages and
Limitations... 216
10.5.12 Other Special CHS Systems................................. 217
10.5.12.1 Many Head Measurements in a CHS..... 217
10.5.12.2 Hybrid pdCHS for Deep Boreholes 217
10.6   Stretch of Liners as Important to FLUTe Methods.............. 218

Chapter 11   FLUTe Vadose Multi-Level Measurements 223

11.1   Pore Gas Sampling 223
11.1.1   The Geometry 223
11.1.2   The Gas Sampling Procedure 223
11.2   Pore Liquid Sampling in the Vadose Zone........................225
11.2.1   The Use...225
11.2.2   The Geometry of Pore Liquid Sampling................225
11.2.3   The Sampling Procedure for Pore Water 225
11.2.3.1 Other FLUTe Measurements in the
Vadose Zone..226
11.2.3.2 In Summary 227

Chapter 12 The TACL (Traveling Acoustic Coupling Liner) 229

    12.1 The TACL Method 229
    12.2 Use of the Blank Liner to Provide Coupling of Fiber
            Optic Cables 232

Chapter 13 Application of Combinations of Liners and Other Methods 233

    13.1 The FLUTe Sequence 233
    13.2 Lahd (Liner Augmentation of Horizontal Drilling) 234
    13.3 Progressive Packers 234
          13.3.1 Purpose of Design 234
          13.3.2 The Method 235
          13.3.3 Emplacement Technique 235
          13.3.4 The Means of Keeping the Liners Pressurized 235
          13.3.5 Other Concepts of Potential Use of the
                    Progressive Packer 236
    13.4 Towing Sondes and Supporting Boreholes for Logging 237
    13.5 Transparent Liner 237
    13.6 Duet Method 238
    13.7 Vertical Conductivity Measurements Using
            FLUTe MLSs 239
    13.8 Liner Pressurization for Shallow Water Tables or
            Artesian Conditions 241
          13.8.1 The Problem Addressed 241
          13.8.2 FLUTe's Weighted Inverted Liner
                    Design (WILD) 242
                13.8.2.1 The Wild Method 242
                13.8.2.2 Advantages and Limitations
                         of the WILD Design 243
          13.8.3 The Submerged Standpipe Design 244
                13.8.3.1 The Function of the Submerged
                         Standpipe Design 245
                13.8.3.2 Details of the Function 245

Chapter 14 CSC (Continuous Screened Casing) Design 247

    14.1 Purpose and Design 247
    14.2 Calculation of Flow in the Interrupted Annulus 249
          14.2.1 The Results of the Calculation 251
                14.2.1.1 Calculation No. 1: Calculation with
                         No Seals in the Annulus 251
                14.2.1.2 Calculation No. 2: Calculation with
                         Grout Seals in the Annulus 253
                14.2.1.3 Calculation No. 3: Calculation with
                         Seals in the Annulus and Allowing
                         Radial Horizontal Flow from the
                         Annulus 253

14.2.1.4 Calculation No. 4: Calculation with
Seals in the Annulus and Allowing
Radial Horizontal Flow from the
Annulus and upon Increasing
Formation Conductivity by Factor of 10... 254
14.2.2 What May Be the Definition of Significant
Vertical Flow? 254
14.2.3 What Steady-State Flow Calculations Show
about Significant Bypass of the Seals 256
14.2.4 Optimizing the Design 259
14.3 The Construction of the CSC Design 261
14.4 Combined Overburden and Bedrock Access......................... 263
14.4.1 Discussion of the Design Function........................... 265
14.4.2 Conclusion of the CSC Design 266
14.5 T Profiles in Continuous Screened Casing 267
14.5.1 Bypass of the Liner in the Sand Pack...................... 267
14.6 Conclusion 272

Chapter 15 Other Applications of Liners... 275
15.1 Use of Liners in Angled, Horizontal, and Tortuous
Boreholes or Pipes 275
15.1.1 The LAHD History 275
15.1.2 The LAHD Method... 277
15.1.3 Advantages of the LAHD Method 280

Chapter 16 FLUTe Calculational Models 281
16.1 The Crooked Pipe Model 281
16.1.1 History and Purpose... 281
16.1.2 The Drag Model in a Crooked Pipe 282
16.1.3 Parameters for a Crooked Pipe Calculation of
Liner Travel 283
16.1.4 Advantages of the Crooked Pipe Model.................. 286
16.2 Transient Correction Model of the T Profile Method........... 286
16.3 Extrapolation to the Equilibrium Asymptote for the RHP.... 289
16.3.1 How to Calculate an Asymptotic Limit for an
Exponential Approach to Equilibrium 290
16.4 Fracture Aperture Calculation Model Using the
T Profile Data 291
16.4.1 The Model 292
16.5 Data Reductions of T Profile 295
16.5.1 Who Does the Data Reduction 295
16.5.2 When Is the Data Reduced to a T Profile................. 295
16.6 Data Reduction of RHP 295

16.7   Data Reduction for the Act 295
16.8   Fact Diffusion Models 296

Chapter 17 Installation Procedures of Many Kinds 297

Chapter 18 The Manufacturing Machines and Facilities Developed for
Liner Fabrication 299

18.1   Specially Designed RF Welding Machines...........................299
18.2   Dye Striping Machine 299
18.3   Compression Wrapping Machine 299
18.4   Air-Driven Canisters 300
18.5   EP Marking Methods 300
18.6   Port Welding Machines and Other Attachments...................300
18.7   Long Trays for Eversions..300

Chapter 19 Conclusion 301

References 303
Index 305

Foreword

I first met Carl Keller when I was a young researcher tasked with developing and field testing new environmental characterization and monitoring technologies for the Department of Energy's Savannah River Integrated Demonstration Project in the early 1990s. At the time, Carl was an employee of Science and Engineering Associates (SEA) and had developed a method he named SEAMIST (Science Engineering Associate Membrane Instrumentation and Sampling Technique) for collecting depth-discrete gas samples from the subsurface using a tubular multi-port, everting membrane that was installed into boreholes. The system was initially everted with at least ten ports at different depths and associated sampling tubing on the outside of the rolled membrane. The flexible liner was everted into our very stable, vadose zone boreholes to a depth of approximately 100 ft. The ports were sampled regularly to help characterize our chlorinated solvent contaminated site and monitor the performance of our horizontal well-based remediation technologies. When Carl left SEA, he renamed his technology FLUTe and began to expand the technology from 1 useful patented method to 30 and counting.

As part of the Integrated Demonstration Project program (both at Savannah River and Hanford), there were regular meetings with the various participants working on developing and deploying the largely prototype technologies. I had the opportunity to reconnect with Carl when he participated. With his white moustache and smiling, relaxed but rugged face (not too different than the way he looks now), Carl confessed that whether it's kayaking, sailing, riding motorcycles, performing applied field research, or in earlier days, alpine climbing, he spends a lot of time outside. What was remarkable to me during those meetings was when reviewing the various technologies that were presented both for their technical design and the practicalities of deploying them in the field, Carl immediately identified the technology's most important issues either with their measurement principles or potential field implementation. In his quiet, measured, yet direct way, Carl was often the first to identify when the laws of physics were being violated. His constructive comments helped several technologies succeed and identified a few that could not. We became friends and later periodic collaborators through those early encounters.

This book is a compendium containing comprehensively developed methods, empirical observations and techniques, anecdotal case studies, and a few potential opportunities to explore. It includes a wealth of information on tools and methods that were developed in response to needs that were not satisfactorily addressed by baseline and "gold standard" technologies. The ability of liners to exploit a significantly larger surface area than other techniques allows for higher resolution sensing and sampling, more effective isolation, and other advantages. However, the flexible liner techniques have not been universally and overwhelmingly adopted yet. This is partly due to a lack of awareness of the FLUTe technology – Carl's focus has always been on the development and proof of the science and methods rather than on aggressive marketing. But it is also partly due to the inherent inertia and overcautiousness in many engineering disciplines that are governed by old standards, reticent regulators, and frugal clients. In addition to overrelying on conservative, generalized techniques and technologies, the environmental field is paradoxically plagued by having to engineer-specific solutions in heterogeneous environment that too often favors personal favorite niche applications that have anecdotally "worked before".

There is no other comprehensive description of the inverting and everting liner techniques and clearly no better person than Carl to describe them. The liner techniques detailed in this book provide distinct advantages over current baseline technologies in several situations including fractured rock and dramatically heterogenous distributions of subsurface materials, phases, and chemicals. The advantages are scientifically defensible. The engineering solutions have been demonstrated and are often quite elegant.

The inverting and everting liner technique is both useful and underused, but it won't be neglected forever. In addition to the techniques described in this book, there are potential applications at both the large scale and the very small scale. It is likely that the small-scale applications will emerge first due to the tremendous recent advances in "nano" technologies, including protein design and construction as well as other chemical and physical engineering at the molecular scale. Large-scale techniques may be implemented in the course of space exploration and expansion as specialized techniques are required that have different constraints than the ones we're familiar with on earth. The conceptual models and practical designs and applications described in this book will provide the foundation for the future designs and applications. As I stated above, at this point in time, Carl Keller is the only person with the breadth and depth of understanding and experience to compile and capture both the fundamental and advanced knowledge on this topic, but I believe both he and I look forward to future champions and advances.

Joe Rossabi

Preface

The main purpose for writing this book is to record what has been learned over 25 years about the use of flexible borehole liners for underground measurements and other applications. As a recording of the art and science accumulated, it is not an historical record of the evolution, but it is primarily a textbook on the theory and practice of the many methods explored in the development of liner applications.

There is a temptation to record the full history behind the evolution but that would include the pursuit of dead ends and extraneous detail. I have included some short historical descriptions of the purpose of each method that guided the development, and have included only some of those people who were particularly helpful in aiding the development with the first tests at their sites, or the FLUTe employees who were major players in the refinements. I have had many capable employees who were craftsmen/craftswomen and who took care of the business aspects essential to running a developmental effort. There are some interesting stories of the various kinds of adventures over the years like the move from Houston and the establishment of the FLUTe current plant, but those are not included.

The main purpose of my founding of FLUTe was to perfect the methods of liner use and to prove that they could be manufactured and installed at reasonable cost. As the founder, I was mainly interested in the science and not the business aspects of the adventure. The fact that the methods addressed the ground water needs of the world was of interest to me. The expectation was that after a few years, a company with deep pockets would purchase the technology and exploit the full commercial potential and I would retire to some other project. That did not happen. In the meantime, the technology was continually improved and new methods invented to increase the overall utility of the technology. Experience was gained and methods were further refined. Now over 25 years later, there is a growing realization in the hydrologic community of the utility of FLUTe methods over several of the traditional hydrologic methods. Sales have improved. I was continually encouraged by the progress. However, it has been interesting to see the devotion to historical practice in the hydrologic community. I suspect part of that skepticism of liner methods is due to the flood of marketing claims in the community and part of it is the concern that the regulators will disapprove of any method not backed by decades of historical practice. Fortunately, there were some individuals bold enough to try the new methods.

Sales of FLUTe systems have served mainly to support the development expenses and the staff needed to manufacture, test, and install liner systems at many sites to prove the feasibility. Some years were very lean depending on factors at times political, unfortunately, and the lack of enforcement of environmental regulations or the great recession. Some applications were not so volatile such as for mining, augmentation of horizontal drilling and relining of piping, but the focus of necessity was on those applications that did not require large development investments and were easiest to apply. FLUTe is the classic example of a company "bootstrapped" within the revenues and funds available and never very well financed as are many startup enterprises. All marketing has been essentially word of mouth after a few publications and a website.

As was inevitable, much has been learned about the mechanics of flexible liners, the chemistry, the limitations, the improvements needed, the manufacturing methods, the extreme variety of field installations, and most of all the customer needs. A better method is not alone sufficient to be accepted. Of necessity, the method must be cost competitive with the alternatives and accepted by the regulatory agencies. However, regulators must also be careful, and acceptance of traditional practice only is always defensible. A common reason I hear for not using a FLUTe method is that the regulators are not familiar with the method and therefore will not accept it. It falls to the contractor to educate the regulator, which does not happen with some regulators. Perhaps this book will help. Patent protection was obtained as needed to prevent a simple copying of the methods painfully developed and tested. Papers were published in peer-reviewed journals to aid the acceptance in the competition with traditional hydrologic practice. Everting borehole liner mechanics was not easily understood or appreciated by some people.

Finally, there was the belief that improvements in hydrologic methods were good for everyone. My wife, Lisa, keeps telling me this. The FLUTe technology is simply applied science and mathematics of many kinds and what has been learned as illustrated in this text. Ideas and methods have been adapted from many fields of my personal professional experience and even from mountain climbing, sailing, and Boy Scout training. My beginning in the theory and mathematical modeling of a wide variety of physical processes, such as the flow and condensation of steam in porous geologic media at nuclear tests while at Los Alamos, was helpful to my understanding of hydrologic concepts and associated measurements. I learned a lot of geology, drilling methods, geophysics, and rock mechanics in my 25 years as a member of the Containment Evaluation Panel, which reviewed all US underground nuclear tests. Of course, my early education in physical processes in general was the basis for the FLUTe designs of machines and many methods. It may not be obvious that mechanics, heat transfer, fluid flow, pneumatics, diffusion processes, strength of materials, chemistry, electronics, welding methods, plumbing, etc. are all included in FLUTe designs and fabrication methods. The ultimate judgment of utility is the benefit of the results.

I hope that this is a useful record of what we have learned about the application of flexible liners to underground measurements of many kinds. All these methods can be improved, but that is another lengthy description. The technology has not yet reached its full potential, and some applications are even patented and sitting on the shelf waiting for the proper funding. Liner applications are useful well beyond the hydrologic assessments. Only a few of those have been done.

During FLUTe's history, none of the FLUTe-patented methods were developed on government contracts. The nearest such contracts of that kind have been several to test the FLUTe methods but not to develop them. Therefore, unlike many new methods, the US government has no paid-up license on any FLUTe methods. I have been asked that question.

This text is to explain the theory, the practice, and the results of flexible liner techniques as developed by FLUTe. Some methods described have only been tested at FLUTe's facility, and others have not been used in field applications. I chose to include them as concepts well developed and interesting to me and part of the state of the art, even if not in general application. In the text, they are identified.

Carl Keller

Acknowledgments

Most important acknowledgements are of my wife, children, and good friends who were understanding of my time spent in development of the science. Additionally important are the FLUTe employees including some family members who have assisted with a variety of talents in the fabrication and installation of liner systems and the machines for fabrication and installation. A third important group of contributors is those individuals willing to try the new methods and to tolerate the imperfections of prototype designs. Without those adventurous customers, the technology would have failed.

A fourth group of contributors are those who were helpful in teaching me the fine points of hydrologic measurements needed and of traditional practice that defined the ultimate objectives of the applications.

A few names need explicit recognition in the first group as follows: Anne Nguyen learned to use a hot air gun to weld many of the early prototype liner systems in the heat of a warehouse in Houston. She built many of the prototype liners. Ian Sharp now with over 16 years at FLUTe installed very successfully many FLUTe systems and fabricated many of the first machines used in fabrication and welding. He also manages all FLUTe fielding activities and solved the installation problems and brought back the information needed for improvements. Sylvia Martinez took care of the contracting and book keeping duties for over 15 difficult years as the business evolved. Steve Martinez with over 15 years at FLUTe has built more liner systems than anyone else. Mark Sanchez did the massive rebuilding of FLUTe's current facility from a dilapidated mica milling plant and built much of the current fabrication facility plus assisting with fabrication of liners for many years. Bill Lowry built and tested the first everting liner system while working for Carl at SEA in 1989. These are but a few of those very helpful to the effort to prove the FLUTe methods.

Among those who dared to do the earliest installations: John Cherry of the University of Waterloo recognized the utility of the FLUTe methods and helped to obtain some of the first serious customers. Beth Parker of the University of Guelph was also helpful in keeping FLUTe in business with support of FLUTe methods with potential users and provided the information needed for refinements. She also has evolved to a regular user and collaborator of FLUTe methods in her programs and courses. Many were the customers who were willing to try the first installations without 10 years of previous experience as some folks required. Were they just bold or possessed of good judgment? Only a few are mentioned in the snippets of this history, but we are grateful to all of them.

Rod Baker, our patent attorney, has been very generous in his help for over 16 years in obtaining most of our 30 liner method patents. Erik and Kurt Sommers, our legal counselors, were especially helpful.

There are many more people who were helpful in the development of FLUTe methods. They are too many to list, but they are appreciated. Even the faith of our bankers is appreciated.

Special recognition for my wife, Lisa V., who supported me these many years and provided valuable editing and suggestions for the organization of the book. She suggested that I should start my own company in 1996.

Author

Carl Keller was born in Indiana in 1941 on a farm. At 18 years, he received a scholarship to Valparaiso University for his National Merit Test scores. He graduated in 1963 with two BS degrees in both physics and mathematics. His first job was at Connecticut Advanced Nuclear Engineering Laboratory operated by Pratt and Whitney. They sent him to Rensselaer Polytechnic Institute for an MS in engineering science, which he earned in 1965. He took a job offer from Los Alamos National Laboratory in 1966 working in the underground nuclear test program. In 1974, the Defense Nuclear Agency (DNA) hired him to a position of Division Leader in charge of all Department of Defense underground nuclear test containment designs and associated research. From 1974 to 2000, he was a member of the Containment Evaluation Panel as the DNA representative, and then as an independent expert after leaving DNA in 1985, reviewing all US underground nuclear test designs. In the time at Los Alamos and DNA, Carl wrote calculational models for the new field of underground nuclear testing and designed experiments for model validation and nuclear test designs. In 1989, he invented his first everting liner system for underground measurements and received the R&D 100 award for the invention. In 1996, he founded his own company, Flexible Liner Underground Technologies, LLC. In 2008, he received the National Groundwater Association Technology Award for his liner designs. He holds, as of 2021, 31 US patents plus foreign patents on flexible liner methods that are in use in many countries.

In the meantime, he constructed an 18-ft. carvel-planked sloop sailboat with hand tools, sailed in many nice places, climbed mountains from Mexico to Alaska, Austria, Nepal, and other places, enjoyed wood carving, motorcycling, and competed successfully in cross-country ski races.

Abbreviations

ACT	Air-Coupled Transducer system
BGS	Below Ground Surface
Blank	The simple borehole liner without attachments
CHS	Cased Hole Sampler
daFACT	Directional Assessment Flute Activated Carbon Technique with three carbon strips
DEIL	Discrete Extraction and Injection Liner
DNAPL	Dense Nonaqueous Phase Liquids, density greater than 1.0 g/cc
EP	Eversion Point, the depth of the end, or the end point, of an everting/inverting liner
FACT	FLUTe-Activated Carbon Technique
FLUTe	Flexible Liner Underground Technology, LLC
GCMS	Gas Chromatograph Mass Spectrometer
LAHD	Liner Augmentation of Horizontal Drilling method
LDPE	Low-Density Poly Ethylene
LFS	Low-Flow Sampling
LNAPL	Low-Density Nonaqueous Phase Liquid, density less than 1.0 g/cc (e.g., gasoline)
MIP	Membrane Interface Probe, a direct push device by Geoprobe
MLS	Multilevel Water Sampling system often with head measurements
NAPL FLUTe	The color reactive covering of a blank liner for NAPL detection
NAPL	Nonaqueous Phase Liquid
NAWC	Naval Air Warfare Center, Trenton, NJ
ND	Non Detect, for analytical results below the resolution limit
Packer	The Inflatable bladder on a pipe used to plug a borehole
PCE	Perchloroethylene, "dry cleaning fluid", a DNAPL
pdCHS	Positive Displacement Cased Hole Sampler
PID	Photo Ionization Detector
Profiler	The name given to the FLUTe transmissivity profiling machine
PVDF	Polyvinylidene Fluoride
RF	Radio frequency
RHP	Reverse Head Profile
RNS	Ribbon NAPL Sampler
SOP	Standard Operating Procedure
Spacer	The permeable surround of a liner that defines the interval from which a fluid sample is to be extracted

Straddle packer	A pair of inflatable bladders for isolation of an interval in a borehole for the purpose of injection, extraction, or head measurement
Stroke	The volume expelled during the pumping procedure or the act of expelling a volume from the Water FLUTe pump system
SWF – Shallow Water FLUTe	The Water FLUTe system for shallow water tables
TACL	Traveling Acoustic Coupling Liner
TCE	Trichloroethylene, a common degreaser, a DNAPL
TOC	Top of Casing
Vadose FLUTe	The multi-level vadose pore fluid sampling system
Water FLUTe	The multi-level ground water sampling system
Well Development	The process of removing mud and cuttings from fractures in the borehole wall
WILD	Weighted Inverted Liner Design
WT	Water Table

FREQUENT PARAMETER TERMS USED

Ti	Tension in the inverted liner at the EP
ΔP	The differential pressure between the inside and outside of the liner
A	Cross-sectional area of the liner, or other area
Z	The depth coordinate
R	The ratio of the borehole radius to the range of ambient head in the formation (but sometimes R is the universal gas constant)
H	"head," in the term pressure, $P = \rho g H$ where ρ is density and g the acceleration due to gravity. H is the depth below a water table usually
ΔH	The excess head in a liner, equivalent usually to DP in terms of force per unit area

1 Introduction/Purpose

This book is a documentation of the science of flexible liner applications as developed by Flexible Liner Underground Technologies (FLUTe). Over a 25-year period, the application of flexible liners for underground measurements has evolved into many methods. Some of those applications are described here in greater detail than in earlier publications. This description of the technology includes the theory, the liner designs, the methods of installation, the machines invented to do the manufacturing, and the machines designed for the installations. The mechanics of everting/inverting liners are described and the application of those mechanisms to even more diverse functions such as installations of cured-in-place liners in the piping in the walls of the Smithsonian National Museum of Natural History in Washington, DC, is also elaborated. As the experience with liner use was accumulated and challenges were encountered, new designs were developed and methods changed to improve the utility of the liner uses.

The main objective of the FLUTe technology is to exploit the unique characteristics of everting flexible liners. Some applications are hydrologic, others are seismic measurements, some are repairs of piping, and augmentation of drilling methods. Once the basic mechanisms were understood and methods of manufacture and installation were obtained, the application to many different uses became possible. This book is to convey the science for a better understanding of the utility. There is a common reluctance to trust anything considered new in the hydrologic community. There is an exceptional allegiance to the "gold standard" of traditional methods. With this description, perhaps there will be less suspicion and more understanding of the utility and the limitations. The actual cost of use of FLUTe methods is less than traditional methods for many measurements, and it provides spatial resolution otherwise not possible. None of these methods are highly technical but are simply the application of mechanisms well-known both inside and outside the hydrologic community. This is an explanation of the applied science of flexible liners to improve the general understanding of the technology.

In general, the advantage of liner measurements has been twofold. The ability to rapidly seal an entire borehole and transport measurement devices while the liner is being installed was one advantage. A second general advantage is to quickly measure with high spatial resolution the common hydrologic characteristics normally obtained with traditional methods.

This description includes the details of the hydrologic measurements and the physics behind the engineering of those measurement methods. How the liners are everted through crooked piping and boreholes under a variety of hydrologic conditions is explained. Installations through direct push rods and how that is done in sediments are described. Examples of results are included for each method. Comparisons with traditional hydrologic measurements are discussed briefly. Calculational models developed for several methods are described.

DOI: 10.1201/9781003268376-1

Also described are devices invented or devised to meet the special needs of liner applications for installation and monitoring.

Some applications that have only been tested in the FLUTe facility are described at the end of the book. They are offered as food for thought as FLUTe addresses the potential for the evolution of the methods.

2 Brief History of Flexible Liner Underground Technologies (FLUTe) Methods

In 1989, I, the author of this text, had a contract with Los Alamos National Laboratory to design an experimental validation of flow models to be used for 10,000-year predictions. The objective does seem absurd. However, that is the design objective for underground nuclear waste storage facilities. In one sense, the task was both physics-driven and a philosophical exercise. I accepted the contract realizing the limitations and first considered those mechanisms not normally included in the flow calculations which I had developed, such as earth tides. With that broad charter, I also considered what the experiment might include. In the unsaturated zone of Yucca Mt, the test must measure flows in fractured rock over very long time periods at large distances. But water as the transporting liquid does not flow into boreholes in the vadose zone unless near full saturation. Water tends to stay in the fractures with the higher capillary tension. Therefore, if the experiment needed to collect water samples distant from the source, the boreholes for interception and collection of the flow must have competing absorbers for collection of the fracture flows near the boreholes.

Installation of absorbers on inflatable packers can be done in long unstable horizontal holes such as drilled at the Yucca Mt. site by installing an absorber covered packer in a protective pipe, removing the pipe, and inflating the packer to press the absorbers against the borehole wall. However, it is not practical to slide the protective pipe back over the absorber covered packer bladder for recovery. I suggested to my engineer, Bill Lowry, that the packer could be of a thin flexible material and inverted from the borehole while under pressure, and could also be emplaced by the reverse procedure of everting the flexible packer. There was some skepticism expressed. But when the tubular liner was built by Bill and his wife, of urethane-coated nylon tent rainfly material (purchased at the local outdoor sports store in Santa Fe, named Base Camp), the procedure worked as conceived. That method was the beginning of what later became FLUTe technology.

There was some doubt on the part of my employer, at the time, that the method should be the subject of a patent. But, when I offered to buy the rights from my employer, the employer, Science and Engineering Associates (SEA), allowed me to write a patent with the help of a local retired Dupont patent lawyer in Santa Fe, NM. Since unknown to me, Ray Wood of the Isle of Mann had already invented an

DOI: 10.1201/9781003268376-2

3

everting flexible liner mechanism for relining sewers, I had to narrow the claims of my first liner patent which was granted in 1993.

In the years between 1989 and 1996, I was hired/purchased along with my patent by Eastman Cherrington Environmental (a horizontal drilling company) in 1993. However, Eastman Cherrington Environmental was closed by its owner in 1995 which orphaned the flexible liner technology that I had established at their plant in Houston. The flexible liner methods had not yet been integrated with the horizontal drilling as planned. I founded my own flexible liner company in 1996 called Flexible Liner Underground Technologies, LLC; *FLUTe* for short (www.flut.com).

I built liners for the next two years in Houston near the Eastman Cherrington facility with the same staff until 1998 when I moved the equipment to Pojoaque, NM, near Santa Fe. In 2013, the plant was moved to its current location, a much larger facility near Velarde, NM. I now hold 30 flexible liner U.S. patents plus foreign patents, with more pending. The technology has grown to worldwide sales from Europe to South Africa, Brazil, Australia, Japan, Canada, and many other countries plus all the 50 U.S. states.

A list of the order of the liner methods in general use is interesting to the history of the evolution of the technology. The flexible liner methods evolved in the following approximate sequence:

	Year	Method
1.	1991	vadose methods of several kinds
2.	1997	Water FLUTe of current design
3.	1997	NAPL FLUTe
4.	1998	Duet
5.	2001	Magic Gland
6.	2003	T profile
7.	2010	FACT
8.	2010	ACT
9.	2010	RHP
10.	2017	Transparent Liner
11.	2018	DEIL
12.	2018	TACL
13.	2018	CHS
14.	2018	CSC
15.	2019	WILD

This is not a complete list of FLUTe methods. Most dates are prior to the date of patent award if patented. FLUTe's 31 patents include more than these methods or multiple patents related to these methods.

Each of these methods is described in this text. Not all of the methods are in use and several are awaiting more aggressive funding or marketing. Some concepts are described as potentially useful concepts and are awaiting actual field applications. For the most part, the FLUTe focus for practical reasons has been on those applications with purchase requests rather than potentially useful methods. Much of the technology has been motivated by customers request for solution to a problem described. Refinements are often based on the field experience. It is obvious that some methods were very slow to gain acceptance. That may be somewhat related to FLUTe's limited marketing investment and the demand for other FLUTe methods as a distraction. It is interesting how many new contacts have never heard of FLUTe methods and regret the lack of earlier information on the option. However, the focus in developing further applications does distract from the marketing effort.

3 The Mechanics of Flexible Liners

This chapter is an explanation of the mechanical aspects of flexible liners. Flexible liners have many uses as illustrated by the many Flexible Liner Underground Technologies (FLUTe) applications. Subsequent chapters will describe the applications of these mechanical aspects of everting flexible liners and how some of those characteristics also apply to liners that are simply lowered into place. Underground applications are also affected by the hydrologic and geologic aspects of the underground installations, thus designs must consider those situations, which are addressed in subsequent chapters of this book. This description may seem too simple, but later, some of the mechanisms involved can be used for interesting measurements of the subsurface environment. Anyone installing liner systems should be well aware of these simple relationships. FLUTe has extended the use of liners to beyond hydrologic methods and more can still be done such as liners traveling on or under lakes or overland for several applications in mind.

3.1 THE FLEXIBLE LINER CHARACTERISTICS

There are many kinds of flexible liners. Some of the most useful are strong and flexible and were originally constructed from urethane-coated nylon fabrics. However, some less useful liners tested have been constructed from plastic films and even silicon rubber for high-temperature situations. The strength of the liner is important to many applications and that is why the first useful liners were made of coated nylon fabrics. The coating found to be the most useful was a tough urethane film which is very well-bonded to the fabric. The quality of the bond to the fabric is important as will be discussed. The inferior liner materials tested are described later. The construction details of the liners are described throughout this book.

The typical liner is a cylindrical tube with a diameter equal to, or greater than, the pipe or borehole into which the liner is to be installed even though some uses don't involve such a passage.

The advantages of the liner depend on its attributes:
1. It must have a good tensile strength.
2. It must be very flexible.
3. It is best somewhat elastic, but not too elastic such as are silicone rubber liners.
4. It must be capable of being formed into the tubular geometry.
5. It must be resistant to tearing.

DOI: 10.1201/9781003268376-3

6. It must have chemical characteristics that are compatible with the application.

7. It must have a friction coefficient that allows the liner to be easily everted.

These characteristics are addressed as they are important to the use of the flexible liner. The most common installation is into a pipe or borehole, so that is the main focus herein, but other applications are also described.

3.2   THE EVERSION OF A FLEXIBLE LINER

The liner characteristics affect the ability to evert the liner into the hole or pipe for particular applications. Figure 3.1 illustrates a simple horizontal liner and the geometry of the eversion process. The liner is driven with a pressurized interior fluid. That fluid might be air, water, mud or some other fluid such as molasses. The fluid is under a pressure, Pl, greater than that of the surrounding environment, Po. In some situations, such as in a borehole in a geologic medium, the pressure in the borehole and in the medium must be sufficiently lower than the interior liner pressure, Pl, such that the liner is inflated and expanded to its full diameter or to the diameter of the opening into which it is being everted. Liners can be everted on the surface of a lake or into a liquid surround without the confinement of a hole or pipe. Liners can be everted across an uneven surface or through a forest as has been done. Guidance mechanisms have been developed to control the direction of propagation if the liner is not constrained in a passage such as pipe or borehole. However, the everted liner under pressure is relatively stiff and tends to propagate nearly straightforward.

The eversion process drives the liner through an extreme deformation of the liner fabric. The deep folds of the inverted liner are collapsed by the liner pressure and then forced by the liner pressure to unfold to the full everted state. Most coatings and thin polymer tubular films during the eversion process will delaminate from the fabric or form a series of perforations in the film that then leaks, violating the need for an airtight liner. Only the high-quality urethane coating is sufficiently strong/tough to not form perforations due to plastic flow in the film nor to separate from the fabric when everted.

3.2.1   THE TOWING FORCE

As illustrated in Figure 3.1, the liner is internally pressurized, the pressure, Pl, expands the liner with the pressure against the "everted" portion of the liner. The

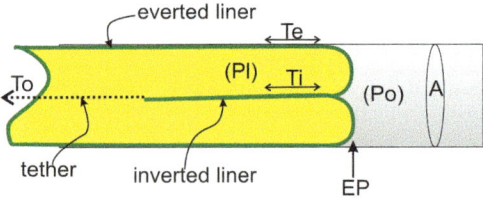

FIGURE 3.1   Geometry and terminology of an everting liner.

pressure is also acting to collapse the "inverted" portion of the liner and the fluid pressure is also against the inverted end of the liner labeled the eversion point (EP) in Figure 3.1. That pressure against the EP develops an end load on the EP of the liner equal to ΔP A where ΔP is the difference between the interior liner pressure, Pl, and the exterior/outside pressure, Po, across the end of the everting liner, and A is the cross-sectional area A of the liner; hence, $\Delta P = (Pl - Po)$. The end load tends to displace the EP toward the lower pressure Po. Resistance to that displacement is of two kinds. The first is a tensile stress per unit circumference of the liner (σ_e) in the everted liner. That stress multiplied by the circumference of the liner C produces a restraining tension, Te = σ_e C. The second restraint of the end load is the tension developed in the inverted portion of the liner (Ti = σ_i C). Ti is called the "towing force" since it tends to drag the liner and any attachments toward, and through, the everting portion of the liner. The sum of the two tensions is equal to the end load on the everting portion of the liner when the liner is stationary. Therefore, Te + Ti = ΔP A. The relative magnitude of each of those tensions has been determined experimentally and is related to the shape of the EP under pressure and to whether the liner is everting or inverting. In the simple eversion process, Te is nearly equal to Ti. Therefore, Ti =Te = ΔP A/2. Since the everted liner is pressed firmly against the hole wall and the outer surface area of the everted liner is large, the everted liner typically does not move under the tension Te. If the everted portion of the liner is not always against a restraining hole wall, the everted and inverted portions of the liner will stretch as much as its tensile strength and elasticity allow.

The inverted liner under its tensile load Ti will stretch, but more significantly, the tension in the inverted liner will cause the inverted liner to move toward the EP, unless restrained. That restraining tension (To) at the entrance of the hole or pipe must be greater than or equal to the Ti. If To is less than Ti, the inverted liner tends to slide toward the EP. In the simple case of Figure 3.1, the EP will then propagate by the process of eversion.

The tension Ti depends on the value of ΔP except for the fact that ΔP must be sufficient to cause the liner to evert. That minimum ΔP required to evert the liner is defined as $\Delta Pmin$. If ΔP is less than $\Delta Pmin$, the liner will not evert even if Ti is near zero. The value of $\Delta Pmin$ depends on the friction of the liner on itself in the folds of the inverted end of the liner at the EP and upon the stiffness of the liner material. Therefore, stiff liners require a greater pressure Pl in the liner to evert. If Ti is not zero, because the inverted portion of the liner is restrained with a tension To at the open end of the liner (entrance of the borehole or pipe or canister outlet), the tension Ti is then $(\Delta P - \Delta Pmin)A/2$ as the liner everts. The tension Ti on the liner can cause the liner to stretch. That mechanism is addressed later in Section 3.2.5.

3.2.2 DRAG (FRICTION EFFECTS)

In a frictionless environment, To = Ti = $(\Delta P - \Delta Pmin)$ A/2. However, the inverted liner is usually sliding against the everted liner producing a drag resistance (D). Therefore, the tension Ti at the EP must exceed the restraint To and D or To + D < $(\Delta P - \Delta Pmin)$ A/2 in order to evert. If Ti is greater than To + D, the liner will evert and the EP will propagate, towing the inverted liner toward the EP.

This eversion can continue as long as the ΔP is sufficient or when the sealed end of inverted liner reaches the end of the liner and the eversion is prevented because there is no more inverted liner to evert to the everted state. This description is for the simple eversion of a liner in a constraining hole (e.g., a borehole) or unconstrained, as along the surface or into a lake, either vertically or horizontally along the bottom of the lake. It is useful to note that if the cross-sectional area of the liner is less, the towing force is less. For this reason, a liner can pass a partial obstruction in the borehole, but the towing force can be reduced. If the obstruction is more than half of the hole diameter, the deformation of the EP can also halt the eversion.

3.2.3 EVERSION INTO A BOREHOLE

If the liner is everting under water into a water-filled hole, the simple eversion is not so simple as in a common vertical borehole. In an open hole above the water table, the value of Pl and therefore ΔP depends on how high the water column ΔH is in the liner. Figure 3.2 shows such a geometry where the liner is on a reel and the towing force of the liner at the reel, To = Ti–D, is sufficient to pull the liner from the reel. As the liner propagates into an air-filled hole, the liner may seal the open hole and cause the air trapped beneath the everting liner to be compressed. The increasing air pressure beneath the liner will decrease the value of ΔP until ΔP no longer exceeds 2(To+D)/A + ΔPmin and the eversion will stop. If the eversion is not in a pipe, the air trapped beneath the liner may vent into the formation outside a drill hole. Or the pipe may be the surface casing which extends below the water level in the formation forming a sealed air volume beneath the liner. Figure 3.2 shows an air vent tube which must be lowered into the hole prior to the liner installation to vent the compressed air below the liner. The everting liner can pass the air vent tube. However,

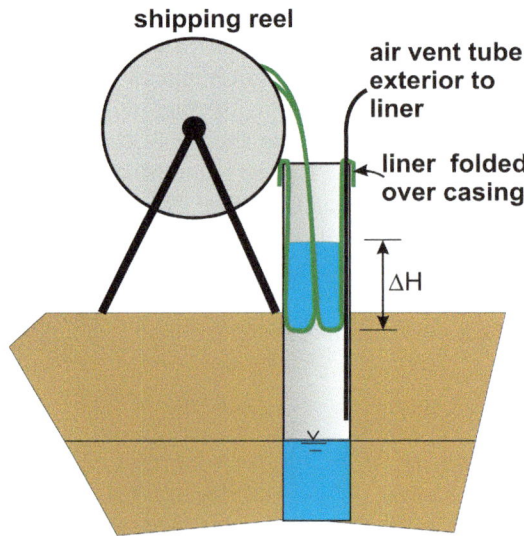

FIGURE 3.2   The installation of an everting liner into a vertical water-filled borehole.

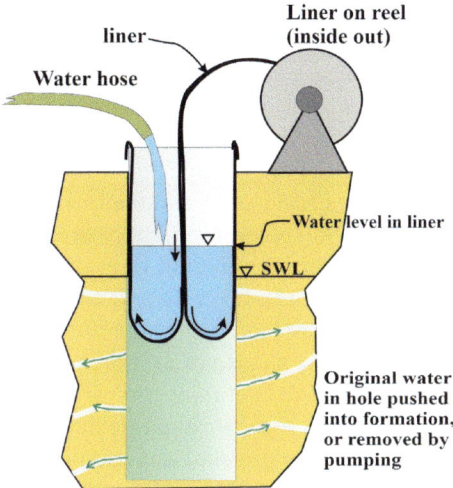

FIGURE 3.3 Eversion of liner below the water table.

the air vent tube does deform the normal EP liner geometry, and the tension Ti may be reduced to less than $\Delta PA/2$ and, with the drag, D, the liner may not provide enough tension, To, to pull the liner from the reel. Adding more water to the interior of the liner to increase ΔH is a simple means of increasing To in order to continue the eversion to the water table in the borehole/pipe.

As the EP enters the water table (Figure 3.3), the water pressure, Po, beneath the EP increases with depth and the value of ΔP decreases. When the excess head ΔH now in the liner (the water column height inside the liner above the water level in the formation) drops to the value of $\Delta Pmin + 2(D + To)/A$, the liner will cease everting. The simple solution is to add more water to the liner to increase the excess head inside the liner. In other words, only the water height inside the liner above the formation water table drives the liner. If the head difference between the liner level and the water table cannot provide a sufficient ΔP, the liner will stop everting.

However, as the liner everts, the water in the borehole must be displaced into the formation or the water pressure, Po, beneath the liner will quickly increase and decrease ΔP, causing the eversion to halt. In a permeable formation, the liner will continue to evert at a rate controlled by the flow rate of the water into the formation. This process is used with the FLUTe transmissivity measurement method to map the transmissivity distribution in the formation, which is described later in Section 10.3. The liner can be installed in impermeable boreholes by pumping the water from beneath the liner as described in Section 3.2.4.

In some situations, the surface casing is larger in diameter than the liner. In that case the air is not compressed beneath the liner and the liner can propagate into a water-filled surface casing. If the water beneath the EP does not easily flow into the formation, the water level will rise in the annulus between the liner and the larger casing. That rise in water level will increase the pressure (head, Po, beneath the liner). The head increase beneath the EP will decrease ΔP and can halt the eversion, or slow it, as the water flows into the formation. Once the EP is stalled by the decreased ΔP, the water will often flow into the formation until ΔP increases and the eversion continues. However, the common effect of the liner propagation into an oversize hole is to cause an oscillation in the annular water level causing an oscillation in the eversion rate with an oscillation in the To at the surface.

Once the liner everts into a smaller borehole below the casing, the oscillation due to propagation into the oversize casing stops and the liner everts at a rate controlled by the flow of water from the borehole into the formation. However, as described later, the liner descent rate is dependent on not only the flow rate out of the hole, but also on variations in To and in ΔH. If the To and ΔH are held constant, the liner descent rate is controlled by the flow rate out of the borehole. An exception to that generalization is if there are variations also in the borehole area, A, discussed in the next section. That effect is addressed later in Section 10.3.

In summary, the liner tends to propagate at a rate dependent on ΔP, the pressure difference between the liner interior pressure and the pressure beyond the EP. The greater the drag of the liner on itself or at the surface, the higher the pressure needed within the liner to cause eversion. The drag of the liner on itself can retard the eversion. There are several causes for variations in the drag term, D. If the hole or pipe is curved, the drag, D, can be very large. Those effects are described later. For the stiffer liners or of higher friction coefficient, a higher ΔP is needed. There are practical limits on the height of the water column that can be developed in the liner. Those are also addressed later in Chapter 8. Some boreholes have a spiral trajectory which can increase the drag because of the effective curvature. In a curved hole, an increase in To also increases the force of the liner against the curve and therefore the drag friction will increase.

3.2.4   OTHER FACTORS INFLUENTIAL ON THE LINER PROPAGATION

3.2.4.1   Hole/Liner Diameter

Clearly the value of A, the cross-sectional area of the liner at the EP, affects the calculated towing force, Ti. Smaller holes have a smaller cross section according to πr^2 where r is the radius of the liner. If the liner is everting into a passage whose radius is less than that of the liner, the effective value of A is less. The result is that it takes a higher ΔP for smaller diameter liners to propagate against a given value for To. It is also an experimental fact that $\Delta Pmin$ is larger for liners with small diameter. The difference is probably due to the fact that the liner thickness is not scalable. Thinner liners have a smaller $\Delta Pmin$ and are less stiff. Double-coated liners are stiffer than those coated only on one side. As will be explained later, adding materials such as spacers and tubing to liners makes them effectively stiffer and can increase the value of $\Delta Pmin$. The value of D will increase, if the smaller liner is "crowded" by additions on the outside of the everting liner. The crowding and resulting wrinkles in a liner when everted into a hole smaller than the liner diameter also increase the value of $\Delta Pmin$.

3.2.4.2   Wet Film Adhesion

An important mechanism that increases the drag, D, is called wet film adhesion. This is due to the fact that when two nominally flat materials have a water film between them, the meniscus at the edge of the film drops the pressure in the film effectively causing them to sort of "bond" which increases the drag of one layer sliding on another. For an air-filled liner wet by water, the everted liner can effectively tend to adhere to the inverted liner and greatly increase the drag of the inverted liner on the everted liner. If the everted liner is not dilated by an interior pressure, as when above the water table, the intimate contact of the inverted and everted liner can stop the eversion with a large increase in the drag term D. A means of reducing the wet film drag above the water table is to inflate the everted liner with an air blower to collapse the inverted liner and to dilate the everted liner to minimize the contact of one on the other. As the inverted liner descends into the water-filled interior of the liner, the wet film adhesion is eliminated because no meniscus is then possible. However, there is still a drag component between the inverted and everted portions of the liner, but not usually significant except for extreme liner lengths or crooked holes. Wet film adhesion is worse for very deep water tables due to the longer wet film adhesion area in contact between the inverted and the everted liner above the water table.

3.2.4.3 The Minimum Tension

The liner and tether should always be under some tension during the eversion installation. This is an important point. The tension should not be so high as to greatly retard the eversion, but the eversion mechanism is aided by some tension in the inverted liner, Ti, at the EP. The tension causes a larger ΔP necessary for eversion to occur and keeps the liner well-inflated. If the liner travels some distance to the water table, the hanging weight of the liner above the water table must also be offset by the tension, To, at the surface. Otherwise, the liner can buckle of its own weight and interfere with the eversion at the EP. The minimum tension also causes the ΔP to be sufficient to cause the liner to seal the borehole as it descends. There are other consequences of insufficient tension associated with stretch of the liner during installation with very deep water tables. That is due to the effect of the wet film adhesion causing a drag on the everted liner. That drag will stretch the everted liner if the everted liner is not installed with sufficient interior pressure to force the everted liner against the hole wall preventing that stretch. The net effect of the stretch is to allow a buckling of the liner at the water table trapped by a later increase of ΔH. This is a more common problem with the installations by those less experienced not maintaining an air inflation of the liner above the deeper water tables.

It is noteworthy that during a low barometric pressure day, a hole when open will allow air which was injected into a permeable vadose zone during a high-pressure day to flow back out of the permeable zone into the hole. That flow into the borehole must be prevented by sufficient air pressure inside the liner to keep it dilated. The dilation of the liner can reduce the wet film adhesion which is more significant in formations with deep water tables and with long uncased intervals in the vadose zone. Obviously if the borehole is cased, this is not a concern. The air vent in Figure 3.2 can be helpful in venting air between the liner and the hole wall by air flow in the interstitial space adjacent to the air vent tube.

3.2.4.4 The Difference between the Eversion and the Inversion of the Liner

The above description applies to the eversion of a liner. If the tension To is increased with a constant internal pressure of the liner, the eversion process can be reversed

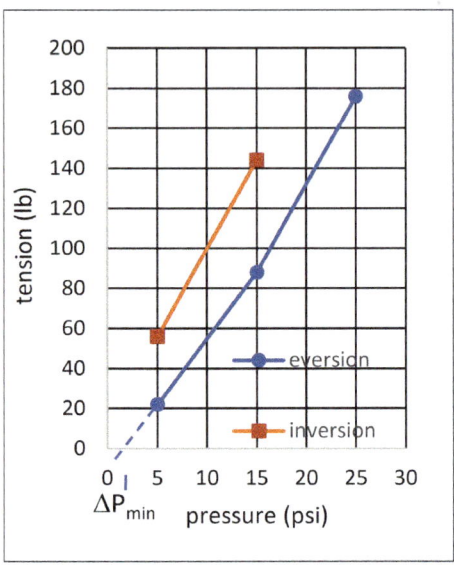

FIGURE 3.4 The plot of the inverted liner tension, Ti, during the eversion and inversion at different internal liner pressures, Pl. Note the inversion tension is higher but the two graphs are nearly parallel. ΔPmin is the minimum pressure needed to drive the eversion process.

to cause the liner to invert. The inversion process is very similar in many respects to the eversion reversed except that the tension, To, needed to invert the liner is now the sum of D and Ti. The tension in the inverted liner at the inversion point, Ti, is somewhat higher than that during eversion. During inversion, Ti is approximately 2/3A ΔP as determined experimentally. The graph of experimental data in Figure 3.4 shows the tension, Ti, during both the eversion and inversion processes. The pressure where the extrapolated data (the blue line) crosses the abscissa during the eversion process is the value of ΔPmin, the minimum eversion pressure. During inversion, the tension is higher, because the inversion deformation of the liner provides an effective resistance to inversion. However, that is not just the negative of the ΔPmin during eversion. Figure 3.5 illustrates a drawing of the liner shape during eversion versus during inversion. It is observed that the inversion of a liner produces a different shape in the end of the liner at the EP. The crown of the inverted liner has a larger diameter, Ai, than the crown, Ae, during eversion. This affects the effective area in the tension calculation, and suggests that the distribution of the end load on the liner is borne more on the inverted liner than the everted portion of the liner during the inversion (i.e., Te < Ti, where Ti is the inversion tension).

Figure 3.5 also shows how the attempted inversion of a liner, with insufficient interior pressure, Pl, to prevent slippage of the everted liner on the hole wall, can buckle the liner. This is discussed more in Section 3.3 on the liner removal procedure. The friction of the everted liner on the hole wall is typically less in a poly vinyl chloride (PVC) casing than for an open borehole wall. The buckling of the liner is to be avoided. It is the main frustration of the inexperienced operator in liner removals.

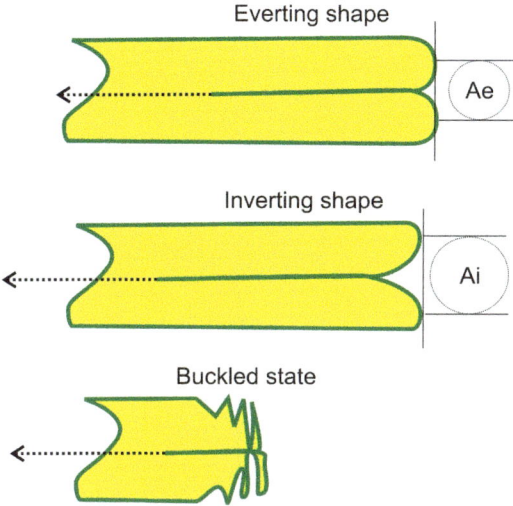

FIGURE 3.5 Difference in liner shape during eversion and inversion. Also shown is the buckling of the liner that can occur during inversion with insufficient interior pressure in the liner and contact with the hole wall.

3.2.4.5 The Air Balloon Drag and the Air Vent

As shown in Figure 3.6 the typical borehole liner installed in a water-filled borehole is closed at one end, has a tether strongly connected to the closed interior end of the liner, and an air vent near the closed end of the liner. Figure 3.7 shows the geometry of the inverted portion of the liner as it is drawn beneath the water-filled interior of

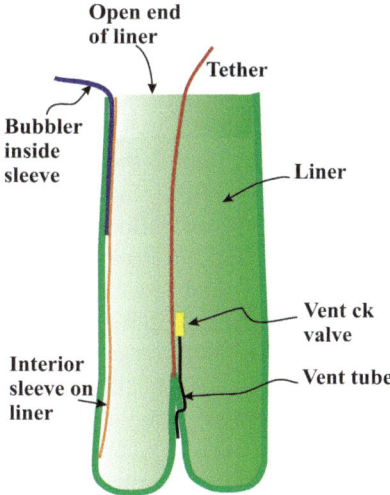

FIGURE 3.6 Typical features of the basic blank liner. The vent check valve is often two valves of different designs to assure the valve does not leak.

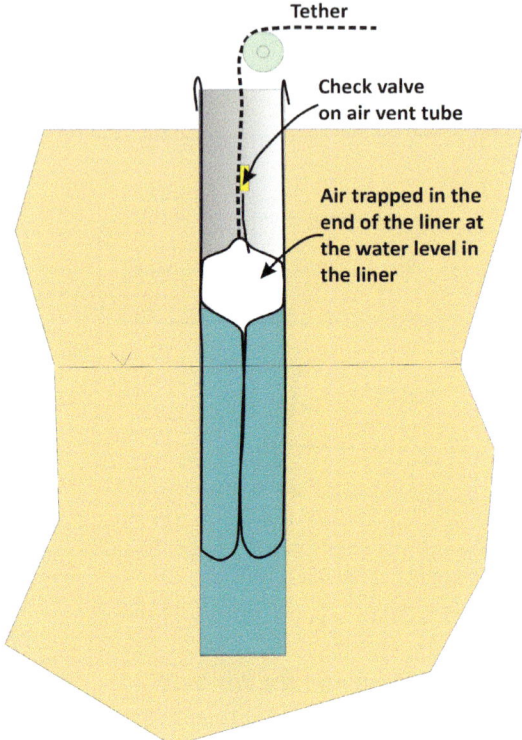

FIGURE 3.7 Geometry of balloon formation in liner above liner water level.

the liner. The water, which is driving the liner eversion, compresses the inverted liner as it descends beneath the water level in the everted liner. That compression forces any air trapped in the folds of the inverted liner upwards toward the descending closed end of the liner. As that trapped air accumulates in the closed end of the liner, the inverted liner is dilated and an "air balloon" is formed in the closed end of the liner. The inverted liner can then dilate to the full diameter of the everted portion of the liner and the dilated inverted liner can drag heavily on the everted liner. Because of the large surface area of the air balloon, the drag can be excessive even for a low air pressure in the balloon and for a small friction coefficient of the liner against itself. As the inverted liner descends beneath the water in the liner, the pressure in the balloon increases and the dilated liner balloon can stop the liner advance. For this reason, an air vent tube is built into the end of the liner with a pair of check valves (see Figure 3.6 which shows a single check valve).

As the pressure increases in the air balloon, the trapped air vents through the air vent tube into the interior of the liner. If more water is added to the liner, the submergence of the air balloon forces the air to vent more quickly. The vent tube containing the check valves is about 10 ft. long so that the vertical gradient in the water column between the inflated end of the liner and the open end of the vent tube produces an even greater pressure differential and greater venting rate for the trapped air. The check valve prevents that water which is interior to the liner from flowing out of the bottom of the liner when it is fully everted. Such flow of water would be an effective leak in the liner. The pair of check valves of different designs is usually used to avoid the possible malfunction of both valves which would allow the liner water to leak into the borehole.

The drag of the air balloon is not the only resistance to eversion, the buoyancy of the balloon adds to the tension, Ti, inhibiting the eversion.

3.2.4.6   Effect of Breakouts on Liner Eversion

Enlargements in the borehole can occur due to fractures or weak cementation of the formation or Karstic features in the formation. If the enlargement is of a diameter greater than the liner, the liner is not supported by the hole wall. The lack of liner support can result in a burst of the liner if the ΔP in the liner is greater than what the liner can withstand. The unsupported liner can also buckle in the enlargement during inversion as shown in Figure 3.5. Also, if the liner is everting through a large opening such as a cavern penetrated by the drill hole, the liner may not align with the borehole in the floor of the cavern or enlargement. The liner advance may be halted if it does not align with the borehole at the bottom of the enlargement. Several methods are available as described in Chapter 12 to address these concerns. A remedy against the burst of the liner is to not overfill the liner or to use a stronger liner fabric.

3.2.4.7   The Impermeable Borehole Installations

As the liner in Figure 3.3 enters the water table, driven by the excess head in the liner, ΔH, which provides the inflating pressure, ΔP, the liner acts like a descending piston well sealed to the borehole wall. The descending liner tends to compress the water in the borehole, but unlike the air trapped beneath the liner, the water is nearly incompressible and will stop the liner eversion unless the water in the borehole flows into the formation. Normally that flow does occur, but the rate of flow depends on the conductivity of the formation or the transmissivity (conductivity times the open saturated length) of the open hole beneath the liner. As the liner descends, it seals off the flow zones as each flow zone is passed. (Note this process is used to measure the conductive intervals of the formation, Keller, 2013). If there are no conductive intervals remaining in the borehole, the liner will stop. Also, as each flow zone is sealed during the liner eversion, the liner descent typically slows down.

In many applications of the everting liner, it is advantageous to speed the liner descent by pumping the water from the borehole beneath the everting liner. In other situations, it is a disadvantage to have the borehole water, often contaminated, to be displaced by the liner into the formation. Pumping the water from beneath the descending liner is a simple concept. The pump and water removal tube to the surface only temporarily violates the seal of the borehole by the liner. The sealing liner is usually a very desirable function of the flexible liner and therefore the water removal tube is usually removed for the long-term seal of the borehole.

3.2.4.7.1   Shallow Water Table Pumping from below the EP

The air vent tube of Figure 3.2 can be extended from the top to the bottom of the borehole as a "water removal tube" to pump the water from beneath the liner using a sufficient pump at the surface. A sufficient pump is able to draw the water from the water table to the surface. The liner installation into a borehole of low transmissivity will raise the water level in the water removal tube to improve the flow rate. Typical pumps for pumping from the surface include a simple bilge pump, double diaphragm pumps, or even a peristaltic pump. But peristaltic pumps are too slow to be practical.

3.2.4.7.2 Air Lift Pumping from beneath the Liner EP for Deep Water Tables

In the early days of liner installations, several devices were used to remove the water from beneath the liner with deep water tables, but a pump beneath the liner was not easy to remove even if the liner was deflated after installation. Leaving the pump in place with the water removal tube to the surface violates the seal of the liner. A very simple device often now used is an airlift pumping system which is easy to remove from the borehole with a liner in the borehole. The airlift system is shown in Figure 3.8. It consists of an outer tube (the water removal tube, often ½″–1″ in diameter) with an interior air injection tube. The outer tube is called a water removal tube by FLUTe, but the outer tube has many other names. It is lowered into the water in the borehole before the liner installation. As air is pumped through the interior air tube, it aerates the water in the water removal tube reducing the effective density

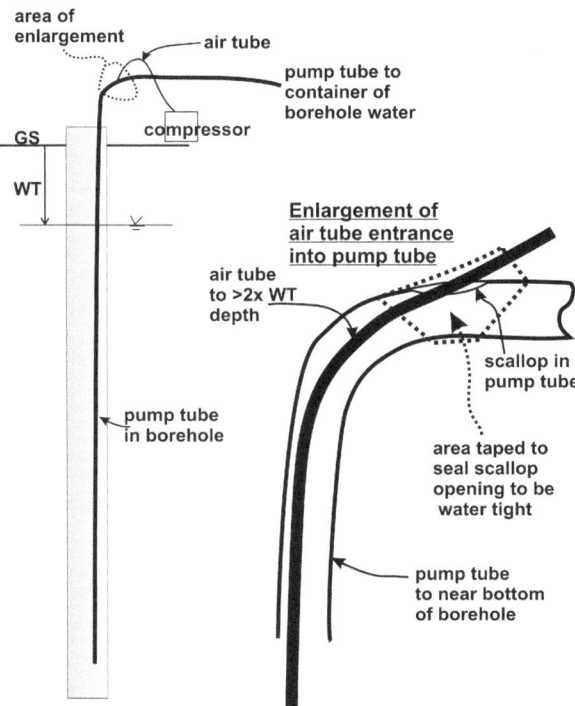

FIGURE 3.8 The lift pump system used for removal of water from beneath the everting liner. The enlargement shows the details of the air addition tube inserted into the larger pump tube.

of the water inside the water removal tube. The water outside of the water removal tube is of normal density and tends to displace the more buoyant aerated water in the water removal tube upward. This is the simple effect of gravity on the water of different densities. The aerated water in the tube may be displaced to the surface if the aerated water column is long enough to provide a sufficient buoyant force. Typically, the submergence of the water removal tube below the water table must be greater than the depth of the water table from the surface or the pump is not effective.

The major advantage of the airlift pumping system is that it is simple with no moving parts (discounting the air compressor unless a gas bottle is used). The water removal tube is also very slender and the removal is simple. When the liner has reached the bottom of the borehole, the water level inside the liner is pumped down until the liner collapses far enough and the water removal tube is pulled from the hole. The liner is then refilled to dilate it to seal the hole. Figure 3.8 includes an insert drawing showing the very simple version of a common air injection tube and water removal tube geometry. The water removal tube after being lowered into place is opened above the surface with a small scallop cut from the water removal tube. The air tube is inserted through the scallop and the scallop is taped closed as much as possible to seal the scallop around the tube. Some leakage is expected. Air is injected at a pressure above the depth of submergence of the air tube and the upper end of the water removal tube is directed into a container. If the water level in the borehole (or in a liner as is sometimes another use) drops due to the pumping, the air lift will cease to function as a pump. That is a limitation of the airlift pump. The pumping rates of the airlift pump are typically 1–3 gal. a minute and not nearly as high as common centrifugal pumps. But the airlift pump is far less expensive, not counting the common compressor needed, and it is much more slender, which is essential for easy removal. A limitation of the air injection is that the compressor must provide a pressure greater than the submerged length of the air tube. Otherwise, air will not flow from out of the bottom of the air tube. For deep installations, the air tube may need to be much shorter than the water removal tube but still a water table (WT) depth of submergence (i.e., 2 WT depths below the surface).

The airlift pumping method is commonly used with much larger water removal tubes/pipes, and larger air flow rates, to remove sediment from the bottom of a drill hole (Driscoll, 1995). The method has the additional advantage of being able to pump sand and gravel without the usual damage to the common mechanical pumps. Fine sediment can be removed from boreholes with water removal tubes of 1″ diameter. However, if the sand being pumped adds too much weight to the water column inside the water removal tube, the flow will stop because the sand has effectively increased the density in the water removal tube.

Pumping the open borehole water from beneath everting liners is useful for some FLUTe liner methods to improve the measurements as described later in Chapter 10 for the FLUTe activated carbon technique (FACT) method.

3.2.5 STRETCH OF THE LINER

The tension on the liner, Ti, will cause some elongation of the typical liner since the liner materials are slightly elastic. This can have an effect on some of the measurement methods described later in Chapter 10. A much more detailed description of stretch is provided in Section 10.6 including the measurements of stretch in different liner materials and how the stretch of liners leads to procedures to adjust when needed for the stretch. The simple estimate of stretch for a liner installation is that the change in length due to a tension, Ti, is: $\Delta L = E\ Ti\ L/C$, where E is the coefficient of elasticity defined from liner tests, $Ti = \Delta PA/2$, L is the liner length under tension, and C is the circumference of the liner. Values for E vary from 0.0005 to 0.0028 for a wide range of liner fabrics with the units of T in lb., L in ft., and C in inches. E is determined from the measurements of stretch with applied tension on liners of known length. This behavior can result from trivial stretch to substantial stretch depending largely on the liner length and the Ti during the installation. The stretch is only important to methods that depend on the depth of the liner in the borehole. After the description of the methods in Chapter 10, the measurement of actual stretch and the accommodations needed are described.

3.3   THE LINER REMOVAL METHODS

3.3.1   THE NORMAL INVERSION FROM A PERMEABLE BOREHOLE

In some situations, the liner removal is simply the reverse of the installation. Instead of allowing a modest tension, To, on the liner or tether, a higher tension will reverse the eversion and cause the liner to invert. That will require a tension sufficient to overcome the effect of ΔP on the end of the liner. If during the eversion of the liner the process is halted with tension on the inverted liner/tether, there is little or no drag, D, due to sliding friction, and the tension, To, is approximately $Ti = 1/2A\ \Delta P$, the tension at the EP. Then increasing To, the liner will tend to invert if To is greater than $2/3\ A(\Delta P + \Delta Pmin) + D$. The 2/3 factor is determined during inversion as approximate depending on the liner stiffness and diameter. It is slightly higher than the factor ½ during eversion. The $\Delta Pmin$ for inversion can be different, but it is not expressed as a negative pressure, but as a tension, Tmin, needed to invert the liner, approximately $3/2\ Tmin/A$ in place of $\Delta Pmin$ in the expression above, or $To > 2/3A\ \Delta P + Tmin + D$ is needed to invert the liner.

It is very important to remember that the liner will only invert if it is still inflated. If the ΔP is too low, the liner will tend to buckle instead of invert as shown in Figure 3.5. This is because the liner is not pressed against the hole wall with the necessary friction to prevent the liner from sliding upward above the EP. With no ΔP, the liner will always buckle. The minimum pressure to prevent buckling depends on the friction coefficient of the liner on the hole wall and the stiffness of the liner. The tension, Tmin, to invert the liner must be greater than the tension to buckle the liner. Due to many variables involved, the minimum ΔP to prevent buckling for all circumstances is not known, but some generalizations have been noted as follows:

1. Low friction between the liner and the hole wall can allow buckling to occur (e.g., a mud lubricated hole wall).
2. If the liner is not supported in the borehole or pipe as in an enlargement or cavern in the formation, the liner may buckle.

3. If the liner diameter is greater than 8″, even when not supported in a bore-hole, the liner will usually invert if ΔP is sufficient to inflate the liner.
4. Small liners (<8″) are more inclined to buckle under any pressure if not supported. It seems to depend on the end load on the EP and the normal buckling failure of a column depending on its diameter. Slender pipes or liners buckle more easily.
5. If the head/pressure in the liner is not high enough, buckling will occur.
6. Liners much larger than the borehole or pipe diameter are more likely to buckle during inversion due to the excessive folds in the everted liner.

For these reasons, most liners will invert if the ΔP is high enough and the liner is constrained in a pipe or borehole of diameter smaller than the liner diameter.

In order to prevent buckling during the liner inversion, it is advised to maintain a water level in the liner at least a minimum level above the highest head in the formation. That is usually a head greater than the ΔPmin and varies with the diameter of the liner. As a general rule, at least 5 ft. of excess head for 6″ or larger diameters and more than 10 ft. of excess head for 4″ or less. Liners in PVC casing will buckle more easily due to the lower friction of the liner on the casing. Those should have a higher excess head during inversion. One precaution is to not pump the liner water level below that critical level. Therefore, it is dangerous to lower a pump inside the liner to deeper than 5–10 ft. <u>above</u> the water table in the formation. When the pump is lowered below the water table, the liner is often partially collapsed by the pumping and buckles when inversion is attempted. This is the most common reason for frustrated liner removals by inexperienced personnel.

The end result of the liner buckling is that it continues to buckle until the buckled portion of the liner is tightly jammed in the hole and a greater tension only compacts the jammed liner with no inversion possible. Releasing the tension and adding water to inflate the liner usually does not extend the liner because the buckled portion of the liner is forced by the increased ΔP hard against the hole wall, thereby resisting extension of the liner to its pre-buckled state.

The only remedy that sometimes works is to deflate the liner and to pull the liner at the surface upward to extend the buckled liner (i.e., to unbuckle the liner). If that is well done, the liner can be refilled sufficiently to allow the inversion to occur.

3.3.2 THE PUMP AND DRAG REMOVAL

If the liner is not to be reused and only to be removed, one can pump the liner empty by lowering a pump to the bottom of the liner and removing all the water. The liner can then be lifted to the surface. However, most pumps will draw the liner against the inlet of the pump and prevent the effective removal of all the water from the liner. To prevent or at least reduce the effect of the liner collapsing on the pump, a perforated tube of ½–3/4″ diameter can be lowered in the liner to the bottom of the liner past the pump to allow the flow of the water in the liner from all portions of the liner while the pump is operating.

When the liner is essentially empty, it can be lifted from the borehole, but as the remaining water inside the liner flows to the bottom end of the liner and dilates it, the water inside the liner can dilate the liner as it is lifted above the water table, preventing removal. The pump can then remove the rest of the water if the pump is at the bottom of the liner. Other "tricks" have been employed to complete the removal with water in the bottom end of the liner.

Fortunately, with the liner inversion from the borehole, the procedure can be easier than the pump-and-drag approach, except for one situation as follows.

3.3.3   THE IMPERMEABLE BOREHOLE REMOVAL

If the liner is inverted as described above, the inversion of the liner must draw water into the borehole below the liner or the drop in pressure below the liner will continue until the ΔP (difference between the pressure inside the liner and outside the liner) increases to prevent the liner inversion under any practical tether tension. Increasing the tension at To can tear the liner or burst the liner due to the excessive ΔP. This resistance to inversion actually required a tension of 300 lb being applied to a liner for three days. The EP during that installation had passed the last flow path by only a foot in an impermeable granite hole. The high differential pressure caused by the pressure drop below the liner produces an impressive seal of the liner against the borehole wall. For small boreholes, the pressure beneath the liner can reach a nearly full vacuum. In that case the ΔP in the liner is the full water column in the liner plus an atmosphere (~33 ft. of head). If the water column in the liner is short, the liner can be inverted with a vacuum beneath the liner without failure, but such an inversion is considered dangerous because if the tension cannot be maintained, the failure to restrain the liner will cause it to instantly evert back down to the current water level in the borehole. The everting liner has a dangerously high tension if it were to become entangled. The towing force of the liner is Ti with the ΔP augmented by the full vacuum below the liner and the full water column inside the liner. In contrast, if the water level in the hole is rising and following the liner removal, the liner will rebound a short distance and then descend at the rate the water can flow back out of the borehole.

A remedy to the impermeable hole liner removal is to pump the liner empty as described above. It is called the "pump-and-drag" removal. However, that is usually damaging to the liner and the inversion is also preferred for some FLUTe methods.

Another solution has been designed and used. That solution is to include in the liner construction the "long vent tube" which extends the air vent tube at the bottom of the liner (Figure 3.6) to the surface with no check valves. With that tube built into the liner, as the inverting liner draws down the head/pressure beneath the inverting liner, water can be added via the long vent tube to the interior of the inverted liner which is at the low pressure beneath the liner. The water addition through the long vent tube dilates the inverted liner and allows flow out through the inverted liner, through the EP, and into the borehole beneath the liner. That raises the low pressure beneath the liner (decreases ΔP) and allows the inversion to continue.

The long vent tube is used for those installations that require the liner to be everted to the bottom of the hole and inverted out of the hole such as the FACT method described later. In many cases, the bottom end of the borehole is of very low conductivity and needs the long vent tube for the liner to be inverted until a significant flow path is uncovered by the inverting liner. A limitation of the long vent tube use is when the closed end of the liner reaches the surface. In that case, water addition via the long vent tube dilates the inverted liner above the surface which prevents further inversion with the water addition method. That happens when the liner has been inverted for half its total length. Further removal of the liner by inversion requires other methods of water addition below the liner. One "last ditch" approach for the rare difficult removal may be to add water through the air vent tube as shown in Figure 3.2. If the water can flow past the liner between the liner and hole wall as in a surface casing, it can increase the head beneath the liner to aid removal.

3.4 THE LINER SEAL

3.4.1 INTERIOR VIEW OF THE SEALING LINER

The purpose of the everting liner is often to seal the borehole. The seal is particularly good because of the thin and flexible nature of the liner in contrast to the relatively stiff rubber of a packer. The packer must be thick and tough to withstand the abrasion of emplacement in ragged boreholes. The liner is only about 15–20 mils in thickness and relatively elastic. The borehole photo of Figure 3.9 shows the inside of the liner in a borehole. The liner conforms very well to the rugosity of the hole wall so that fractures, vugs, and breakouts are obvious. Because of the eversion installation procedure, even the large ledge, in the upper left-hand corner of Figure 3.9, is draped with the liner such that the liner appears to be painted on the borehole wall. As long as the interior pressure is maintained in the liner with the excess head in the liner above even the highest head in the formation, the borehole is exceptionally well sealed.

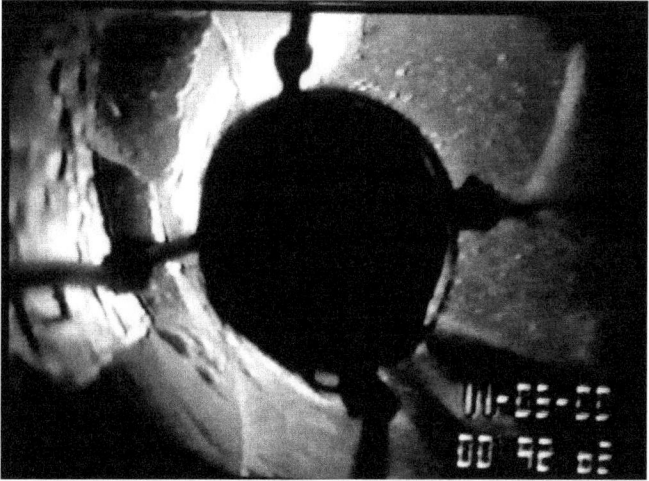

FIGURE 3.9 Snapshot of the interior of a liner showing how well the liner conforms to the vugs in the borehole wall and the ledge in upper left of photo. (Large black central object is the back of the light source.)

If the formation pressure is above the liner interior pressure, the liner is not pressed firmly against the hole wall. Since the actual highest head in the formation is not usually known prior to the liner installation, a simple procedure has been developed that can determine the highest head in the formation once the liner is in place. It is important to remember that the water level measured in the open borehole is between the highest head in the formation and the lowest head and is not at the highest head in the formation. The method for assessing the highest head in the formation, "the highest head measurement," is described in the next section.

3.4.2 THE HIGHEST HEAD MEASUREMENT METHOD

A very simple technique for determination of the highest head in the formation uses the flexible liner emplaced in the borehole. There are two conditions addressed with the procedure with slight variations. The first is the condition of the downward gradient, the most common situation. An upper aquifer is penetrated by the borehole connecting a lower aquifer with a lower water table than the upper aquifer. In some situations, cascading water is observed in the borehole above the water table in the borehole. Since inflow to the borehole occurs from the higher head aquifer and out of the borehole into the lower head aquifer, the actual water table in the borehole is the net result of the inflow and outflow. The water table in the open borehole is often called the "blended head," Quinn et al 2016, or equilibrium head and it is somewhere between the highest head penetrated by the borehole and the lowest head. The transmissivity of the formation determines whether the water table in the borehole is nearer the highest or lowest water table. If the highest head has a higher transmissivity, it dominates the blended head observed in the open borehole. If the lowest head has the higher transmissivity, the blended head is near the lowest head interval.

The downward gradient situation is shown in Figure 3.10. The upper aquifer at higher head is flowing into the lower aquifer when the borehole is open. The result is the blended head is below the water table in the upper aquifer. The sealing liner interrupts the initial connection of the aquifers. Initially, the liner is filled with water above the blended head since there is no knowledge of the actual head distribution in the formation. That causes the liner to seal the lower aquifer of lower head. The upper aquifer may continue to flow into the borehole depending on the water level in the liner. If the water level in the liner is less than the head in the upper aquifer, the liner does not seal the upper aquifer and water continues to flow from the upper aquifer into the borehole collapsing the liner until the water level in the liner rises to that of the water table in the upper aquifer.

If it is noted that the water level in the liner is rising, more water must be added to the liner to seal the upper aquifer. Figure 3.10 shows how that can seal the upper aquifer by causing the liner to dilate to seal the borehole. How does one know that the water level in the liner is well above the highest head in the formation? When the liner interior head exceeds the formation head everywhere, the water level inside the liner is stable and will not be rising. By pumping a measured amount of water from the liner (e.g., one foot or ~1.5 gal. in a 6″ hole), the water level will drop one foot. Using an electric tag line, it is possible to determine if the water level inside the liner is stable. If stable, another foot of water can be removed and the water level

Installation of liner with unknown highest head

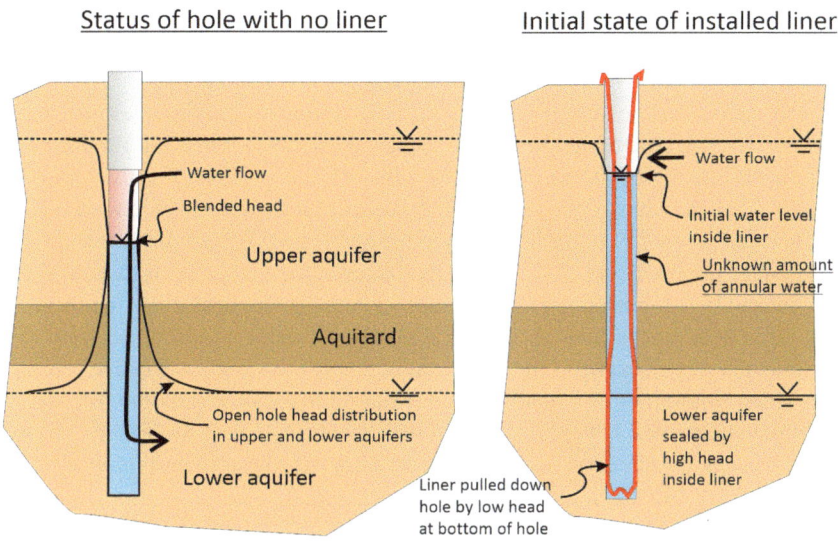

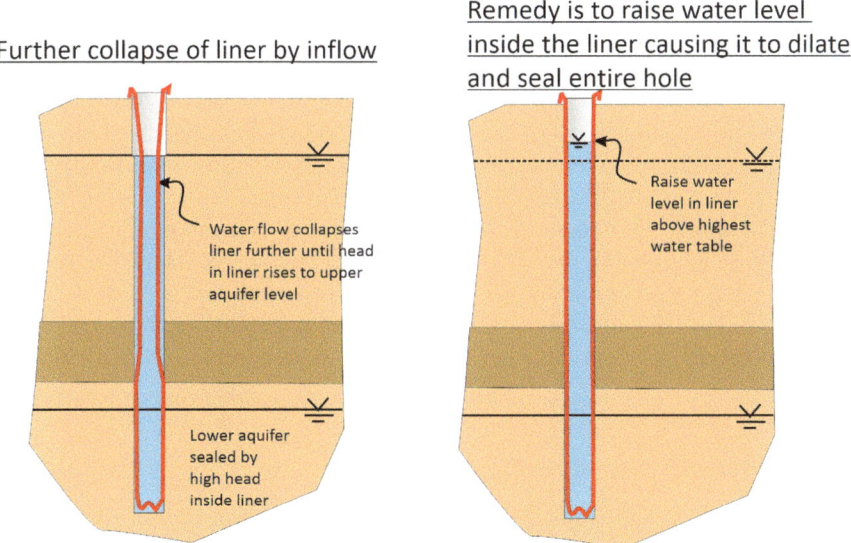

FIGURE 3.10 Water table sequence in open borehole as changed by the liner installation. The liner installation seals the lower aquifer and responds to the upper aquifer with a water level rise in the liner.

monitored again. Each time the water level is lowered and the water level in the liner is stable, the head in the liner is higher than the head in the formation. However, if the water level after a water removal is below the highest head in the formation, wherever it is located, the water level in the liner will rise because the higher head in the formation will collapse the liner, causing the water level in the liner to rise. At that point, when it equilibrates, the new water level in the liner is approximately equal to the highest head in the formation. The water level in the liner will cease rising when the water level in the liner is equal to the highest head in the formation or until the collapsing of the liner exposes a lower head interval. For the latter reason, the stepwise head changes should be small so as to not allow an extended collapse of the liner. In this manner, the highest head in the formation can be determined. This reduction of the liner seal could allow the water at the higher head to flow to a lower head interval, if the collapse is extensive. Once the highest head has been determined, the head in the liner should be returned by water addition to a level of 10 ft. or so above that highest head in the formation. This is a standard FLUTe procedure if there is any possibility of a head in the formation that can violate the liner seal. Any rise of the water level in the liner after a water addition is a strong indication that the liner level is not sufficient to seal the borehole in the highest head interval.

For a Water FLUTe multi-level system described in Section 10.5.1, one has the additional advantage of head measurements at discrete intervals in the formation. However, it is still possible that there is an aquifer that is not detected by the discrete head measurements at higher head than any interval measured. This has been erroneously offered as a limitation of the Water FLUTe system, but the highest head measurement determines the highest head wherever it exists in the formation intersected by the borehole. The procedure is to increase the head in the liner well above any head measured at each port. By performing the stepwise reduction of the head in the liner, one can determine if there is a higher head somewhere in the formation not seen in the discrete measurement intervals of the FLUTe multi-level systems described in Section 10.5. It is also true that adding water to the liner should raise the level the appropriate distance. If it does not, more water should be added until the rise is the expected amount. Then the water level in the liner should be increased enough to obtain a good seal everywhere. The appropriate rise is determined from the borehole cross section and the amount of water added. For example, in a 6-inch diameter borehole, 1.47 gal. of water addition produces approximately a 1 ft. rise in the liner.

3.4.3 Artesian Conditions

An example of an extreme formation head is the artesian head condition. An "artesian head" is defined as a head which if connected to a cased hole/pipe will rise above the ground surface. The illustration in Figure 3.10 shows that the inflow to an open borehole may well exit the borehole at the aquifers that are at lower head than the artesian head when the borehole is not sealed. Therefore, an artesian upward flow can exist in an open hole without overflow of the open hole. In many cases, if the open borehole is overflowing at the surface or rises a short distance in a surface casing, it is probable that the artesian condition exists in subsurface with a head well above the surface. The liner does not seal a borehole with an artesian condition unless there is a means of increasing the head in the liner to above the artesian head.

In some situations, the potential artesian condition may not be obvious unless local pumping is stopped or there is infiltration from melt or rain at a higher elevation causing overflow to the top of the borehole. In that case, the liner may have been sealing the borehole, but when the artesian condition occurs, the water level in the liner can rise by collapse of the liner until it overflows the surface casing. This has occurred in a liner installed in a subsurface vault. The artesian condition collapsed the liner and forced nearly half of the liner water fill to overflow the surface. When the artesian condition ceased, the water level in the liner descended to the new "blended head" in the liner which is probably above the blended head in the open hole since the liner may be sealing some of the lowest head intervals. The liner was no longer sealing the borehole except at those intervals where the liner was dilated at those intervals with a formation head less than the blended head in the liner. However, in a partially sealed borehole, the head in the liner will not be that of the highest head, but below that value.

Several methods have been developed to seal the artesian intervals. Those methods are as follows:

1. Extend the surface casing and liner to allow a fill of the liner above the highest head. This is not always practical.
2. Fill the liner with a heavy mud. However, that has limitations as addressed below.
3. Seal the top of the liner and pressurize an air volume above the water level in the liner.
4. Use a sealed wellhead and what is called a submerged extension which is a mud-filled pipe that pressurizes the water in the liner above the formation pressure.
5. Use a Weighted Inverted Liner Design (WILD).
6. Fill the liner with grout to prevent liner collapse by the artesian head.

Figure 3.20 addressed in Section 3.5.7 shows the effect on the vertical pressure gradient for each method as compared to the typical fill of a liner to a level below the surface.

Method 1 above is easy, but not practical in many situations where a tall surface conductor is not acceptable. Method 2 has a very steep gradient which adds to the liner interior pressure ΔP with increasing depth. In deep boreholes, the excess head ΔP deep in the borehole may be so high as to burst the liner. And near the surface, the excess head in the liner due to the mud may not exceed the artesian head of a confined aquifer. Another concern with a mud fill of the entire liner is that the weighting material may settle in the mud leading to a lower mud pressure at the top of the hole. This does not occur in a PVC pipe, but tests show that it occurs inside a liner for some reason. Method 3 is faced with the difficulty of sealing the top of the liner airtight and the air pressure may decay due to diffusion through the liner, like a helium filled balloon, or due to leakage past the top seal. An eternal airtight seal is difficult to achieve.

Method 4 has a short heavy mud fill at the bottom of the liner. By pressurizing the sealed liner with a water, the small diameter "submerged standpipe" which extends into the mud fill experiences a rise of the mud level in the small pipe. In that case, most of the liner has the normal water-filled liner gradient, but it is shifted upward to provide a relatively uniform differential pressure in the liner with depth. Method no. 4 is described in Section 13.8.2.

Method 5 (WILD) uses a heavy wellhead of smaller diameter which is sealed to an extension of the liner (see Section 13.8.1). As the heavy wellhead is allowed to descend into the water-filled liner, the liner pressure increases throughout its entire length. The increase in pressure depends on the weight and therefore length of the heavy wellhead. The weight of the heavy wellhead is similar to the tension, Ti, at the EP of the liner described above. Except in this case, the top end of the liner is also inverted. A refinement includes a sheave at the bottom of the heavy wellhead through which the tether is routed with an additional weight on the tether end which enhances the downward force of the heavy wellhead and also maintains each access to the tether which is normally secured at the top of the surface casing for the typical liner. See Section 13.8.1 for details of the WILD method. The WILD method has not been used as of 2021. The spreadsheet for the calculation of the appropriate weight for the wellhead has been done. A disadvantage of the method is that the weight used must be substantially greater as the diameter of the surface casing is increased.

The advantage of methods referenced above, 1, 3, 4, and 5, is that the liner still has the normal hydraulic head differential ΔP along its full length.

Method 6 has the disadvantage of making the liner a permanent installation which is more acceptable for a FLUTe multi-level system rather than other liner methods. Because of the typical grout density of 1.35 g/cc, the excess head obtained, like that of a mud fill, may not be sufficient for a shallow artesian condition. Also, the grout fill must be installed in reasonable lifts to not overpressure a deep liner. The grout fill is attractive for liners that must survive the high pressure of some remediation fluid injections. Tests have been done to determine that the heat of hydration is not a concern for liner materials. The grout fill can be weighted with a common barite compound if needed.

A general problem with dealing with an artesian condition is that it often varies with the season and nearby pumping and can frustrate the methods 1–5. Methods 3–5 can be easily adjusted. Method 6 can be a better long-term solution, but can still allow leakage to the surface under artesian heads greater than the grout pressure during the grout emplacement.

A design of a method has been developed which does allow the liner to be pressurized in the casing above any threatening leakage from a shallow artesian aquifer. That method does require a grout fill of the liner. When that method has been tested, information will be available from FLUTe. A possible Method 7 could be called "the entrapped fluid layer" which is enclosed between two cast grout layers in the casing as a seal of the casing against leakage to the surface. The fluid layer can be pressurized by several means to dilate the liner more strongly against the casing. It does not enhance the grout seal of the borehole below the casing. That seal would need to be enhanced below the casing by increasing the pressure on the grout column during its cure or use of an expanding grout. This method has been conceived to solve the problem of relentless, but rare, seepage of shallow artesian water out of the top of boreholes sealed with grout-filled liners.

More details are provided on the devices used with Methods 4 and 5 in Section 13.8.

3.4.4 LINER SEAL COMPARISON WITH PACKERS

Packers have been the traditional means of temporarily sealing boreholes. Figure 3.11 shows the head distribution contours in the formation near a typical packer emplacement with a vertical downward gradient in the borehole and vertical conductivity in the formation. It is easy to see that, unless the packer is located in an aquitard, some bypass of the packer will occur to the open borehole below the packer. The formation vertical conductivity and the head difference across the packer control the amount of bypass in the formation. The borehole roughness controls the seal quality of the packer against the borehole wall. For a rough wall, that packer seal is often poor. Since the packers are emplaced in open boreholes filled with contaminated water, any bypass of the packer will violate a sample drawn from the interval between the packers. This is especially important when packer testing is done for the deepest contamination in the borehole. A comparison of packer contaminant sampling results with Water FLUTe flexible liner results in the borehole OSW-2 at the Chemsol superfund site showed 19 µg/l contamination in the deepest port of

FIGURE 3.11   Drawing on the right of the excess head contours near a packer with a higher head above the packer. Bypass flow (dashed lines) can occur both in the formation and at the packer contact with the borehole wall. In contrast, on the left, there is no bypass to an open hole with an everting liner.

the Water FLUTe as compared to a far higher level of 430 μg/l of contamination in the packer water sample drawn from the same interval. The trace of contaminants in the Water FLUTe measurement could be due to the open hole cross connection. The Water FLUTe results in the 200 ft. interval varied by 50-fold. All the packer measurements varied by only 16 percent in the same 200 ft. interval with seven measurement intervals. One might suspect that uniformity of the packer results is more like the variation in the open borehole water which may be that uniform with depth.

The liner as a thin flexible coated fabric conforms well to the borehole rugosity. The long liner does not have an open borehole above the liner, so the bypass to an open hole is not possible (Figure 3.11). In general, a sufficient seal of the borehole is achieved if the vertical flow at the contact of the liner with the borehole wall is less than the flow that would occur through the native material before the borehole was drilled. A perfect seal is not likely. In a conductive formation, the borehole sealed with a liner will have less vertical flow than would occur through the material removed by the drilling process. The same would be true of a packer, except the open hole above and below the packer offers no resistance to vertical flow. Furthermore, the gradient across a packer is usually much larger than the gradient in the formation. Because the borehole is completely sealed by a liner, the vertical flow in the formation is essentially the natural flow without the influence of the borehole.

If the liner leaks and loses the excess head for sealing, then it may not seal the borehole. It is important not to damage the coating of the liner and to maintain the excess head needed for a good seal. A video is available from FLUTe which describes the liner installation procedure in detail. The liner unlike a packer is not at risk of entrapment by slough of the hole wall as supported by liner experience.

3.5   LINER INSTALLATION DEVICES

Sections 3.1–3.4 describe the basic liner mechanism. However, installations of liners are performed with a wide variety of devices. The installation using the natural head of a water fill of liners as shown in Figure 3.3 is the most simple, but it was not the first method used. The first installations were in the vadose zone and used air pressure for everting the liner into boreholes. These were done with air pressure canisters.

3.5.1   AIR PRESSURE CANISTERS

The first installations were for the purpose of wicking pore fluids from the vadose zone. Some air-driven liners carried tubing to extract gas samples or to draw pore gas samples through silica gel to capture tritiated water vapor from beneath radwaste storage pits.

As shown in Figure 3.12, the air pressure canister deploys the liner from a reel inside the canister. The liner is loaded on the reel by rolling the tether, followed by the liner, onto the reel. The open end of the liner is then clamped to the exit pipe of the canister. A compressor or blower is then used to increase the air pressure inside the canister. The higher pressure inside the canister compared to the atmospheric pressure develops the ΔP needed to evert the liner. The liner can then propagate horizontally across a surface, or into a borehole or pipe in any direction from vertically

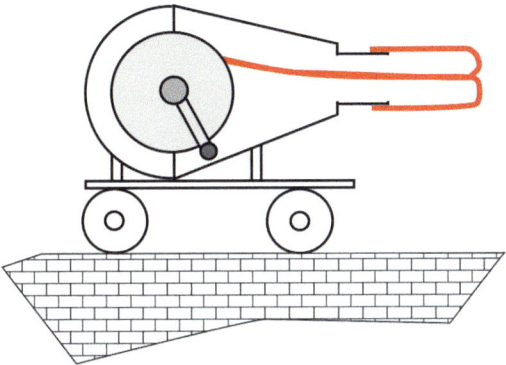

FIGURE 3.12 Air canister on cart. Air pressure supplied to the canister drives the red liner eversion pulling the liner from the interior reel.

downward, horizontally, or to vertically upward. The air-driven canister installation has been used for a wide variety of installations such as installing cure-in-place liners in piping in the walls of the Smithsonian Museum of Natural History, or for vadose zone measurements. An air-driven liner is relatively lightweight. A simulation was demonstrated of a borehole from the back of a tunnel as shown in Figures 3.13 and 3.14. A pipe was secured to a ladder to simulate the borehole upward from a tunnel.

FIGURE 3.13 2″ liner everting from air pressure canister. Liner is guided to enter the pipe simulation of a borehole through the back of a tunnel. Note how rigid is the liner under pressure.

FIGURE 3.14   Liner has propagated through a "borehole" and beyond as in an enlargement.

As the liner everted, relatively rigid, it was guided from the surface into the elevated pipe/borehole. Once the liner enters the pipe, it continues to evert and as shown in Figure 3.14 with the liner extending beyond the pipe as through a breakout in the borehole. This same mechanism was used to demonstrate the towing of a sonde vertically upward from a deep nickel mine in Sudbury, Ontario. The normal tether attachment to the liner was used to tow the sonde through the interior of the liner as the liner everted.

The everting liner was also used to tow logging tools in horizontal passages beneath landfills as shown in Figure 3.15 and described by Keller, 1996, as a method for monitoring for leakage. The liner was everted into the open hole for half of its length and a sliding valve was closed on the tether. The sonde was attached to the tether through an air lock. The sliding valve of the airlock was then opened and the liner continued to evert while towing the sonde. The electronic cable to the neutron moisture sonde was fed through a sealing gland at the near end of the air lock as the liner everted towing the sonde at the closed end of the liner, as the liner everted toward the propagating EP.

The installation under a landfill in Indiana is shown in Figure 3.16. In some cases, such as under landfills, the liner was driven with water in the liner which is pressurized by air in the canister. In Figure 3.16, the liner is driven with water added in the pipe beyond the canister as the liner is deployed from the large canister through a roller elbow to align the eversion with a horizontal drill hole. The canister was

FIGURE 3.15 Towing of neutron moisture log beneath experimental radioactive waste landfill for the University of Texas. A 6″ liner deployed from 24″ canister driven horizontally with air. Neutron sonde cable fed through gland on the opposite side of trailer.

pressurized with air to maintain the water pressure sufficient to evert the liner. As the liner driven by water was everting, more water was added above the roller elbow through a valve. The procedure was part of the method called a "liner augmentation of horizontal drilling" (LAHD) described in Chapter 15.

A major advantage of air-driven installations is that liners can be driven vertically upward as from tunnels or from the basement of the Smithsonian Museum.

FIGURE 3.16 A 48″ air canister liner installation beneath the landfill following horizontal drilling reamer. Installed eight sampling ports. Water-filled liner driven with air.

The air-driven reel canister is used nearly daily to evert liners as part of the liner fabrication procedure of liner systems.

3.5.2 Hose Canisters

Another air-driven method of liner eversion uses a long canister called hose canister. The hose canister geometry is shown in Figure 3.17. The advantage of the hose canister is that it has a much higher pressure capacity than the typical air canister of Figure 3.12. Another advantage is that the liner being everted can contain a rigid pipe or tubing attachments that cannot be rolled on the small reels of Figures 3.14 and 3.15. This has been done for carrying casing into horizontal drill holes. The tether is fed through a sealing gland at the closed end of the hose. The tether or electronic cable, if used, can be deployed from a reel outside of the hose. A disadvantage of the hose is that the hose must be the full length of the liner and requires much more space for deployment near the borehole. If a rigid pipe is to be towed by the liner, the hose canister must be the full length of the pipe and the liner.

In one of the installations in New Mexico, the hose was extended through a forest near the borehole for a relatively deep installation of absorbers for wicking tritiated water from the vadose zone (described in Chapter 11). The liner can be deployed from the hose using air or water. A disadvantage of the hose canister installation is the drag of the long liner on the long hose as a relatively higher driving pressure is needed for hose canister installations. However, the hose pressure capacity is well above that of the liner or the air canister. The hose can be shipped on a reel with separate end pieces and a short clear pipe at the outlet end to view the advance of the liner markings.

For both kinds of canisters, a roller elbow can be used to redirect the liner after it exits the canister. The roller reduces the additional drag of the liner in the turn traversed to a new direction. The roller housing must be able to withstand the liner driving pressure. A calculational model was developed to determine the driving pressure of the liner in its travel through crooked piping. The model includes the effects of the direction of the eversion (gravity effects on the liner and on drag), the number and angle of elbows in the passage, and the weight and diameter of the liner. That model is described in Section 16.1.

A significant advantage of a hose canister is to allow a higher driving pressure in the liner when the overhead space is too limited for a gravity-driven installation or

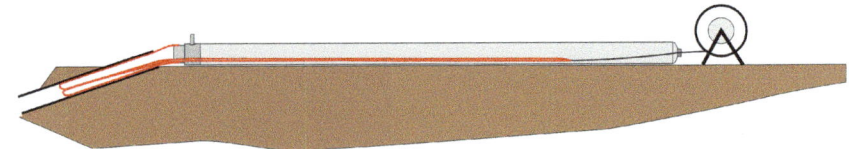

FIGURE 3.17   The hose canister can be rolled on a reel for shipment and extended on the ground. The liner can then be pulled into the hose, clamped to the open end of the hose, and then the hose can be inflated with air or water. The tether can be contained on a reel with a braking system or crank to control the extension.

where the scaffolding for higher pressure installations cannot be erected (e.g., installations in the galleries of a dam). The scaffolding erection and its use to extend the water column in the liner is often considered a safety hazard and therefore requires certain precautions.

3.5.3 Gravity-Driven Installations

The installation shown in Figure 3.3 uses the head of a column of water to provide the ΔP needed to evert the liner. By increasing the excess head above the formation's highest head, the value of ΔP can be increased. The heavy mud installation described in Section 3.5.7 uses the higher density of a partial fill of the liner with mud to increase the driving pressure in the liner. There are other advantages of a short mud fill of the liner, such as increasing the liner tendency to propagate vertically downward. The heavy mud acts like a plumb bob directing the liner more nearly vertically. The mud has less advantage as a plumb bob in an angled hole. However, the mud may mix with the water fill of a liner to provide a distributed increase in the liner differential pressure along its length. A disadvantage of a mud fill is that the temporary water removal tube used, as described in Section 3.2.4.7, cannot be easily removed by pumping down the water level in the liner to collapse the liner. The mud-filled section of the liner is usually at a higher head than the exterior formation head and the liner may not collapse without the removal of the mud in order to remove the water removal tube. Note, one cannot pump a liner head below the formation head because the formation head will fill the hole and collapse the liner to raise the liner water level back up to the blended head.

The partial mud fill of the liner to enhance the driving pressure is very convenient for shallow water tables where the excess head desired may not be available without an extension of the surface casing and associated scaffolding.

The ability to achieve a sufficiently high driving pressure with shallow water tables is an advantage of the air canister or hose canister. However, the additional apparatus is not needed with the gravity-driven pressurization. One limitation of the gravity-driven system is the scaffolding height that may be needed to achieve a sufficient driving pressure as for artesian conditions. A method to overcome that limitation is described in Section 10.3.9 (a bulbous wellhead) as it was used for artesian head installations.

3.5.4 Magic Gland

The "magic gland" design was developed for installations of liners too long for a hose canister or reel canister, yet needed to be driven with a substantial pressure. This was designed for the LAHD use under rivers and large buildings. The geometry of the magic gland as shown in Figure 3.18 allows the liner to be fed between two roller belts which provide a significant reduction of leakage of the liner driving fluid from the housing interior past the liner at the entrance. The liner is fed into the pressurized volume inside the housing and out through a suitable pipe exit of the housing. A roller can be incorporated in the pressure housing to redirect the liner and to monitor the travel of the liner. The liner is clamped to the housing exit much like the outlet of the hose or air canisters. When pressure is applied to the interior

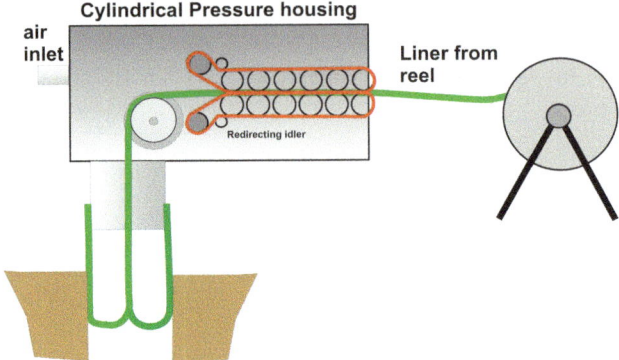

FIGURE 3.18 "The magic gland," a feed-through device for liner travel from a reel into a pressurized volume with minimum pressure loss in the everting liner. Designed for very long bulky liners too large for the air canister designs.

of the housing, the liner everts beyond the pipe exit as desired and pulls the liner from the exterior reel through the sealing belts at the entrance. The belts above and below the liner are pressed by the housing interior pressure against the liner to minimize leakage past the liner. The rollers at the edge of the belts keep them in close contact. The advantage is that the reel containing the liner is exterior to the housing of the magic gland and the liner can have an essentially unlimited length and bulk. This design has only been tested in a prototype and not refined yet as preferred, but it works well as conceived. This design is also patented.

3.5.5   THE DROP-IN-PLACE LINER INSTALLATION

In a more recent evolution of liner technology, some newly designed multi-level liner sampling systems have been emplaced in cased holes (e.g., cased holes with multiple screens) by simply lowering the liner with a weight at the bottom end of the liner into the casing. Whereas abrasion of the liner is not a concern associated with the travel of the inverted liner through the everted liner, the drop-in-place method in a PVC casing is not likely to abrade the liner to cause it to leak. It has also been found that a drop-in-place liner can be installed in short (less the 150 ft.) open uncased vertical holes. Clearly the drop-in-place liner would be more of a concern for abrasion in deeper uncased boreholes. In deeper uncased holes, a protective sheath is used on the outside of the collapsed liner to be lowered into the borehole. When in place, a water balloon made of the same liner material is inflated at the bottom end of the liner to anchor it firmly in the borehole. A short mud fill can also be used to anchor the liner against the hole wall instead of the water balloon. Then the sheath is lifted from off the liner, leaving the liner in place. The liner is then inflated with water through a central tube built into the liner. Mud or other fluids can be injected to seal the borehole as for the everted liner installation. Removal of the liner requires pumping the water fill from the liner and lifting the liner from the borehole as for the pump-and-drag procedure described in Section 3.3.2. A mud fill of the liner must be diluted and pumped from the liner before lifting the liner is possible. This installation technique has also been patented for installations of a variety of FLUTe measurements. The multi-level water sampling systems using the drop-in-place installations are now in regular use at many sites. They are described in detail as CHS systems in Sections 10.5.3 and 10.5.4.

3.5.6 THE BULBOUS WELLHEAD FOR ARTESIAN INSTALLATIONS

Both the air and the hose canisters can be used to overcome the artesian head while installing an everting liner. However, neither is suitable for the transmissivity profiler measurement method described in Section 10.3 nor for the Water FLUTe installation described in Section 10.5.1. Both installations require access to the liner interior during the installation. A special design was developed for performing transmissivity profiling installations in boreholes with high artesian heads and high flow rates out of the top of the casing. It was named the "bulbous wellhead design." It is described in more detail under the T profile description Section 10.3.9. However, it has the following features:

1. Overcomes the artesian head during the liner installation allowing liner eversion (e.g., with 30 ft. of artesian head in New Brunswick, Canada).
2. Allows temporary flow from the top of the well at very high flow rates (e.g., 150 gal./min at one site).
3. It is essentially a gravity-driven installation with unique features.

This design was developed for use at a potash mine site in New Brunswick, Canada, for artesian conditions produced by heavy brine infiltration from surface ponds. The same site is remembered for successful liner installations off scaffolding in a howling blizzard several days before Christmas.

3.5.7 MUD-FILLED LINERS

3.5.7.1 Purpose of Mud Fill

Often it is difficult to achieve a ΔP that is much greater than $\Delta Pmin$ and sufficient to overcome the drag and hence the liner will not evert. The situation is that the ΔH to advance the liner is more than a shallow water table depth. If the borehole is artesian and water is overflowing the surface casing, that is an extreme example.

In these situations, a means of increasing the pressure in the liner is required. Since heavy water (deuterium) or liquid mercury are not available for use in everting liners, it is easier to employ the drillers' common heavy mud as the driving fluid inside the liner.

Heavy mud is usually a bentonite mixture with barium sulphate or barite. Barite is powdered and used for mixing with bentonite. It is also called baroid as a commercial name for Halliburton's various additives of mud mixtures. Mud is the common term for bentonite with various amounts of barite in water because it is a mud. Barite has a density of 4.5 g/cc in crystalline form as compared to water at 1 g/cc. So, a barite and bentonite mixture is called a heavy mud. Simply pouring barite into water, it sinks to the bottom at a rate depending on the particle size. Bentonite is a thixotropic material which means that it forms a gel if stationary and not being sheared. Upon mixing or shearing, it turns into a viscous liquid. The gel strength and viscosity depend on the amount of bentonite in the water. In drilling, the mixture of bentonite is used for the purpose of suspending the cuttings if the drilling is stopped, in which case the gel may support the cuttings in the mud. If the cuttings settle too much, they can lock in the drill string and prevent further drilling. The thicker the mud, that is the more bentonite in the water, the better the ability to support the cuttings. Likewise, the more bentonite, the better it is able to support the barite additive. The heavy mud provides a stabilization of the hole wall due to the higher pressure with depth in the mud and the mud also coats the hole wall to further prevent slough of unstable formations with heavy mud in the hole. Another characteristic of bentonite mud is that the more bentonite added, the more viscous the mixture. If more than about 0.7 lb/gal. of bentonite is added to the water, the viscosity is so high that the mixture is difficult to pump through less than 1″ ID tubing with typical pumps. Bentonite comes in various quality ratings. The more pure the bentonite, the more the gel strength, and also the higher the cost. Drillers use a mixture called high solids bentonite grout that can be used to seal the annulus between the borehole and the casing. That mixture is too thick for use in liners. For high solids bentonite, drillers do not need high-quality bentonite. FLUTe uses high-quality bentonite because the gel strength is more predictable. It is also important to have the right pH in the water when mixing bentonite. Mud chemistry is important to well drilling, but not covered here. A pH of >7 for the water is preferred. The "hardness" of the water is also important to the mud behavior. Hard water has a high mineral content and is bad for gel strength. FLUTe uses ion exchange filters to reduce the "hardness" of the water used for mud mixtures.

The overpressure produced by the mud is adjusted by the amount of barite added. The mud density is important to the stabilization of the borehole or in use in liners. The pressure of the mud at the bottom of a column of mud is $P = (\rho - 1)gH$, where ρ is the density of the mud and 1 is the density of the water in the formation. The term g is the acceleration due to gravity. As is addressed hereafter, the overpressure must be sufficient, but not too high. The term ρgH is also the basis for hydraulic head, H, of a column of water. The recipe used for mud in liners is shown in Figure 3.19 as a function of the density desired and the height of the mud column. The table values are the overpressure (in psi) at the bottom of a column of mud. The overpressure is the pressure above a similar column of water as in the formation. The graph in Figure 3.20 shows the pressure of a typical column of water below the water table labeled saturated formation water pressure. A second curve (liner filled with water to ΔH) is the pressure with depth of the water column in a liner with the water level in the liner a distance ΔH above the water table. The difference between the liner pressure with ΔH and the formation water pressure below the water table is the pressure that seals the borehole (the blue shaded area). For comparison is shown the pressure in a column of mud (the coarse dashed line) with the mud fill of the liner to a height ΔH above the water table. The actual mud density determines the slope of the mud curve. A fourth curve plotted is the pressure with depth of an artesian condition, with a water table. Ha, above the surface. It is interesting to note that the water

Mud mixture for Liner fill

Density gm/cc	Bentonite #/gal	Baroid #/gal	20	40	60	80	100	120	140	160	180	200	250	300	350 ft of mud
			pressure with depth of mud (PSI) if filled to the water table.												
1	0	0.00	0	0	0	0	0	0	0	0	0	0	0	0	0
1.05	0.3	0.32	0	1	1	2	2	3	3	3	4	4	5	6	8
1.1	0.6	0.66	1	2	3	3	4	5	6	7	8	9	11	13	15
1.15	0.6	1.24	1	3	4	5	6	8	9	10	12	13	16	19	23
1.2	0.6	1.85	2	3	5	7	9	10	12	14	16	17	22	26	30
1.3	0.6	3.11	3	5	8	10	13	16	18	21	23	26	32	39	45
1.5	0.6	5.89	4	9	13	17	22	26	30	35	39	43	54	65	76
1.8	0.6	10.83	7	14	21	28	35	42	48	55	62		87	104	121

Add half a teaspoon of soda ash per gallon. (three teaspoon: Blowouts have occurred at 25 psi for 400d DC
(16 table spoons per cup, or 48 teaspoons per cup)   However, that is ~1/2 normal burst pressure
(i.e., one cup of soda ash is good for ~96 gallons) relatively comfortable level of high pressure
7-10 ft. preferred , 5 ft minimum (~2 psi)
Too low for sealing need excess head above water table

FIGURE 3.19 The overpressure of mixtures as a function of density and height of the column. The second and third columns indicate the amount of bentonite and baroid (barite) added per gal. for the indicated density in column 1. The pink-shaded cells are overpressures in ft. of water considered a risk to a 400-denier nylon liner.

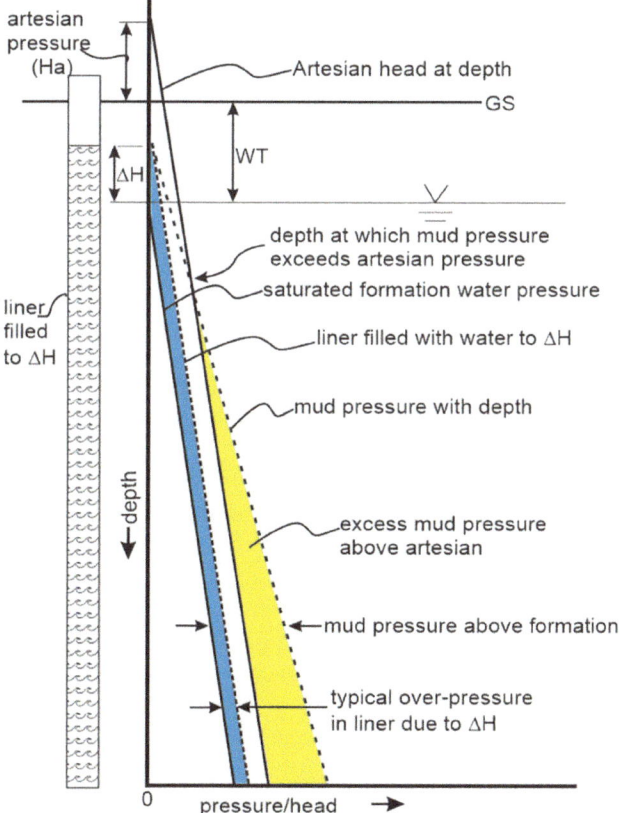

FIGURE 3.20 Comparison of normal hydraulic gradient to that in a liner with excess head, that with a mud fill and compared to the artesian gradient. A mud fill can exceed the artesian head but the mud pressure can be dangerously high.

overpressure in the liner (i.e., the difference between the liner pressure and formation pressure) is less than the artesian pressure everywhere, so a water fill would not seal the liner against the hole wall. However, below a critical depth the mud pressure exceeds the artesian pressure, shown as the yellow-shaded area. The mud is sometimes used to overcome the artesian pressure. However, it is important that the mud overpressure at the top of the mud column is zero because $\Delta P = (\rho - 1)gH$ and the value H is zero at the top of the mud column at the distance ΔH above the water table. The mud column would seal the liner for its entire depth below the water table. However, at the full hole depth, TD, the $\Delta P = (\rho - 1)g$ TD may be so high as to burst the liner where it is not supported, as in a breakout of the hole wall.

The blue-shaded area is the overpressure dilating the liner in a lake where the normal pressure curve for water applies. But the saturated pressure curve may be different for a real geologic environment and the formation pressure may actually be on the artesian curve so that the normal water fill is not sufficient for a seal. That is one situation where the mud fill can be used.

The other situation where a mud fill is used is when the overpressure ΔH is not sufficient to evert the liner as first mentioned above. In that case a short column of mud in the bottom of the liner will provide an elevated pressure at the EP sufficient to cause the liner to evert. This method is often used when performing a transmissivity profile as described in Section 10.3 where a higher pressure at the EP will drive the liner faster completing the T profile in less time or thereby reducing the significance of variations of water tables in permeable intervals of the formation.

A precaution is that the barite is not supported by the bentonite gel strength in lined boreholes indefinitely, apparently due to the chemical effect of the liner. That makes the grout fill more reliable over the long term. A grout fill acts like a mud pressure, but the grout can be installed in lifts which then prevents the high head at the bottom of the hole, since the cured grout supports the uncured grout head above.

Section 3.4.2 above describes a method using the flexible liner to determine the highest head in the formation, which is very useful to know, although without any indication of where that highest head is occurring in the formation sealed by the liner. Unfortunately, that information is not available when the mud is added.

3.5.7.2   An Example of the Mud Pressure Calculation

If the water table depth is 3 ft. below ground surface (BGS), filling the liner to the surface is not sufficient to evert the liner or provide a good seal. The borehole is 240 depth, TD = 240 ft. The mud density of 1.2 provides a lateral overpressure of 48 ft. of water with an overpressure of ~21 psi at the bottom of the borehole. The liner burst pressure is about 60 psi or 138 ft. of head. The 21 psi should be safe for this mud density. A safe excess head is about one-third of the burst pressure in many engineering applications.

If the objective is to speed the liner descent and the water table is sufficiently deep (e.g., 20 ft.) to seal the entire liner in the borehole, a mud slug 15 ft. tall and density of 1.8 g/cc may be added to the liner with a mud overpressure of $15 \times (1.8-1) = 12$ ft. of water head in addition to ΔH. If the liner is filled to the surface, $\Delta H = 20'$ and the net head on the EP due to water and the mud is 32 ft. At the top of the mud slug, the mud overpressure is zero and the excess head on the liner is 20 ft. When the mud slug reaches the bottom of the hole, even if diluted to be distributed over 30 ft. or more, the net excess head on the bottom end of the liner is the same 32 ft. It makes no difference how the mud might be distributed in the liner, as the pressure at the bottom end of the liner is the same. Unfortunately, if the mud is diluted, the bentonite may not gel and support the barite, which thnan can settle to the bottom of the liner. In that state, the powdered barite may act like a sand fill and, due to a high shear strength, may not allow the liner inversion for removal. It was noted that the use of barite alone to increase the head in the liner does settle and impede the initial liner inversion.

An interesting observation when using a mud slug in order to speed the T profile and to overcome variations in the formation head was that the liner was descending, but the water level in the liner was not dropping. This is related to two factors. The head in the borehole beneath the mud slug is greater than the head in the liner above the mud slug. The other factor was that the cable to the transducer in the borehole provided a small, but not zero flow path past the mud slug. The effect was that flow past the cable to the lower head in the water-filled liner was collapsing the liner above the mud slug. The collapsing liner caused the water level to rise inside the liner. This was only noted with a very tight borehole with almost no eversion of the liner except as that needed to displace the borehole water past the cable. In other words, there was no flow out of the borehole because it was nearly impermeable. Therefore, the liner advance was not related to flow out of the borehole and the transmissivity profile was not achieving the basic flow into the formation to perform a transmissivity profile. The attempt was abandoned. From that experience, it was clear that the smallest possible communication cable diameter should be used or that the transducer should be suspended on an extremely slender steel cable near the bottom of the hole after the recording of the transducer had been initiated. Fortunately, the flow past the liner at the cable is negligible for most borehole measurements. The use of a long mud slug might be considered helpful, but the longer flow path past the mud slug is only offset by the increased head difference due to the mud.

Another caution in the use of mud is that when a water removal tube is lowered into the borehole to remove the water beneath an everting liner, as for a WF installation, the mud can make it difficult to remove the water removal tube. That is because the usual procedure of deflating the liner to remove the tube does not reduce the extra lateral pressure of the mud-filled liner interval against the hole wall and against the water removal tube. The mud fill can entrap the water removal tube, preventing its removal when the water is pumped out of the liner to otherwise reduce the friction on the water removal tube. A precaution is to place the water removal tube at a final depth that is above the mud slug final depth. However, if the mud has been diluted, it is not certain where the diluted mud is pressurizing the liner above the formation pressure. The same problem can be encountered if there is a very low head at the bottom of the borehole but a higher head above. In that case, the liner can be nearly emptied to release the water removal tube, but that is not a practical solution for the mud filled liner.

3.5.7.3 In Summary, How the Heavy Mud Is Used

These conditions lead to a partial, or full, fill of mud in the liner:

1. Shallow water table where filling to the surface does not provide a sufficient overpressure.
2. A stiff liner material that requires a higher pressure only at the EP.

3. Artesian conditions at known or unknown depths.
4. When an occasional rise in the water table such as when nearby pumping is halted and the new water level excess head is not sufficient to seal the borehole.
5. Adding to the driving pressure of a T profile for shallow water tables.

It is useful to address how each of these conditions is met with a mud addition and what other methods can be employed to achieve the same or better advantage as mud to address the situation.

3.5.7.3.1   Limitation of Heavy Mud

An important limitation of a heavy mud is that in time, in a liner, but not in a PVC pipe for some reason, the weighting material will settle. It is not known to what extent and for which mixture yet, but the settlement observed makes the use of mud for long-term pressurization less attractive. Early tests in tall PVC pipes showed that the baroid did not settle over several years, but later observations show that in time the baroid does settle in flexible liner lined pipes. The chemistry of the water is important to the mud gel and apparently the liner affects the chemistry of the mud.

3.5.7.3.2   A Limitation of a Grout Fill

It has been observed that a grout fill of the liner will often raise the pH of water outside of the liner and can affect the pH of water samples. In time that effect decreases and in some situations a grout fill has had little effect on the pH of water samples. It is not understood why the variable experience. An extensive purge as recommended prior to sample collection helps to reduce that effect.

4 Chemistry of the Liners

Over time, various questions arose about the chemical characteristics of the liner materials and the effect on water samples. The several questions were as follows:

1. Arsenic content of the liners?
2. Toluene in the liner material?
3. 1,4-dioxane in the liner material?
4. PFAS compounds in the liner materials?
5. NDMA in liner materials?

4.1 ARSENIC

Arsenic was detected in water samples and the question was to what extent the liner may have provided some of that arsenic. The early liner materials of urethane-coated nylon were of the standard military specification, which required a fungicide to prevent mildew growth on the coated fabric as used for many applications from life vests to awnings and pool covers. Leach tests showed that low levels of the arsenic in that fungicide could be leached from the liner material. As of 2014, all FLUTe liner stock is specially ordered to be free of arsenic. Each new lot of liner material is sent for a lab analysis to establish that it is free of arsenic. FLUTe liner materials no longer contain arsenic. Some liner materials are still available with the fungicide for special applications. They are of a different color (red and orange) to assure they are not misused.

4.2 TOLUENE

Toluene is used in the fabrication of the urethane-coated liner fabric. Toluene has also been seen in the water samples obtained from FLUTe multi-level sampler (MLS) systems. The same water samples from the field showed that the toluene concentrations were limited to levels well below the drinking water limit and also the toluene seen in the samples was much less or undetectable after a typical time of three months. The longest presence of low levels of toluene in samples was seen to be one year in very low-flow formations. Samples of liner stock used for FDA applications, manufactured without toluene, were examined and found to be mechanically inferior to the standard material manufactured with toluene. Because of the limited lifetime and low levels of toluene leaching from the fabric and the superior characteristics with the standard process, FLUTe elected to live with the presence of the limited toluene leached from the liner material. More recently, toluene has been recognized as a useful tracer in the liner material to identify actively flowing fractures in the daFACT design (Section 10.2.2). We noted that elevated levels of toluene in the FACT-activated carbon also correlated with elevated levels of contaminants

DOI: 10.1201/9781003268376-4

43

when contaminants were in the groundwater. We deduced that the leached toluene was being carried to the carbon by fracture flows since the carbon is isolated from the liner with a diffusion barrier and not in direct contact with the liner. The FACT measurement is performed when the borehole is first drilled and before the toluene would dissipate. Hence, the toluene can serve as a temporary useful tracer.

4.3   1,4-DIOXANE

1,4-Dioxane is a common contaminant in water samples and the question arose of whether the 1,4-Dioxane in water samples was coming from the liner components. It was also noted that 1,4-Dioxane was in the soapy water used to do the leak check of the liner after being welded into a tubular form. Samples of the soap used and of the liner materials were sent for analysis. Only the soap had measurable levels of 1,4-Dioxane. Using the levels of 1,4-Dioxane in the soap, and the quantity of soap used for leak checking, it was discovered that the MLS water samples in question contained more 1,4-Dioxane in each of the sequential water samples from the Water FLUTe system than was contained in the entire quantity of soap water used in the leak check of the entire liner. Therefore, the 1,4-Dioxane could not be coming from the liner leak checking. The leak check procedure was modified to use even less soap solution. Because 1,4-Dioxane is ubiquitous in the groundwater from many sources such as soaps, and none of the liner materials showed 1,4-Dioxane, it was concluded that sampled water is not affected by the liner materials. There was some suspicion at this site that 1,4-Dioxane came from the drill water source that was hydrant water reclaimed from wastewater treatment.

When the same question arose again in 2019, the materials that had not been tested, such as the talc used to lubricate tubing pulled into the interior sleeves of the liner, were tested and also found free of 1,4-Dioxane. Subsequent samples at the site in question were collected after extreme purging and also found to contain higher levels of 1,4-Dioxane at one site in Florida. This demonstrates that the large purge volume of FLUTe recommended prior to the sample collection can avoid the concern about groundwater samples being affected by the contents of the borehole and perhaps can also reduce the impact of poor drilling practice such as recirculating the drilling fluid while drilling through contaminated intervals.

Ironically, one project manager in a regulatory agency decided to alert all of their project managers regarding the presence of 1,4-Dioxane from FLUTe liners before the testing and calculation above were done. We still get questions about that memo six years later. No correction was ever provided.

4.4   POLYFLUORONATED ALKYL SUBSTANCES (PFAS)

The effect of PFAS derived from the sampling equipment on water samples is a recent concern. FLUTe liner materials were again lab tested with a weeklong leaching of the samples in 250 ml and in 1 liter of water. Most of the materials were seen to show non-detects (NDs). However, Tyvek does show several PFAS compounds. The PFAS compounds otherwise detected were few and very near the detection limits.

Tyvek is not included in FLUTe systems associated with water sampling. A subsequent test of several multilevel systems by drawing deionized (DI) water through the typical system for the prescribed purge volume prior to the sample collection showed undetectable levels of PFAS in the water samples. The earlier Water FLUTe systems that used Teflon tape in the fittings and a Teflon check ball did show a very low level of PFNA (0.301 ng in 250 ml, RL = 0.25 ng). The absence of PFAS from current FLUTe MLS systems was confirmed by using the very low leach rates measured from the polyvinylidene fluoride (PVDF) tubing, with the typical residence times in the tubing, to calculate the level of PFAS expected in the sample water. The calculation produced undetectable levels in the sample water. Again, it is useful to avoid such sample perturbations by purging the recommended volume prior to sampling. Sampling with other procedures, which do not, or cannot, purge the water adjacent to the borehole are more likely to be impacted. It is also noteworthy that PVDF tubing in the FLUTe MLSs did not produce PFAS in water samples tested. However, HDPE tubing is available in most of the FLUTe systems in place of the standard PVDF tubing, if desired. PVDF tubing has been tested to be far superior to the other common tubing materials (such as LDPE, HDPE, and Nylon; ref. Parker and Raney tests at CREEL SR96 03 and SR97 02). Trichloroethylene (TCE) and other common VOCs (volatile organic compounds), sampled through PVDF tubing, had much lower absorption and leaching than occurred for other more common tubing materials.

4.5 *N*-NITROSODIMETHYLAMINE (NDMA)

Only two sites have expressed an interest in NDMA. The NDMA concern is at such a low level, ng/l, that few people test water samples to such low levels. However, one organization does wish to measure NDMA in water samples to such low levels. According to the EPA (2017), *"NDMA is not currently produced in pure form or commercially used in the United States, except for research purposes. It was formerly used in production of liquid rocket fuel, antioxidants, additives for lubricants and softeners for copolymers (ATSDR 1989; HSDB 2013)"*. Because of that interest, the Water FLUTe components were tested by that interested organization for NDMA by pulverizing the samples and leaching the pulverized samples for two weeks. Some NDMAs were detected in some components. We are not free to report those levels at the customer's request. However, it is interesting that while some water samples from the Water FLUTe at the site showed NDMA, others from different intervals showed ND (none detect). The Water FLUTe was removed and, after two months, extensive straddle packer tests were done on each screen interval in the well. After extensive packer purging (over 300 gal. from each interval), the packer samples were seen to be converging to the levels seen in the Water FLUTe samples. The Water FLUTe system had been in place for several years, sealing the well. The Water FLUTe has been reinstalled. As with many such issues (e.g., FACT assessments described in Section 10.2.2), the important question is about actual significant effect, not just the possible effect. It was also concluded that purging is important to obtaining representative samples. Another conclusion from those tests was that cross-connection in open boreholes can prevent representative water samples.

5 Kinds of Blank Liners

5.1 DIFFERENT DIAMETERS

Borehole tubular liners are welded to many different diameters because it is difficult to evert overly large-diameter liners into smaller boreholes. It is also difficult to obtain a better seal desired with many folds in oversized liners. Liners with welded attachments to the exterior of oversize liners are exceptionally bulky and more difficult to evert. In order to obtain a good seal, the liner must be equal to the borehole diameter, but many boreholes have variations in diameter throughout the borehole due to raveling of soft layers or bit wear as the hole is drilled. For that reason, most liners are provided at a slightly larger diameter than the nominal hole diameter. Cased holes are sealed with liners of the same diameter as the casing. The liners dilate slightly under pressure.

Large-diameter liners unsupported by the borehole or casing have a lower burst pressure than smaller diameter liners of the same material. Most boreholes support the liner, which would allow very high differential pressure, ΔP. However, many boreholes have enlargements that do not support the liner. The liner must be able to withstand the pressure differential, ΔP, without the borehole support. Cased holes support the entire liner, and liners in cased holes can withstand very high differential pressures, which assures the optimum seal. For very large boreholes, it is prudent to use stronger fabrics such as the 840-denier liners.

Some boreholes are drilled with different diameter bits, usually changing to smaller diameter at greater depths. For liners everting through an oversized casing into a borehole of a much smaller diameter, the liner fitting the smaller borehole will propagate through the oversized casing independent of the casing diameter. At the bottom of the casing, the usual step change in diameter can impede the liner. Since most boreholes are not perfectly plumb, the drill string will leave the casing on the lower side of the casing. The everting liner is nearly weightless under water when driven with water and may not align well with the borehole entrance. The liner is self-centering for a small step change, but for large casing to borehole diameter differences, a small volume of heavy mud added to the bottom end of the liner can better align the liner with the entrance into the borehole.

The significance of the oversized casing is greater when inverting the liner. Without the casing support of the liner and the associated friction, the liner may buckle as described in Section 3.2. The solution available is to build the liner of two different diameters with a tapered transition from one size to the next. Such a taper is frequently done during the radio frequency (RF) welding procedure into a tubular form, but the taper adds an additional cost and the taper must be located precisely at the transition. Extreme diameter differences are best avoided. Common tapers are 6-4″ and 8-6″. FLUTe has built inversion aids that can be located in the liner once the liner is inverted into a larger diameter. These can help to prevent the buckling of a 4″ liner in a 6″ casing during removal, but these are usually only available for

DOI: 10.1201/9781003268376-5

blank liners without welded attachments on the exterior of the liner. Another remedy to avoid buckling is to pump the liner empty upon inversion into the large diameter casing and lift the liner out of the casing. In an uncased hole, this could result in abrasion of the liner and may prevent its reuse.

5.2 FABRICS

5.2.1 NYLON LINERS

Nylon liners are especially useful for everting liners because they are strong and elastic and take the urethane coating well with a good bond of the urethane to the fabric. Nylon fabrics are readily available and have a melting temperature that is well above the urethane temperature, which allows good RF and conduction welding without melting the fabric.

5.2.2 POLYESTER LINERS

Polyester liners were attractive because the polyester was more resistant to the attack of remediation fluids injected to neutralize some DNAPLs (dense none aqueous phase liquids) like TCE (trichloroethylene). However, they do not seem to be resistant forever. Polyester liners are useful if the greater elasticity of nylon is objectionable.

5.2.3 SILICON RUBBER LINERS

Silicon rubber liners have exceptionally high melting temperatures and are useful in high-temperature environments as used in Yucca Mountain near heater tests of nuclear waste storage effects on the formation. That application required special fabrication methods. A serious problem with silicone rubber is low tensile strength. Under pressure, silicone rubber liners undergo excessive dilation and can develop an aneurism of the liner at the EP (eversion point), which deflects the EP and stalls the eversion. Silicone rubber liners were only able to be everted into the horizontal holes of the Yucca Mt. site at relatively low pressures from air canisters. Silicone rubber also has a high friction coefficient that was moderated with the use of talc inside the liner. The fabrication of silicone rubber liners required the liner to be manufactured as a round extrusion of the proper diameter, since they cannot be welded as is done for the urethane-coated liner fabric. The addition of interior sleeves for tubing in silicone rubber liners was done with a high-temperature silicone rubber adhesive. The sleeves were made of tubular fiberglass. The tubing in the sleeves was of Teflon. End seals of the liners were of Teflon plugs whipped with copper wire. The tethers were of Teflon-coated wire or cable. Feedthroughs were brass fittings. Silicone rubber liners are rarely used but it was an interesting application and worked well with careful emplacement of the liners for both gas sampling and absorber applications.

5.2.4 TRANSPARENT LINERS AND GEOPHYSICAL LOGGING

Transparent liners are of use in boreholes with the advent of high-resolution borehole cameras. However, the suitably strong and transparent liner was not available until recently. The ability to see through a liner has some advantages at the same time that the liner is supporting the borehole wall and sealing the hole against the cross-connecting flow for contaminant migration.

Liners have always been "transparent" to several kinds of logging sondes such as gamma, neutron moisture, induction-coupled resistance measurements, and acoustic tele-viewers. Figure 5.1 shows the acoustic tele-viewer image of the borehole wall with and without the liner in place. The image also shows the tether inside of the

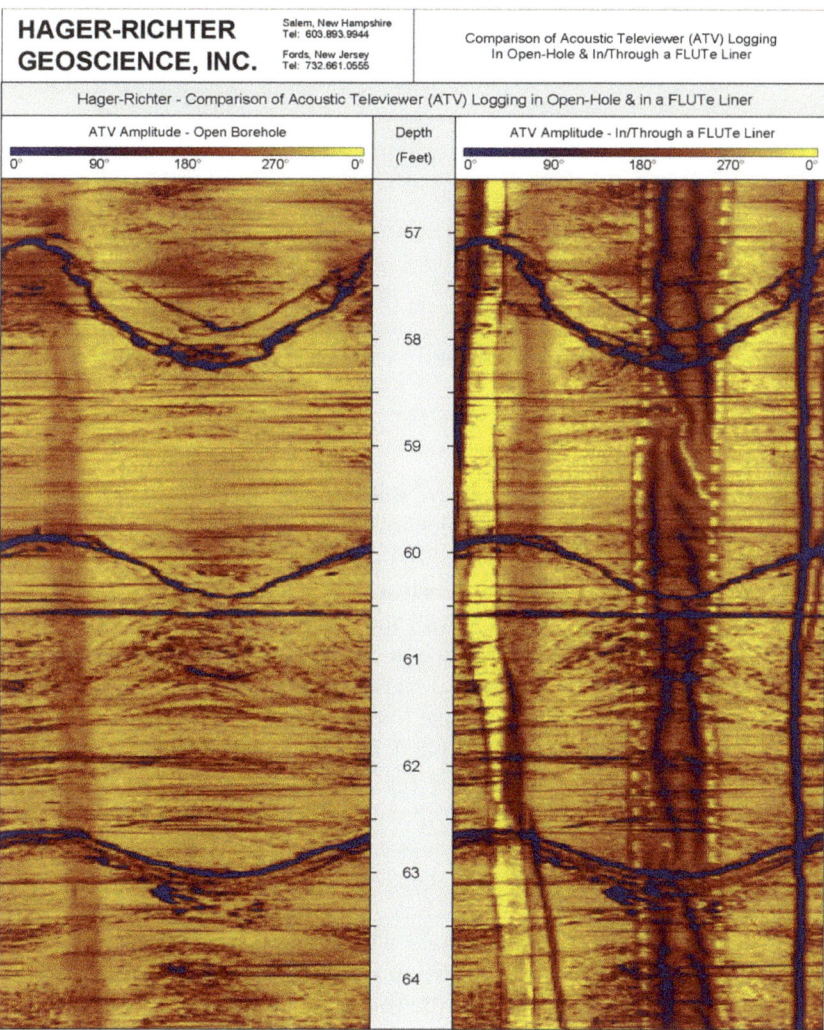

FIGURE 5.1 Acoustic tele-viewer image of the borehole wall on the left without the FLUTe liner in place. The right-hand image is with the liner in place. Fractures are clearly visible through the liner. The black line on the right is the tether in the liner and the central dark shadow is the sleeve welded to the inside surface of the liner. This liner is not optically transparent.

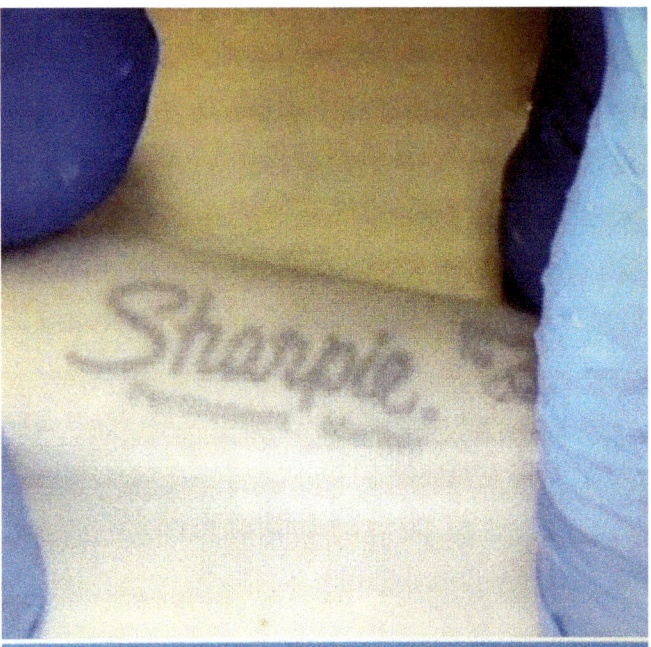

FIGURE 5.2   Print on Sharpie™ marker as seen through liner under water.

blank liner and the 2″ wide interior sleeve welded to the liner with a stitch-like weld of the sleeve to the liner. The double thickness of the sleeve produces some obscuration of the formation structure.

A strong transparent nylon liner is newly available from FLUTe, which allows the optical tele-viewer to see the borehole wall and detect color changes behind the liner. Figure 5.2 is a snapshot of a video view of the print on a marking pen as seen under water behind the nylon liner. Figure 5.3 shows the stain of a TCE contact with the NAPL FLUTe cover on the outside of the liner against the borehole wall. Normally, such stains are viewed by inverting the blank liner from the borehole. However, since the NAPL/FACT system is left in place for 2 weeks, the NAPL stains, which can develop in less than 1 hour, can be viewed with a borehole camera before the liner is removed for the analysis of the FACT carbon.

Another interesting application is the detection of the arrival of a potassium permanganate injection from a nearby hole. In fact, such an injection in a central hole surrounded by other boreholes lined with transparent liners would allow the mapping of connecting fractures in the surrounding boreholes. A camera run in the borehole at later times can detect the first arrivals and the development of the permanganate stains over time. Figure 5.4 shows the potassium permanganate stain as the permanganate flows from a small tube beneath the liner. The stain is immediately visible.

There is also the potential for detection of the arrival of tracers and fluorescent dyes at the borehole wall. The big advantage is that the borehole is sealed and the

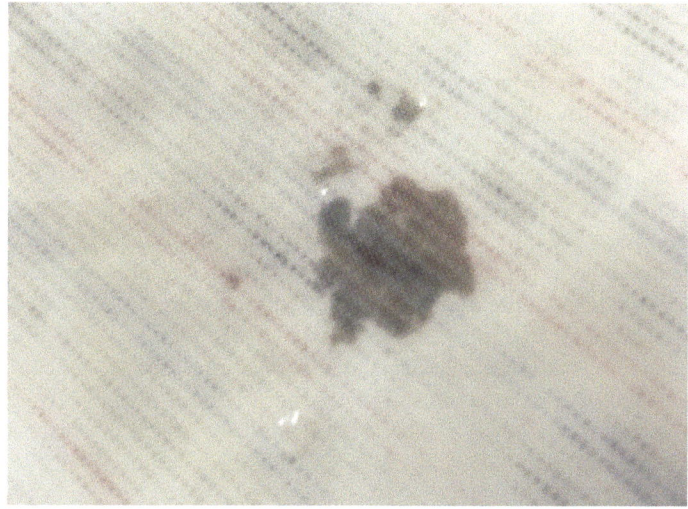

FIGURE 5.3   DNAPL stain on NAPL FLUTe cover as seen through transparent liner.

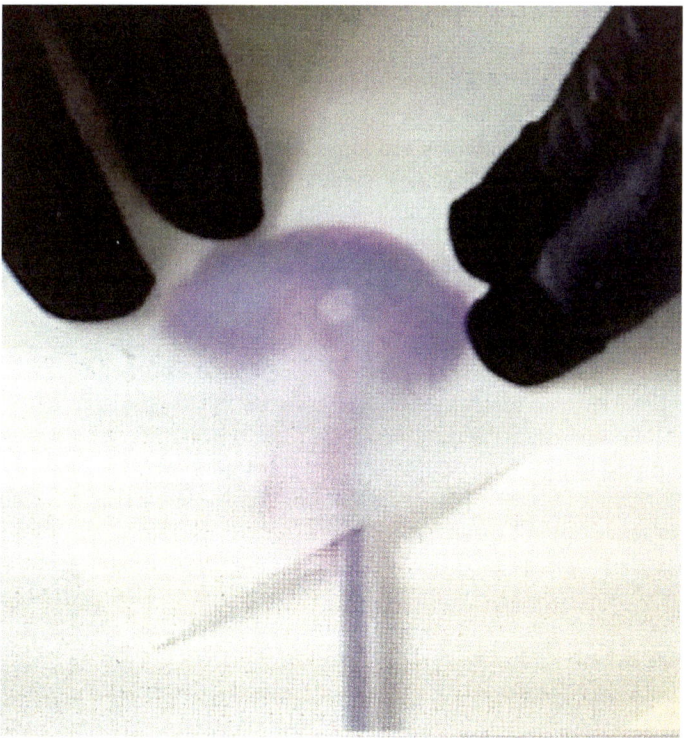

FIGURE 5.4   Potassium permanganate stain as seen flowing from a tube under the liner under water.

tracers are not mixing with the open borehole fluids. The ability to fill the liner with clean water also avoids the obscuration of the turbidity of the open borehole water. This is a new FLUTe development and has not been fully utilized with a camera yet. A limitation is that the liner is only relatively optically transparent to objects in contact, or near contact, with the liner under water. One cannot view the borehole wall in an enlargement greater than the liner diameter. Nowadays (2021), all FLUTe standard 400-denier single-coated blank liners are fabricated using the transparent material. The liners are only nearly transparent when submerged in water, which has a refraction index nearer that of the yarn as compared to that of air on the yarn in the liner. For this reason, the double-coated liner materials are not usually so nearly transparent because air is trapped in the fabric between the two layers of urethane. If the water outside of the borehole is very muddy, one may not see much detail of the borehole wall features in breakouts. Transparent liners do not use dyed fabric. An interesting potential use being explored is the ability to view fluorescent hydrocarbons at the hole wall with a special UV lamp and special camera.

5.2.5 DIFFERENT FABRIC WEIGHT LINERS

All of the liner materials are manufactured in different fabric weights identified by the yarn weight per unit length in terms of "denier". The most common FLUTe liners are manufactured of 400-denier nylon fabric with a urethane coating on one or both sides/surfaces. The urethane coating of the fabric is required for the impermeability to air or water necessary for everting liners. The liners that require welded attachments to the exterior of the liner are coated on both sides. The double coating also provides insurance against damage of the impermeable coating as can occur with the use of various geophysical sondes inside the liner or damage to the coating from abrasion. A disadvantage of the double-coated liner is that it is stiffer and therefore more resistant to eversion leading to a higher ΔPmin for the liner. It is also more expensive.

Less expensive PVC (polyvinylchloride)-coated fabrics often do not have the high-quality bond of the coating to the fabric necessary for everting liners.

Other liners are made of higher and lower denier. The denier (the fabric yarn weight) of the liner also leads to more or less stiff fabric and affects the ΔPmin of the liner. FLUTe uses fabrics ranging from 70 to 840 denier with tensile strengths ranging from ~50 to 400 lb/in., depending on the denier and the fabric construction. The strongest fabrics are of 840-denier ballistic nylon. The ballistic nylon also has a much higher tear strength as well as tensile strength. Both are important characteristics of everting liners as used in the field. The 840 ballistic nylon can withstand a higher ΔP and more abrasion than the lower denier liners. The denier is also related to the cost with the higher deniers being more expensive. On rare occasions, FLUTe has used 1000-denier fabrics that are more common in inflatable boats. However, most such boat applications use PVC-coated fabrics that are less expensive, but lack the bond strength of the coating to the fabric as compared to urethane-coated fabrics.

Because of the strength differences, higher deniers are used for larger diameter boreholes since the burst pressure is inversely related to the liner diameter. FLUTe uses only high-quality urethane-coated fabrics that are relatively free of pinholes in the urethane. On the rare occasion when pinhole leaks are discovered in leak testing, the pinholes are sealed. The highest quality coated fabrics are more expensive and are used in such applications as inflatable life vests.

5.2.6 TUBULAR PLASTIC FILM LINERS

One of the first materials tested for potential use as everting liners was made of plastic film, especially low-density polyethylene (LDPE) films. They are very inexpensive. However, upon eversion testing, it was seen that the plastic film suffered plastic yielding in the folds of the inverted liner as the liner everted. The plastic yielding leads to a row of perforations at each fold with extreme leakage of air from the tested liner. The row of perforations looks like a zipper and hence termed "zippering" for the failure mode. Weakly bonded coatings on fabrics also show the zippering failure of the coating as well as delamination.

Another disadvantage of the thin films available in tubular form (mainly used for packaging) is the low tensile strength of the very thin film even though the film material has a high tensile strength. The thicker films are also stiffer, zipper, and are not useful for everting liners.

5.3 CARRIER LINERS FOR COVERINGS

Carrier liners are usually blank liners with special features to allow attachment of special coverings as described later in Sections 10.2.1 and 10.2.2. In some cases, absorbent coverings are used on blank liners. If the covering is continuous throughout the length of the liner, it is usually attached to a simple device built into the end of the liner called a "pig tail", which is attached to the cover after the carrier liner is everted into the cover. The pig tail must be included with the end seal and tether attachment when the carrier liner is constructed. For short discrete coverings, the classic button-on-a-tab is welded to the liner at several surrounding points at a given location. The cover material is provided with buttonholes for attachment to the buttons. As the liner is everted, the short coverings can be "buttoned" to the liner or "unbuttoned" for removal. Such attachments are labor-intensive and relatively costly. If the liner is to be used many times with different short absorbers, the button attachments are useful.

FLUTe has often in presentations on the everting liner suggested methods to deploy special coverings of the customers' design. Those can be absorbers or reactive covers that react upon contact with the formation such as ore bodies. Or the coverings can be a kind of photographic film that is sensitive to radiation and can be developed upon recovery. This kind of radiation detection was done by Science Engineering Associates for the DOE (Department of Energy). However, that group has disbanded. Such activities required special radiation worker training and special installation devices.

In general, the offer still exists to build special coverings for carrier liners. FLUTe has developed two such coverings as the NAPL FLUTe and FACT described in Sections 10.2.1 and 10.2.2. A new covering much more absorbent of TCE has been tested, which can attract and hold 8–20 times the TCE as a normal NAPL FLUTe covering. This covering may be very useful for removing pure NAPL fluids from connecting fractures as compared to pumping 100,000 gal. of water to recover a few grams of dissolved DNAPL.

5.4 LAY FLAT HOSE LINERS

FLUTe has everted common lay flat hose into pipes to test that capability. It is much easier to evert hose material of larger diameter and not so stiff as some hose constructions. The advantage of a liner constructed of hose material is that it is readily available. However, it is also stiff, and offers a good seal only if the exact fit with the diameter of the borehole. Any wrinkle in such material is a major leak path unless the interior pressure is very high. The main advantage of hose material is that it has a very high burst pressure far above that of coated nylon fabrics. Most hose materials of PVC or rubber have a very high friction coefficient and provide exceptional drag while everting. Pressurizing an inverted hose can delaminate the impermeable lining. That also limits the applications. Lay flat hose is also used as a pressure canister for deploying normal liners as described in Section 3.5.2.

6 Novel Applications of Blank Liners

6.1 SURFACE EXTENSIONS

Liner installations are normally done into boreholes or pipes. However, liners will evert across a flat or inclined surface from an air reel canister or hose canister. The eversion can be guided with the device as shown in Figure 6.1. The narrow tow strap is assembled inside the inverted liner and attached to the same fitting as the open end of the liner (e.g., the pipe exit of the air canister in Figure 3.12). The other end of the strap is attached to the closed end of the liner at the tether attachment. A sheave can be attached to the tow strap and can be used to guide the direction of the liner eversion. It is also useful as an aid to the liner eversion because the inverted portion of the liner is compressed onto the strap inside the everted liner. Tension on the tow strap will then be transmitted to the inverted liner and aid in overcoming any drag on the inverted liner, such as when being deployed horizontally, or if the liner is bending around a turn in a pipe or past a post or tree on the surface, or if the sheave is attached to a powerboat on the water. An earlier version of this device is used for the liner augmentation of horizontal drilling (LAHD) described in Section 13.2. A flat leader (tow strap) of high friction coefficient is beneficial in transferring tension to the inverted liner without slippage.

6.2 EVERSIONS ON OR UNDER WATER

If the liner is everted over the top of a water surface such as a lake or a river when air driven, the liner will propagate over the surface. A long weight (e.g., a chain in an interior sleeve of the liner) will tend to keep the liner oriented on the water surface if that is desired. However, the interior sleeve must be parallel to the axis of the liner since the liner when inflated strongly resists twisting of the liner.

Another variation is to propagate the liner with water as the driving fluid into a lake. The same chain will cause the liner to propagate along the lake bottom and the chain will keep the liner oriented such that tubing and ports in the liner can draw lake bottom samples of mud or water. When filled with water and weighted, the liner is not buoyant. Various coverings as described in Section 10.2 can be used on the liner against the lake bottom (e.g., detection of dense nonaqueous phase liquid (DNAPL) on the bottom mud surface of a lake). The eversion of the liner causes it to be deflected rather than halted by the encounter of smaller objects as it propagates.

6.3 VERTICAL UPWARD UNSUPPORTED EXTENSIONS

The liner can be everted directly upward in the air for some distance before buckling. Larger diameter liners will propagate further as self-standing structures/towers. If guylines are attached to the exterior of the liner and inverted with the liner, those

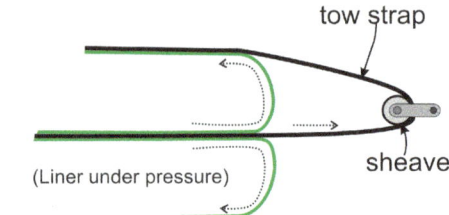

FIGURE 6.1   Tow strap clamped by the inverted liner. Tension on the sheave aids the liner eversion and guides the liner if not constrained in a passage. The tension also aids the liner eversion (Patented).

can be used to enhance the vertical liner stability by anchoring the guylines as the guylines and attachments are everted. Theoretically, such a tower could be erected quickly if needed from a small air canister with an interior reel. Section 3.5.1 illustrates the extension of a small diameter air-driven liner at an angle to enter a borehole in the back of a tunnel. The vertical extension is easier, especially for larger diameter liners less prone to buckling.

6.4   EVERSIONS THROUGH CROOKED PIPING SYSTEMS

Everting liners are often installed in piping with elbows, especially with sweep elbows which are easier to traverse than short radius 90-degree elbows. Those installations are described in Section 16.1 and were used to carry a multi-level sampling system into vertical wells at the end of crooked horizontal piping under buildings. This ability to propagate through crooked piping was an advantage applied to installing cure-in-place liners for leaking pipes. Figure 6.2 shows a resin-soaked diagonally woven fiberglass liner being everted through several elbows. The interior liner drives

FIGURE 6.2   Photo of a resin-soaked liner being everted through crooked piping. This kind of cure-in-place liner was used to line rusted cast iron piping in the walls of the Smithsonian Museum of Natural History in Washington, DC in order to prevent leakage of water into the walls of the building.

the resin-soaked fiberglass liner though elbows as was done in the walls for the Smithsonian Museum of Natural History in Washington, DC. The resin-soaked liner is then heated via the black carbon fibers woven with the fiberglass to initiate a cure of the resin and then the liner was inverted from the cured fiberglass pipe lining. The piping was badly rusted cast iron. The process avoided the need to remove and rebuild the masonry walls of the historic building to replace the old piping.

This emplacement is designed with a calculational model developed by FLUTe (see Section 16.1) to determine the driving pressure required while driving liners of known weight and friction through a tortuous passage with many turns upward and downward including gravity and drag. This model is also used for installing liners under pavement, through turns and into vertical boreholes as demonstrated in Chapter 16. The model calculates the minimum driving pressure to evert the liner through the crooked passage.

6.5   LINING BOREHOLES TO PREVENT GROUT LOSS OR GROUT SHRINKAGE OUTSIDE OF A CASING

A design was presented at a fracking conference in Ohio to use a blank liner to seal a borehole to prevent loss of grout during grouting of casing in boreholes, to prevent grout flowing between the casing and the borehole wall from entering solution channels. This is more of a concern with high-pressure reservoir gas as in oil drilling where the sealing of casing is critical. The loss of the sealing grout can be extreme and expensive in zones of lost circulation where there is no return of the drilling fluids due to loss into solution channels or karst caves intersected by the drill string. The design suggested is very simple as shown in Figures 6.3 and 6.4. The liner is everted into the borehole for half its length (Figure 6.3). The bottom end of the liner is not sealed and when it arrives at the wellhead during the eversion, the casing is

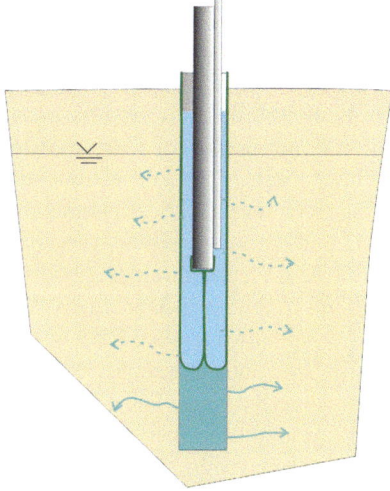

FIGURE 6.3   Installation of casing connected to the liner. Note grout tube inside liner.

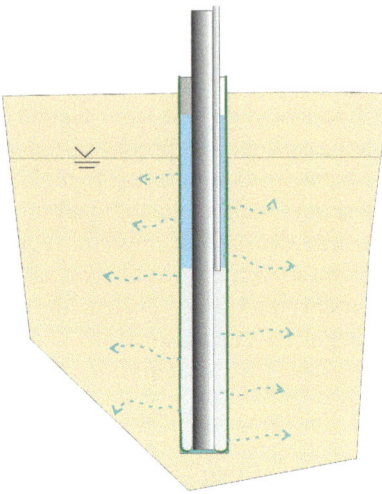

FIGURE 6.4   Grouting of liner outside of casing. Grout fill must be done in lifts to not burst liner.

attached to the open end of the liner. This is easiest if the casing is suspended in a drill rig. The casing sections can be added as the liner everts down the borehole. The liner is then everted for the full length of the casing as shown in Figure 6.4. The casing serves as the tether and follows the liner down the borehole. When the liner is fully extended, a grout line is inserted into the liner outside the casing and the liner is filled with grout outside the casing in reasonable stages so as to not burst the liner with the grout pressure as shown in Figure 6.4. This design has never been used. But liners are often grout filled after emplacement.

The same design was suggested for an inquiry from South Africa of how to seal the upper part of a water well to prevent contaminated water flowing into the well. In that case, the liner is to be simply filled with water to seal the borehole and the well pump can be lowered through the casing without damage to the liner. The alternative of simply grouting a casing into the hole may be more attractive, but the usual grout shoe for that operation is not useful for a pre-existing borehole with an open hole beneath the casing. Other kinds of packer designs at the bottom of the casing may be more attractive. The liner seal has the advantage of being a very thin outer seal allowing a larger casing and the grout will not flow into karst solution channels.

It is noteworthy that the LAHD technique described in Chapter 15 was used to carry a casing into a horizontal borehole in Mustang, OK to test a method patented for such an operation. A third patent was issued for a test of a design that allows the liner to be inverted from the horizontal borehole after the liner exits the borehole with the casing interior to the inverted liner. The casing is then left in place after the liner is removed. That design is also described in Chapter 15 with the LAHD technique. Without the special inversion design, the liner cannot be inverted from off the casing after the casing has been carried into place as may be done using the LAHD method. This method would be most useful for a casing with screens.

A blank liner use has been suggested to prevent shrinkage and cracking of grout surrounding surface casings in the vadose zone. Above the water table, the formation can draw moisture out of the grout or any bentonite backfill material. A flexible liner can prevent that moisture loss to the formation. This use of a liner has not been tried. However, several Midwestern States have done tests (Lackey et al 2009) that show the cracking of the grout fill, due to shrinkage, can allow surface water infiltration to the groundwater violating the purpose of the surface seal. This use of a liner would be a major change to the water well drilling practice and is expected to be slow to be accepted. FLUTe has not pushed that application for that reason.

7 General Advantages of Flexible Blank Liners

Liners can usually be installed with the following advantages:

- No large and heavy equipment is needed (e.g., no crane truck or drill rig).
- Can be installed from small confined spaces even in tunnels and galleries in dams.
- Seal quality is very good.
- Speed of installation.
- Support of the open borehole when everted into place and the low risk of entrapment during removal by inversion.
- Sufficiently simple installation as to be done by other than FLUTe personnel for only sealing of boreholes.
- High spatial resolutions of several measurement methods described later.
- Removable by inversion or pump-and-drag.
- Long lived in the underground environment (e.g., >17 years to date).

The blank liner is the most common device produced by FLUTe. The many measurements that are made using only the blank liner are described in Chapter 10 as the T profile and the reverse head profile (RHP) as well in subsequent chapters such as the liner augmentation of horizontal drilling (LAHD) method. As a carrier of the coverings described in Section 10 and then subsequent reuse for various measurements, the blank liner provides an especially economical function. The rest of this document explains the many applications beyond simple sealing the borehole.

The low head developed beneath the inverting liner can be a borehole development advantage, but the low head beneath a liner can also be a frustration of inversion from a borehole of low transmissivity when pumped into place. Some of the earliest installations by the United States Geological Survey (USGS) in Connecticut (CT) where frustrating to them because of the difficulty of inverting a liner from low transmissivity boreholes. The lesson was that the liner does not need to be everted at the bottom of the borehole to provide an effective seal against cross-connection.

An interesting fact is that the borehole water injected into the formation during a blank liner installation by eversion is essentially withdrawn from the same intervals in the same amount by the inversion of the blank liner.

DOI: 10.1201/9781003268376-7

8 Hazards to the Liner and Precautions

Liner damage has been minimal. The evolution has been to the stronger liners to avoid potential damage. Sharp rocks in boreholes have been less hazardous than might be expected. The urethane-coated nylon fabrics are strong at 200–400 lb/in. tensile strength. That gives a 6-in. 400-denier liner in theory more than 3600-lb longitudinal tensile strength. The 840-denier ballistic nylon liners have an eightfold higher tear resistance than the standard 400-denier liner material. Liners can be damaged in several ways:

- Abrasion from dragging on the ground surface or out of boreholes
- High interior pressure, ΔP, exceeding the burst strength of the liner as in breakouts in open boreholes with deep water tables. The breakouts do not support the liner and can contain sharp rock projections. The high ΔP may be due to a very large gradient in the formation or due to overfilling the liner.
- A large ΔP can be generated by nearby pumping, dropping the formation head to well below the natural level. This is less likely to be anticipated.
- Traversing the interior of the liner with unpadded logging sondes with sharp features to cut the liner, or more commonly, the sonde hitting ledges in the borehole with the liner draped over a ledge. The impact of a hard metal sonde on the liner on a rock ledge can cut the liner.
- Excessive tension on the tether during liner inversion. This is not common unless, contrary to FLUTe recommendations, removing the liner with the tether attached to a vehicle provides extreme tensions on the liner and tether, or, as demonstrated, pulling the liner with a drill rig cable without a tension monitor. Excessive tension on the tether during removal will draw the head beneath the liner to far below the head in the liner. The tension will produce a pressure drop beneath the liner according to $\Delta P = 2T/A$, where T is the tension on the EP and A is the cross section of the borehole. The hoop stress in the wall of a pressurized closed cylinder is twice the longitudinal stress. A high ΔP developed by the tether tension on the EP, communicated via a fracture connecting a breakout adjacent to the everted liner is twice the hazard of bursting the liner with a longitudinal tear. In other words, as the EP under high tension approaches the breakout, it is a threat to the liner in the breakout where it is unsupported and the high ΔP is the greater threat of a hoop stress failure. It has been seen that a liner pulled too hard can suffer a longitudinal tear in a breakout. It is also important that the longitudinal tension in a liner is further reduced by the end load support of the inverted liner, reducing the axial stress by a factor of 4 compared

to the hoop stress. Therefore, the dangerous level of ΔP is the hoop stress limit, not the tension in the inverted liner or tether. Since the tether tensile strength is 4,000–8,000 lb for the typical 3/16 and 1/4″ tethers used, tether failure has never occurred. A poorly tied knot is an obvious exception.

- Sharp rock projections in the borehole. These are more common in old wells in which the soft layers have raveled out leaving thin hard layers which are then broken by the lowering of well pumps in the hole. This can produce broken sharp projections known to be damaging to liners. Occasionally, a sharp chert projection in a limestone formation has punctured a liner, but not in recent years. The chert arrowhead-shaped rock was actually found inside the liner.

9 Special Devices Designed for Use with Liners

Herein are described special machines that have been designed for emplacements of liners. Some of the measurement methods required the invention and fabrication of devices which are not normally available. Several devices were developed for the measurement of water table depths through slender tubing as required for FLUTe multi-level systems (MLSs). Some eversion aids are also described.

9.1 GREEN MACHINE

The green machine (GM) as shown in Figure 9.1 was developed by FLUTe for the manual removal of blank liners from boreholes. The main features of the machine are the wellhead roller of large diameter to minimize the stress of a high tension applied to the liner or components of the liner such as tubing. The GM is also used for the removal of tubing bundles from the borehole. The winch is a sailboat winch used for trimming the sheets of the sails. The small roller behind and beneath the large roller is to route the tether so as to not "foul the winch", that is to direct the tether to the bottom of the winch capstan. The tether is wound clockwise (looking down on the capstan) for several turns, 4–5 wraps. The winch handle is inserted into the top of the winch. Most GM winches are two-speed and have different mechanical reduction when turned in one direction versus the opposite direction. This allows variable tension or speed with torque on the winch handle applied to the capstan. The tether where exiting the top of the capstan must be kept under tension to assure sufficient friction on the capstan so as to not slip.

Since the liner cannot be wrapped on the winch, when the bottom closed end of the liner is lifted above the surface, a different geometry is needed to further remove the liner. Before the liner extends over the large roller, a strap called kellum strap is wound diagonally on the liner at the original tether attachment. A carabiner is connected to the top of the kellum as shown in Figure 9.2. A separate "second tether" cord is connected to the carabiner and routed over the roller through a "first snatch block", which is anchored some distance from the GM and the second tether is redirected back to a "second snatch block/roller and load cell" onto the winch (see Figure 9.2). Tension on the second tether will draw the liner over the large roller until the kellum reaches the first snatch block. At that point, a third short "safety tether" (not shown) is attached to the carabiner and anchored to the same anchor as the first snatch block. The second tether is then allowed to slip on the capstan to provide slack in the second tether to transfer the tension on the kellum to the third "safety tether". A second kellum is attached to the liner at the large roller. The second tether is then reattached to the second kellum and the second tether is tightened with the capstan through the first snatch block to pull the liner further from the borehole. As the liner

DOI: 10.1201/9781003268376-9

FIGURE 9.1   Green machine.

tension is released on the safety tether, the original tether and liner are rolled onto the nearby liner shipping reel to avoid any dragging of the liner on the ground. The procedure is repeated until the entire liner has been removed from the borehole. The sequence of kellum attachment and second tether winching pulls the liner from the borehole in a sort of hand-over-hand manner.

During the liner removal, the liner must be dilated with a sufficient ΔH to cause the liner to invert as it is being removed. As the water level rises with the liner inversion, a pump can be used to maintain the ΔH and prevent liner water overflow of the casing. A tension measurement can be done when a load cell is connected to the second snatch block and anchored at the same point on the GM. The load cell reading is corrected for the angle of the tether through the second snatch block. It is always useful to monitor the tension being applied to the tether. When the liner is stationary, the tension measured can be used to estimate the head beneath the EP in

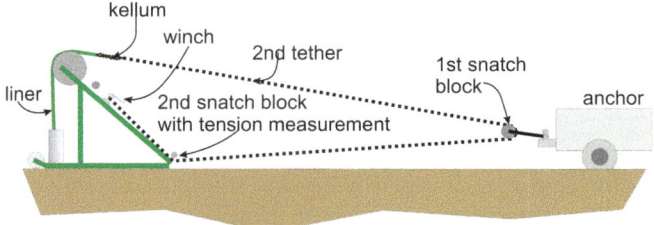

FIGURE 9.2   Routing of the second tether to remove the liner after the original tether has pulled the liner's closed end over the green machine roller. The liner is pulled incrementally from the borehole causing it to invert.

the borehole. This was done at a Maine site to discover a very low formation head at the bottom of a deep hole. The tension relationship to the value of Po is described in Section 3.2.

If the first snatch block is anchored further from the borehole, the length of liner drawn with each new kellum attachment is greater for a more efficient procedure. Pulling a tether with a drill rig or vehicle usually damages the liner and is not recommended by FLUTe.

9.2 LINEAR CAPSTAN

9.2.1 BACKGROUND

The liner cannot be wrapped on the capstan of the GM and hence the kellum attachments and reattachments are described above. The linear capstan shown in the photo in Figure 9.3 is a FLUTe design intended to reduce the manual effort of winching

FIGURE 9.3 Linear capstan removing blank liner from borehole in NJ site.

a blank liner from the borehole. The linear capstan draws the tether and the liner directly out of the borehole in a continuous manner without any damage to the liner.

The liner tether is threaded from the borehole, over the first roller, under the second roller, over the third roller, etc. The rollers are connected with a chain and sprockets on the roller axles to rotate in a manner to move the liner through the machine. A modest amount of tension on the tether, and then on the liner, at the exit from the linear capstan assures sufficient friction on the several rollers to pull the tether/liner from the borehole. The chained rollers are driven with a variable-speed electric motor through appropriate gearing. The tension on the liner or tether is monitored at the first roller as the liner system is pulled from the borehole. As the tether and liner emerge from the machine, they are rolled onto a nearby shipping reel. With 5–10 lb of tension on the tether where it exits the linear capstan, several hundred pounds of tension can be applied to the liner rising from the borehole. If the tension at the exit is relaxed, the liner or tether will slip on the rollers, reducing the tension.

The linear capstan was first designed and built by FLUTe for liner removals at the Santa Susana Field Lab. in California in 2003. The first linear capstan weight was nearly 3,000 lb and carried on a trailer (photo in Figure 9.4). However, the design evolved to the much smaller linear capstan as shown in the photos in Figures 9.3 and 9.5. Ian Sharp, the operator in Figure 9.3, built the machine he is using in Figure 9.3. The modern linear capstan allows the recording and monitoring of tension on the liner, depth of liner in the borehole and is speed-controlled to control the

FIGURE 9.4   First linear capstan used in California to install water FLUTe systems with tubing bundles. It was trailer-mounted and weighed over 3,000 lb.

FIGURE 9.5 Smaller linear capstan that fits in the bed of a pickup truck. Used mainly to remove blank liners. Refinements include tension recording, depth of liner encoders, and speed-controlled motor drive.

tension. The linear capstan is much more expensive than the GM and is normally operated only by trained FLUTe personnel. However, a few regular users of FLUTe liners have purchased a linear capstan. The first monster linear capstan in Figure 9.4 had very large rollers to allow tubing bundles to be routed through the machine. That function is no longer in use. Large tubing bundles are removed with the procedure described in Figure 9.2.

9.2.2 THE LINEAR CAPSTAN DESIGN

The tension relationship at the exit of the linear capstan can be determined from the relation for a cord draped over a cylinder lifting a weight of To lb with a tension T:

$$T = To \ e^{F\alpha},$$

where To is the tension on one side of the cylinder. F is the friction coefficient of the cord on the cylinder and α is the angular drape over the cylinder in steradians. For the routing of the liner in Figure 9.3, and with the n rollers rotating in opposite directions, the tension on a liner at the wellhead is:

$$Tn = To \ e^{nF\alpha},$$

where each wrap on the rollers of the linear capstan is ~180 degrees or π steradians. In this case, Tn is the tension on the liner above the borehole and To is the manual tension applied at the exit of the linear capstan (see Figure 9.3).

The same relationship applies to the tension developed on the capstan wraps on the GM. FLUTe therefore calls this a "linear capstan" that does not damage the liner as is certain were it to be wrapped on the single capstan of the GM. The tension, applied by each roller which is chained to move in the correct direction, is amplified from roller to roller. The angular wrap on each roller is added to that of the previous roller. To a first approximation, the diameter of the cylinder/roller is not mathematically important. A more common experience of this relationship is when pulling a garden hose past/around a tree.

The small linear capstan of Figure 9.3 has gained a number of advantages in its evolution.

It weighs only 650 lb. It has a four-wheeled wagon-like transport system for alignment on the borehole and can even be loaded up a pair of ramps into a pickup by pulling itself on a rope attached to the inside front of the pickup bed. Also, the liner tension measurement uses a pair of recorded load cells that don't require a centering of the liner on the rollers. The first roller has an encoder to allow recording of the travel distance of the liner out of the borehole (i.e., the depth of the liner in the borehole). The power driving the chain system is a variable speed motor allowing a constant tension removal of a liner, until the liner is inverting too quickly. Alternate rollers pivot upward to allow the easy insertion of a liner without the need to thread the liner or tether through the machine.

The pair of FLUTe linear capstans gets a lot of use with, unfortunately, a great reduction of the body-building labor of the GM for pulling blank liners. Currently, only one customer owns a linear capstan of modern design.

9.3   T PROFILER

The transmissivity profiler (T profiler, for short) is a FLUTe design used to measure the tension, control the tension, and measure the depth/speed of the liner installation for the method described in Section 10.3. The basic machine is shown in the drawing of Figure 10.37. The liner is deployed from a shipping reel mounted on a hollow axle reel stand some distance from the wellhead. A braking system on the reel stand controls the tension as the liner is everted down the borehole. The liner passes over a wellhead roller, which is instrumented to monitor both the tension on the liner and the travel of the liner. The roller encoder rotation is recorded in a laptop along with the tension on the roller as measured by a pair of load cells. The depth of the water level in the liner is also measured separately using a bubbler system in which the bubbler pressure is also recorded on the laptop. (A recording transducer in the borehole is not shown as part of this system.) These systems are only operated by trained FLUTe personnel who also, at the same time, control the water level to be essentially constant inside the everting liner. The actual function of this device for the measurement of the transmissivity distribution of the borehole is described in detail in Section 10.3.

9.4 BRAKING DEVICES OF SEVERAL KINDS

The photo in Figure 10.53 (Chapter 10) shows a braking system attached to a shipping reel as a liner system is being deployed into a borehole. The braking system is required to support the hanging load of the liner system (usually a heavy tubing bundle as part of an MLS system). The braking systems consist of a brake caliper connected to the reel stand, which contacts a disk connected to the shipping reel. The tension of the braking system is monitored with a load cell and the braking force is adjusted with the caliper on the disk brake. This system allows the safe suspension of several hundred pounds of tubing and valves in an MLS system being installed into boreholes with deep water tables (400- to 600-ft. bgs).

A braking device (more similar to that described in Section 9.3) is attached to the T profiler shipping reel to maintain a constant tension on the liner during the transmissivity profile.

9.5 THE AIR-COUPLED WATER-LEVEL METER SYSTEMS

9.5.1 THE ACT (AIR-COUPLED TRANSDUCER)

9.5.1.1 ACT Purpose

Water table measurements are an essential factor in understanding groundwater flow subsurface. Water table measurements in a single open borehole are not sufficient to determine groundwater flow in several aquifers connected by an open borehole. It is much better to know the head at any time in the several aquifers. The air-coupled transducer (ACT) method brings an inexpensive approach to the measurement of the history of head variations at many elevations when coupled with the ability to continuously seal a borehole with a flexible liner. FLUTe MLSs allow electric water-level measurements but do not have a large diameter casing needed to lower transducers below the water table. The ability of the ACT system (patented) to monitor water-level changes in small-diameter tubes in FLUTe multi-level water sampling systems at each port in sealed boreholes makes the method especially useful.

9.5.1.2 Background/Comparisons

Water-level measurements in open boreholes have been done with a variety of methods, but the most common are electric water-level meters lowered into the borehole. Water-level variations in time are easily performed with pressure transducers submerged in the open borehole, but they are subject to borehole storage effects as the head changes. However, it is well understood that the water level in an open borehole lies between the highest and the lowest head in any aquifer intersected by the borehole and depends on the transmissivity of each aquifer. The "blended head", as the water level in the borehole is sometimes described, tends to be nearer that of the most transmissive aquifer or conductive feature intersected by the borehole. By sealing the borehole with a continuous liner, the several aquifers are isolated but not available for measurement except through the tubing or casing of the water sampling system connected to each port.

A method developed by FLUTe uses a simple device called vacuum water-level meter (VWLM) for head measurements when the water table is less than 25 ft. below the surface. The function of that system is quite simple. The advantage is measurement through a very slender tube (less than 1/4 in. OD). However, it is not useful for recording the history of water-level changes. The method is described in Section 9.5.2 as a simple means of measuring a shallow water table through a slender tube.

An MLS was developed by FLUTe, which uses a flexible sealing liner to isolate discrete sampling intervals in a borehole for water sampling and head measurements. The device developed by FLUTe called water FLUTe (WF) is described in a paper by Cherry et al (2007). The head at each sampling interval can be measured with an ordinary electric water-level meter. However, the history of the head at each port was measured and recorded by transducers built into the WF system and connected to the surface by transducer cables for downloading the data. When those transducers fail, the replacement is difficult requiring the removal of the liner system and reinstallation.

Butler et al (1999) describe an ACT method for following the water-level changes using a slender tube connected to multiple elevations in a multi-level sampling system. The method of Butler measures the air pressure in a tube above the water table as the water level changes in formation in order to infer the formation head changes. A limiting assumption of the method as described by Butler is that the pressure change is small compared to the original pressure. A more rigorous method for an ACT developed by FLUTe 16 years later, without prior knowledge of the Butler design, is described hereafter. Butler et al (1999) do not describe how the initial water level was measured at each port in the system. One must assume that the same initial water table was used for each port. The focus is on the changes in head and the use of a linear fit for small pressure changes. Another difficulty of the Butler method was the temperature effects on the air tube and the air pressure inside the tube. An important feature not described in the Butler paper is how the MLS was constructed and how the sampling intervals were isolated or sampled. FLUTe was not aware of the Butler method until 2015, and developed the ACT method separately, but recognizes Butler's earlier use of the ACT. FLUTe added the temperature correction that makes the method more generally useful and also has the ability to measure large changes in the water levels.

Another approach to measuring head variations in time was developed recently by the Univ. of Guelph, which uses individual recording pressure transducers that are suspended in a borehole on a slender cable. The transducers are isolated at discrete elevations using a FLUTe flexible borehole liner to seal the borehole. The transducers are recovered by the removal of the sealing liner for downloading the data. Direct connection to the transducers in place is not possible due to the necessary transducer cables violating the borehole seal of the liner. The limitation of the method is that the transducer data is not accessible unless the liner is removed to read the history from the transducer memory; therefore, that approach does not have a continuous recording capability, and there is no sampling capability with the blank liner seal during the transducer measurement as with the ACT.

The Westbay MLS uses multi-level packers to isolate intervals of an otherwise open borehole, but the normal practice is to connect to each interval with a single transducer suspended on a cable inside the central casing. This system is not well suited to the connection of many transducers to many intervals simultaneously, although a very expensive multi-transducer system is offered. Another limitation of the method is that straddle packers are known to be bypassed by flow in the formation or at the contact of the packer with the borehole wall. Such bypass of the packers can frustrate the pressure isolation of the sampling interval. If the packers were long and the isolated intervals were short, the isolation would be better, but that is not the typical design.

Other variations of multi-level head history measurements use individual nested wells in a single borehole or clusters of ordinary wells with short screens. A transducer can be lowered into each well for the head history if the well casing is of sufficient diameter. However, these systems are usually limited to a few wells located in a single or several boreholes and are vulnerable to the seal quality of the backfill materials surrounding the casings. In one variation (Chapman et al 2014), only two nested wells are included in the same borehole with an MLS with slender sampling tubing. The two slender wells use submerged transducers for head histories. The limitation is the crowding of the several systems in the borehole and only two history measurement locations.

Several MLS systems invented by FLUTe provide the multi-level sampling capability and the addition of the FLUTe ACT method provides the head history measurement capability simultaneously at each port. An advantage of the ACT method is that it allows reuse of the transducers at other boreholes since they are not built into the MLS systems as for the earlier WF designs.

9.5.1.3 The ACT Design and Theory

The ACT method is not radically different from that devised by Butler et al (1999). However, the method described here and developed independently has several refinements over the Butler design, which is as follows:

1. It is not assumed that the pressure change is small compared to atmospheric pressure and so large water table changes are measured with better resolution.
2. It is not assumed that temperature changes of the air tube are insignificant (since they are often very significant).
3. A correction of temperature changes is included in the calculation.
4. The measurement of the initial water level at each port is addressed and not assumed to be the same for all levels, as might be expected only in permeable sediments.

The initial geometry of the calculation is shown in Figure 9.6(a). The bottom of the tube is connected to the water in the formation below the water table through a port in the liner at each sampling interval. The initial water table in the formation fills the tube to the same level as in the formation. When the transducer is connected to the tube, it seals the air column in the tube. When the water level rises by the amount shown in Figure 9.6(b) as ΔWT, the air column, of initial length Lo in the tube, decreases by an amount ΔL and compresses the air in the tube. The increased pressure in the air

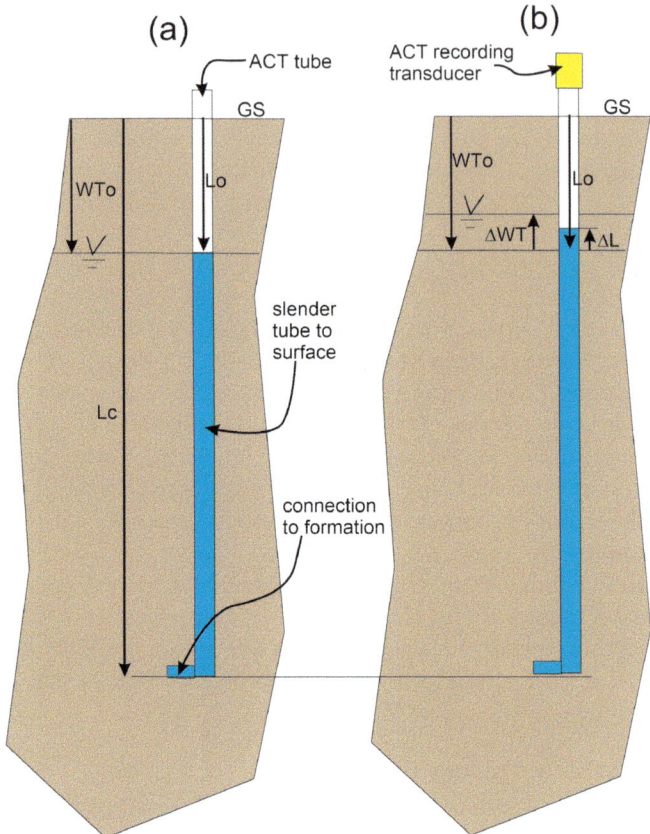

FIGURE 9.6   Geometry of ACT system WT history calculation. (a) The initial geometry. (b) The ACT transducer seals the slender tube. A water level change, ΔWT, in the formation produces a somewhat smaller water level change, ΔL, in ACT tube.

column resists the water-level rise ΔL in the tube. Therefore, the change in the tube ΔL is less than the change in the water table ΔWT. The initial atmospheric pressure on the water table is Po. The initial pressure on the water in the tube is also Po when the transducer is attached. The initial air column length in the tube is Lo between the water level in the tube and the top of the tube. The initial water table depth is WTo below the surface. In summary, the initial conditions are as follows:

- $WT = WTo$
- Pressure on the WT is Po
- Pressure on the water in the tube is Po, the atmospheric pressure
- The distance from the top of tube to the water level in the tube is Lo
- The tube connection to the formation is at the depth Lc below the surface

In this situation, the total pressure in the tube at the connection Lc is $Po + (Lc - WTo)\, \rho g$, which equals the pressure in the formation, which is $Po + (Lc - Lo)\, \rho g$, where ρg is the density of water times the acceleration due to gravity. Then the top of the tube is sealed with a pressure transducer.

Upon the change of the water table:

The water table depth is WT = WTo − ΔWT (for a water table rise). The pressure on the WT is still Po. The pressure on the water in the tube is Po + ΔP (for a water table rise, ΔP is positive). The distance to the water level in the tube is L = Lo − ΔL, ΔL is positive upward. The tube connection to the formation is still the same depth below the surface Lc.

After the water table change, the total pressure on the connection at Lc is equal and Po + (Lc − WTo + ΔWT)ρg = Po + ΔP + (Lc − Lo + ΔL)ρg.

If the pressures are expressed in terms of equivalent water column (i.e., head) then dividing through by ρg and subtracting Lc and Po from both sides of the equation gives:

$$-WTo + \Delta WT = \Delta P - Lo + \Delta L \text{ or the new WT depth is } WTo - \Delta WT$$
$$= Lo - \Delta L - \Delta P \text{ but } WTo = Lo$$

Or, ΔWT = ΔL + ΔP. This is similar to Butler. But hereafter are the differences from Butler.

If we know the level change ΔL and the pressure change, we know the water-level change. The term ΔL is related to the perfect gas equation of state for the closed volume of the air column of length Lo. The volume Vo is Lo A where A is the cross section of the tube inside diameter. The perfect gas law for the air sealed in the tube is:

$$PoVo = nRTo = PV, \text{ if the temperature does not change} \tag{9.1}$$

where Po is the initial pressure, Vo the initial volume, n is the number of moles of air in the column of length Lo, R is the universal gas constant, and T is the temperature. It is noteworthy that P and T are in absolute units of pressure and temperature. PoVo = PV simply says that as the volume decreases then P increases and the product PV is constant. It is useful to note that nR = PoVo/To in Equation (9.1)

$$\text{Therefore, } \Delta P = P - Po = nRT/(Vo - \Delta V) - nRTo/Vo, \rightarrow \text{ where } \Delta V = \Delta L \ A. \tag{9.2}$$

$$\text{With some algebra, } \Delta L = -Lo(PoT/(\Delta PTo + PoTo) - 1). \tag{9.3}$$

Note Lo is the total initial air column length.

$$\text{If } T = To, \Delta L = -Lo\big(PoT/To/(\Delta P + Po) - 1\big) = Lo\Delta P/(Po + \Delta P). \text{ Note,}$$
$$(Po + \Delta P)/(Po + \Delta P) = 1$$

ΔWT = ΔL + ΔP = ΔP(Lo/(Po + ΔP) + 1). If ΔP is small relative to Po, this reduces to Butler's relationship.

9.5.1.4   The Range of Pressure Changes for the ACT Transducers

In theory, the method is very accurate and limited only by the accuracy of the pressure measurement and in fact it usually is accurate. However, experience has taught that there are some practical resolution limits in practice. If the water table is at the depth Lo when the transducer is connected, the air pressure will rise as the water table rises. If the water table rise is such that ΔL is Lo/2, the pressure will rise to double the initial air pressure when the tube was connected. That implies that ΔWT is Lo/2 and ΔP is Po. Or the $\Delta WT = Lo/2 + Po$. If the initial pressure in atmospheric (~33 ft. of water) and Lo is 100 ft., the water table rise would be 83 ft. to increase the air pressure by Po, a very large rise. But if the water table were to rise more than 83 ft., the pressure transducer must be of a range to allow such a large pressure increase. The normal high-resolution transducer preferred has a range of 30 psia with Po near one atmosphere. In rare circumstances, a higher range transducer is needed. In most cases, Lo is essentially the water table depth; therefore, if Lo is 30 ft., the maximum water table rise until the pressure rise is one atmosphere would be 48 ft. or above the ground surface. However, some water tables do rise above the surface after the FLUTe MLS is installed (i.e., the aquifer becomes artesian). That causes other problems, but not with the ACT method.

An extreme water-level change does occur more often when a large pump is turned on in a nearby well. In that case, the ΔP is negative. The theoretical lower limit for an absolute pressure transducer is 0, which is not achieved for a variety of reasons. Therefore, one need not be concerned about over-ranging the transducers. However, there are other limitations to the method for extremely low pressures. One is the effect of the low pressure on the water in the ACT tube. As with peristaltic pumping, the low pressure applied to the water will cause the dissolved gas to effervesce and form bubbles in the water. With the ACT tube usually vertical, in time such bubbles can rise to the surface of the water in the tube and add the gas in the bubble to the air originally in the tube. The effect of a gas addition to a partial vacuum is to raise the pressure and reduce the magnitude of the vacuum. The effect on the calculated water table change is then to decrease the calculated change or produce a calculated water table value below the surface less than the actual result with no gas addition to the air in the tube. That has occurred for a water table drop of over 85 ft. from an initial water table of 30 ft. The first evidence of the error in the ACT result was when comparing the ACT to the manual water-level measurement in the WF for the same port. Another clue was the many spikes in pressure seen in the data as apparently caused by the bubbles expanding into the low-pressure air column above the water level in the tube. Fortunately, that kind of extreme drop is unusual except near production wells and the comparison with manual measurements is very good. In this extreme case, the large differential pressure caused by the water table drop damaged the liner and it was replaced by a stronger one.

9.5.1.5   The Temperature Effect

If in Equation (9.2), the temperature does change, then T does not equal To, and the volume change is

$$\Delta L = -Lo((PoT/(To(\Delta P + Po))) - 1) \tag{9.4}$$

The reality of this equation is that if ΔP = 0 and T = To, there is no volume change. If there is no pressure change and the temperature doubles, the volume change must equal Vo, a realistic value.

The difficulty is how can one determine a temperature effect when the top part of the air-filled tube sees the diurnal changes and the lower portion is sheltered more or less from diurnal changes inside the well of a relatively constant temperature. As Butler described, it is difficult to protect the top of the well and the tubing inside from diurnal temperature changes. FLUTe found the same difficulty. If the surface casing extends above the surface, extreme temperature changes are likely inside the casing depending on exposure to the sun. Twenty-four hours and seasonal temperature changes were not significantly reduced by insulation of the wellhead as proposed by Butler. We have found a useful correction for the temperature changes. Useful is defined as when the obvious diurnal changes are measured and the water table calculation is corrected, the diurnal changes in the water table that is calculated fade to unimportant.

The method that works well is to consider that a portion of the tube sees the temperature change. It is not necessary or expected that the temperature change is uniform over any segment of the tube. If the change is distributed in some manner, it is useful to assign the change to a fraction of the tube. If a known portion of the tube is heated, the gas will expand into the unheated portion until there is pressure equilibrium in the tube. The pressure is the same as if the entire tube was heated to a lower temperature. Hence, the concept is of an effective uniform temperature due to the heating of a portion of the tube. The effective temperature was defined as a weighted average of the temperature in the well and the temperature recorded at the top of the surface casing. The fraction of the tube Lo exposed to a temperature change was defined as "f". The effective temperature change of the entire tube above the water table was defined as Teff = f T(t) + (1 − f) To, where T(t) is the time-dependent temperature measured inside the wellhead and To is the typical downhole temperature or the effective initial temperature of the air in the tube above the water table. It is reasonable that if the distance to the water table is very long and the exposure to temperature changes is only near the top, the factor f would be small to near zero. The term f varies from 0 to 1.

Using this approach, the temperature T is defined as Teff. Because the geometry of the tube exposure probably does not change significantly in time, the f factor can be adjusted until the obvious temperature effects approach zero and that factor can be used for different times of the year or at least near the time of the installation and first measurements.

It is important to note that in the formulation above, all pressures and temperatures are in absolute units as in the perfect gas law, PV = nRT.

9.5.1.6 First Result of the ACT Measurement

The first test of the method in 2010 is shown in Figure 9.7. A tube was lowered into a domestic well located inside a building with limited temperature change. A transducer was submerged adjacent to the submerged ACT tube. An insulating cylinder was positioned inside the top of the well to minimize temperature changes in the well. The water table was 48 ft. below the surface and the transducer was connected

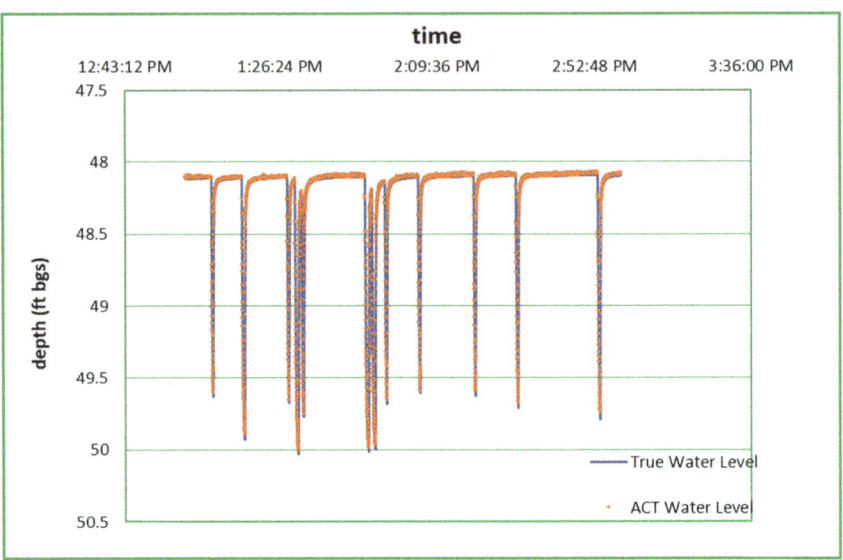

FIGURE 9.7 The first test of the ACT system in a relatively constant temperature environment. The blue curve is the submerged transducer water table result. The orange dots are the water table depths from the ACT measurement. The ACT results are within 1/4″ of the submerged transducer results on the 1 second time scale.

to the top of the tube below the insulating cylinder. When the domestic pump starts, there is an abrupt drop in the water table and slower recovery when the pump stops. The comparison in Figure 9.7 of the water table depth calculated from the ACT transducer data (orange dots) with the head measured by the submerged transducer (blue curve) was within 1/4″ of each measurement on the 1 second time scale. This was a surprising sensitivity of the ACT measurement.

9.5.1.7 The Field Measurements

In the field measurement situation, there are significant differences from the first test. The temperature changes are often extreme (e.g., 50°C) and the duration of the measurement is days to months. In that case, the temperature correction of the calculated water table depth is essential. An installation at the NAWC site near Trenton, NJ, was done with the ACT system connected to a slender tube to one of the ports in a Shallow WF MLS (multi-level sampling) system. The water table depth calculated, the orange curve in Figure 9.8, is without the temperature correction. The temperature history (gray curve) is shown with the right-hand scale. The lower graph, the blue curve, is the water table with the temperature correction. The blue curve with the temperature correction has been displaced by 0.5 ft. to allow the comparison. The initial water level was measured with the VWLM. A final measurement of the water table is shown at the latest time showing that after 90 days the ACT measurement was within 0.4 in. of the actual measured

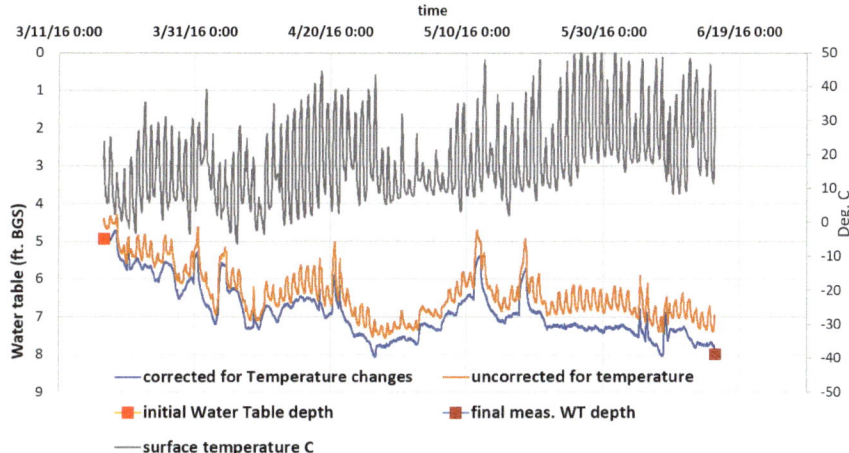

FIGURE 9.8 Comparison of calculated water table without temperature correction (orange curve) and with temperature correction (blue curve). The blue curve was shifted by 1/2 ft. for easier comparison. The final ACT water level is within a half inch of the measured water level after 3 months. The gray curve is the temperature variation of 50°C at the surface. The f factor was 0.3. For this shallow water table, the temperature effects were large. No barometric correction was applied. The port is 50-ft. bgs.

water table. The sharp rises and more abrupt drops are due to stops of the pump-and-treat pump.

9.5.1.8   Input Data and Apparatus

The ACT method uses a transducer of high resolution sealed to the top of the tube. Figure 9.9 shows the ACT location in the top of the WF system (Section 10.5). While that ACT tube can be of variable diameters, it is necessary that the portion of the tube in which the water level is changing is of uniform diameter for the calculation of the cross section as needed. An amplification of the pressure change is possible as will be described later using different tube diameters. However, the basic information needed is the length of the air-filled tube when sealed, and the initial depth to the water table (Lo and WTo in the calculation above). The ACT transducer is typically connected to the tube at the time the water table for that port is determined. The transducer is best located inside the well readily available for removal. This allows the ACTs to be reused on other boreholes or applications where the connection to the water table is a convenient slender tube. There is no disadvantage in using the system for remote inaccessible water table locations using the tube connection. An advantage even in an open hole is that there is no need for an expensive transducer cable to a deep water table. And the transducer is not subjected to corrosive environments below the water table. The transducer preferred is a high-resolution absolute pressure transducer with only the range necessary to cover the changes expected. Most installations use a 30 psi (~2 bar range) absolute transducer. If the water level change is expected to rise more than three-fourths of the original water table depth, a higher range may be needed.

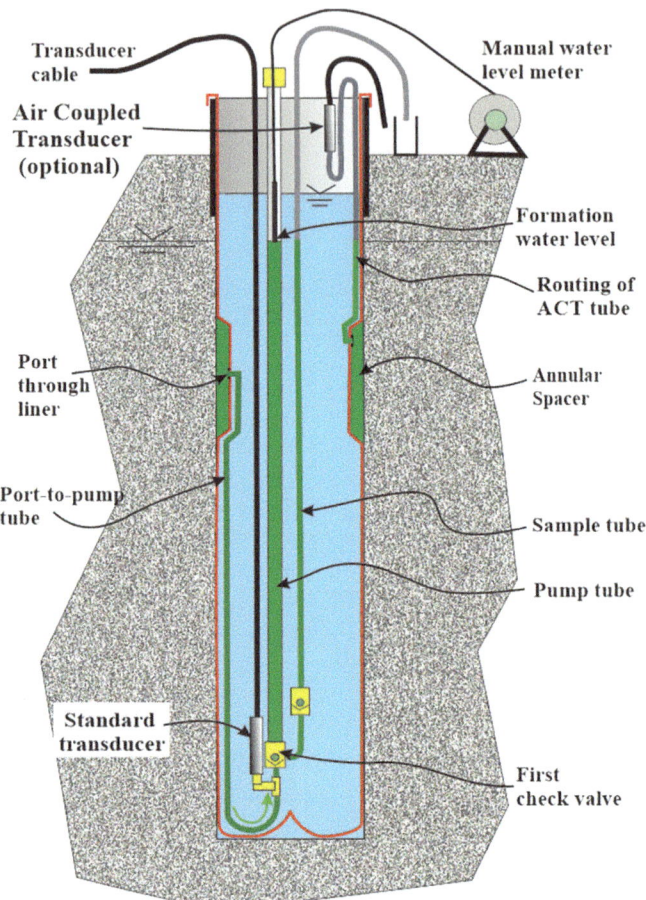

FIGURE 9.9   Water FLUTe MLS system components showing the ACT tube to the spacer and the use of an alternative downhole standard transducer. The ACT transducers are now recommended as more accessible at the surface. The water table in this system can be measured with the manual water-level meter shown through the pump tube.

The data needed to perform an ACT calculation is as follows:

1. Length of the air-filled tube between the water table and the transducer
2. Diameter of the tube at the water table
3. Temperature history of the exposed portion of the tube (a nearby temperature is sufficient if proportional to the temperature of the exposed tube)
4. Initial atmospheric pressure when the tube is sealed
5. The initial depth of the water table
6. ACT pressure history
7. A barometric pressure and temperature history are desired

For the WF and CHS systems, the initial water table is available using an electric water-level meter; unless the water table is so deep, it cannot be measured with an electric meter in the tubing. The VWLM and the air-coupled water-level meter (ACWLM) are available for initial water table measurements. Note that the water table in the ACT tube must be in equilibrium with the water table in the WF pump tube when the pump tube measurement is made. In boreholes of very low conductivity, the water table reaches equilibrium very slowly and the pump tube level has been known to lag behind the rise in the ACT tube due to tube length and diameter in the pump tube. It is important that the pump tube has equilibrated for measurement of the initial water level for the ACT system. This is more of a concern when the WF system is first installed.

9.5.1.9 Usual Applications of the ACT System

ACTs are often used with WF MLS installations (Figure 9.9). Some WFs are as deep as 1400 ft. with the water table at 90-ft. bgs. The ACT results are read continuously in that 1400 ft. installation. For WF installations, the initial water table is easily measured with an electric water-level meter as shown in Figure 9.9. The separate ACT tube connection is to the same spacer as is used for the water sample collection and the head measurement.

The FLUTe Shallow WF (SWF) MLS for water tables less than 25 ft. (Figure 9.10) is convenient for ACT measurements. Such an installation was made at the NAWC

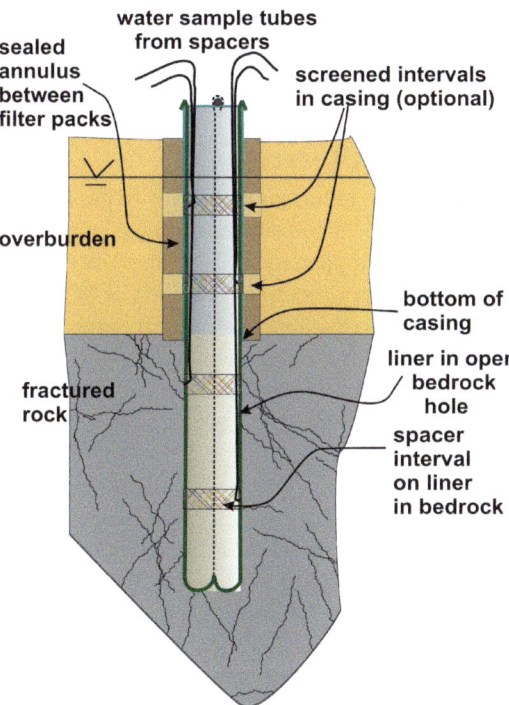

Shallow Water FLUTe system

FIGURE 9.10 The Shallow Water FLUTe system is well suited for use with the ACT by connecting the sample tubes to ACT systems and storing them in the liner between sampling episodes. A VWLM is used to obtain the initial water table when the ACT is reconnected to a sample tube.

site near Trenton, NJ, for the measurement as shown in Figure 9.8. The sample tubes of 1/4″ OD extend directly from the ports to the surface. The ACT transducers must be disconnected from the SWF sample tubes for the water sampling procedure.

Whenever the water table history must be measured and only a single slender tube is available, the ACT is available under license with FLUTe. In one series of WF installations with very deep water tables, the regulatory agency required that there should be an alternative to the downhole transducers (Figure 9.9) built into the WF systems for use after the downhole transducers are expected to fail. Separate tubes to each sampling interval were included in the design as shown in Figure 9.9 and when the first transducer did fail, the ACT transducer was added to the ACT tube for that port.

A recent addition to the FLUTe MLS systems (called CHS for Cased Hole Sampler, which is only useful for shallow water tables for sampling) is an MLS capable of being installed with numerous ports (up to 8) in a 2-in. screened casing. Each of those ports is connected to the surface for sampling and is also available for use with the ACT system. In that situation, the sample tube must be disconnected from the ACT for sampling. In the WF, the water level can be measured in the pump tube at the same time the ACT is connected.

Since the ACT results in Figure 9.8 were obtained through a 3/16″ OD tube, it is feasible to install a flexible liner system with many such tubes to monitor at many elevations. With the CHS system, it is possible to install over 20 head measurement ports in a single borehole, each connected to an ACT suspended in the liner and available for reuse at other boreholes with the same kind of system. This should make the identification of aquitards and aquicludes especially reliable.

9.5.1.10 Resolution of the ACT Method

The resolution depends on the resolution of changes in pressure measured at the surface and other factors. The typical transducer used has a resolution of 0.04″ of head change, which would be the change associated with a 1-in. water-level change in the air tube at a depth of 350 ft. That leads to a resolution of 0.035″ water-level change at a 30 ft. water table. According to the data as shown in Figure 9.7, the difference of the ACT calculation from the water table change measured by the submerged transducer is small but very much larger than 0.035″ based on the pressure resolution alone. The difference between the ACT and the submerged transducer is plotted in Figure 9.11 versus the water table measured. One probable cause for that linear error with water-level change is the measurement of the diameter of the tube used. A change in the tube diameter of 0.005 in. essentially eliminates the error in Figure 9.11 in the ACT water level calculated. The error is in the tube diameter used at the water level to convert the volume change to a level change. This is a very small error for small water-level changes. At a 1/2-ft. water-level change, the probable error in the tube diameter is near the pressure resolution limit of the transducer. It is expected that temperature effects as in Figure 9.8 are larger than the tube diameter error effect.

Occasionally, the ACT result is seen to drift as a result of an air leak from the system. This occurred for two transducers out of ten located in a vault in Guelph, Ont. One tube was cracked while handling the transducer in extremely cold winter conditions. A second tube was damaged when trapped beneath the vault lid upon closing

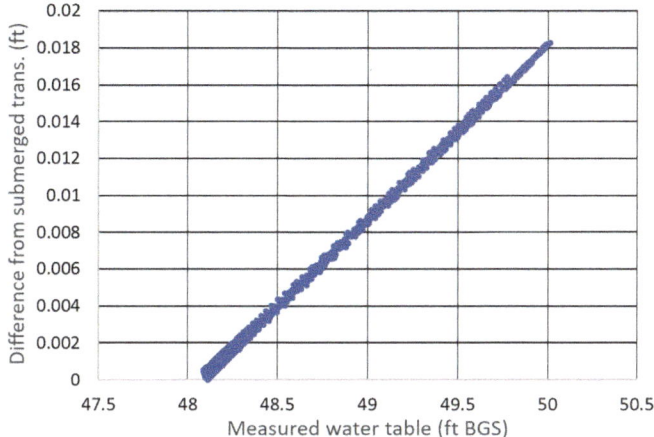

FIGURE 9.11   Difference between water table of submerged transducer and the ACT system for the data set of Figure 9.1. The difference/error is linear with the change in water table suggesting a geometric factor error such as the tube diameter.

the vault. Transducers located inside the liner have only the transducer cable and its connection at the surface. The Guelph borehole had ten water sampling systems in a small 4″ hole without room for the ACT transducers in the liner, so they were placed in the vault. Clearly, air-leak-tight fittings are needed.

The deeper the water table, the less the temperature correction that is needed due to the smaller fraction of the tube exposed to temperature changes.

9.5.1.11   Barometric Corrections

The correction of the effect of a barometric pressure increase is to increase the pressure in the ACT air tube to provide the same effect as a water table rise. Therefore, a correction to the ACT results due to a barometric increase is to reduce the calculated rise of the water table by the amount of the barometric change. That correction is available if desired and included in the data reduction spreadsheet. However, for deep aquifers, the barometric changes are not so well communicated to the aquifer. The ACT data reduction spreadsheet allows a complete or partial correction for barometric changes. The absolute pressure measurement is immune to the error due to water clogging of the capillary tube in the cable used to compensate gauge pressure measurements for barometric pressure changes.

9.5.1.12   How Is the Raw Data Used?

FLUTe provides the MLS systems, the pressure transducers, and the data reduction systems for ACT measurements. Typically, the customer orders ACT systems with FLUTe MLS systems. But they can be installed later. FLUTe installs the ACT tubing in WF systems during the construction of the WF. The SWF and CHS systems need no special construction. The ACTs can be installed at the time of the MLS installation unless FLUTe or the customer does the ACT connection later. The customer provides the first data set to FLUTe and FLUTe inserts the data set into the data reduction spreadsheet using the appropriate parameters for the installation. The completed spreadsheet is then provided to the customer with the first data reduction results and the customer inserts subsequent ACT pressure data sets for the same installation. If the ACTs are moved to other FLUTe systems, the data for the new location is provided. FLUTe offers rentals of ACT systems and the VWLM if needed for shallow water tables. For deep water tables (>25 ft.), the initial water table can be measured through the pump tube of the WF or the CHS systems. ACT tubes can be used in the ACWLM mode as described later in this chapter to obtain the initial deep water table measurement for water tables below the depth of the electric water-level meters. However, the ACWLM is not as accurate as the VWLM.

9.5.1.13   Advantages and Limitations of the Method

The advantages are as follows:

1. The ability to monitor and record water-level histories in a continuously sealed borehole.
2. The ability to measure the history of water-level changes at each port through very slender tubes without the need to lower a transducer into the water. A large diameter tube or casing for transducer access is not needed with the limitations imposed by many large tubes in the well.
3. The tubing used is much less expensive than the cables for transducers in deep water tables.
4. The transducers are not dedicated to the MLS system but can be reused on other MLS systems, or rented.
5. The ability to monitor all ports simultaneously even during water sampling events or pumping tests in nearby boreholes.
6. The measurement has essentially no lag due to borehole storage as for an open borehole, which requires significant flow to the open hole volume for measurement of formation head changes.
7. The system can be used with flexible tubing in FLUTe everting liner systems.

The limitations are as follows:

1. For tight formations, during water sampling events, the water levels in the ACT tube can change abruptly with the potential loss of some air in the ACT tube, requiring a new water table measurement and a disconnect and reconnect of the transducer after sampling.
2. The resolution of water-level changes is good, but not as good as a submerged transducer.
3. If the downward water-level changes are more than half the water table depth, the low pressure in the air column can cause the water to outgas and limits the resolution of the water table change.
4. A water table rise of more than half the depth to the water table when connected requires a higher pressure transducer than the typical 30 psia used for the best resolution of small changes.
5. It is essential that the initial water table depth be measured when the water table is in equilibrium in the initial measurement (e.g., when WTo is measured in the WF sample system and the water level has equilibrated with the formation).

Note the ACT method and data reduction technique are patented. A license for use is provided for use with FLUTe systems. However, a separate license is needed for use with non-FLUTe systems.

9.5.2 THE VACUUM WATER-LEVEL METER (VWLM)

The initial depth to the water table is not always readily available, especially if the only access to the water table is a tube too slender for the electric water-level meter. A convenient device offered by FLUTe is a VWLM. The concept is very simple if the water table is within ~25 ft. of the surface. The VWLM is connected to the tube to the water table (see Figure 9.12) such as the sample tube.

The vacuum is applied with a variety of devices such as a piston vacuum pump or peristaltic pump (the more expensive methods) or a venturi vacuum pump (the preferred pump) connected to a compressor. As the vacuum is applied to the slender tube, the water surface in the tube is raised until visible in the sight glass. The valve is closed to prevent a further increase of the vacuum. The digital vacuum meter provides the vacuum applied in feet or other water column length. The height of the meniscus above the ground surface is measured (H). The water table depth below the surface is calculated as the vacuum applied minus the height H. Since the height of the meniscus can be measured quite accurately, the accuracy of the measurement depends on the precision of the vacuum meter. The vacuum meter supplied is accurate to <1 in. of water over 30 ft. This water level is used for the initial water level for

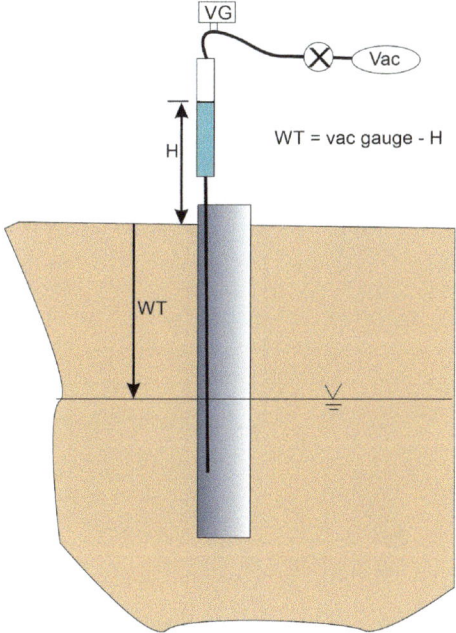

FIGURE 9.12 The vacuum water-level meter geometry. The water table is easily measured with a variety of vacuum sources. The water table must be within ~25 ft. of the surface in order to use this device.

the ACT data reduction if the level cannot be measured with an electric water-level meter. For WF systems, the initial water table is measured with an electric water-level meter in the pump tube of the WF.

A more accurate vertical gradient measurement uses two sight tubes side by side to compare the meniscus level with a common vacuum on each tube. The gradient is as accurate as the measurement of the difference in elevation of the two menisci.

9.5.3   THE AIR-COUPLED WATER-LEVEL METER (ACWLM)

The ACWLM is an extension of the ACT theory in order to allow the measurement of extremely deep water levels through tubing too slender or too deep for an electric water-level meter. The device is a simple temporary addition of air with a syringe to the ACT (see Figure 9.13). The ACT transducer is connected to the downhole tube and the syringe, with the piston open, is connected to the tee. The transducer recording is activated. The syringe is then closed. The resulting pressure change is used to deduce the volume of the trapped air column in the tube to the water table using the same principles as for the ACT monitoring of water table changes. If the cross section of the air-filled tube is constant, and known, the length of the air column can then be calculated. Using the tube volume of air deduced from the pressure change, and dividing that volume by the cross section of the tube, gives the length

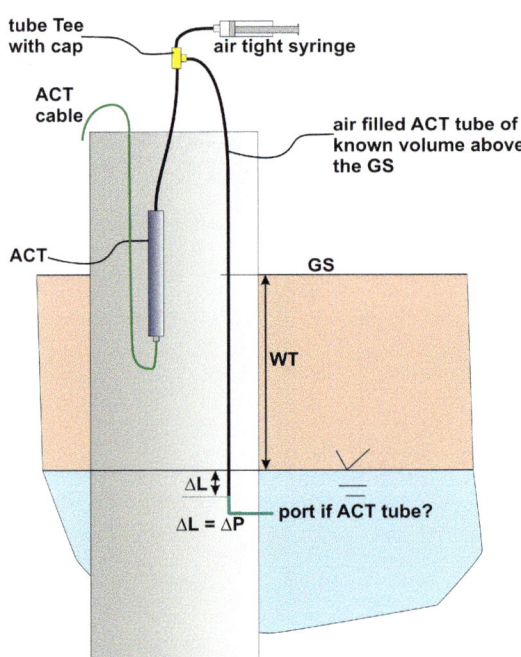

FIGURE 9.13   ACWLM uses a temporary added syringe to the normal ACT system. The closure of the syringe reduces the volume of the system with an associated pressure increase. The magnitude of the pressure change allows the water table depth to be calculated. Estimated resolution is about 1% of WT depth.

of the air column, and therefore the depth to the water table. A basic assumption is that the temperature of the system does not change during the brief closure of the syringe. The method is most useful in determining the initial water table in the use of the ACT system for monitoring water-level changes. The method is sensitive to the precision of the syringe volume change and actual depth to the water table. The deep water tables have about the same percentage of depth resolution as more shallow water tables. While that may be important to the actual water table depth, it is less significant for the monitoring of water-level changes with the ACT method. This is because any error inherent in the tube volume calculation for the water table depth is corrected in the deduced change in the water table. The method is most useful for FLUTe systems that have access only through slender tubes too small for typical water-level meters. For extreme water tables or crooked holes, even the electric tag line measurement of the water level can be difficult.

9.6 EVERSION/INVERSION AIDS

Eversion aids are to allow eversion of a liner without sufficient internal pressure, ΔP, or sufficient friction against the borehole wall or pipe to prevent buckling of the liner during inversion from an oversize hole as described in Section 3.3. The eversion aids shown in Figure 9.14 are several designs that all provide the end pressure against

FIGURE 9.14 Different eversion aids to cause the liner to evert with insufficient internal pressure or aid the inversion of the liner in passages larger than the liner diameter. The two leftmost devices are for different liner diameters and use a push rod that connects to apply the eversion force. The other devices use the weight of the device to aid the eversion or inversion of the liner.

the EP to cause the liner to evert or invert. The two designs on the left side of the photo in Figure 9.14 use a PVC pipe connection to the small PVC pipe to force the device downward against the EP of the liner (see Figure 3.1). The liner is threaded downward through the interior of the PVC ring and back upward on the exterior of the ring. Rollers embedded in the edge of the PVC ring minimize the friction of the liner as it everts through the ring. These are primarily for everting the liner downward although the device can also urge the eversion horizontally or encourage the inversion upward.

The two devices on the right side of the photo in Figure 9.14 are formed of long heavyweights with rollers on the ends. The weights are attached to a flexible tubular member, which surrounds the inverted liner. The inverted liner extends downward through the interior of the assembly and the everted liner passes on the outside of the device. The weight of the assembly aids the eversion of the liner. Rollers on the end of the weights minimize the friction on the EP of the liner as they feed through eversion aid. These are also useful to prevent buckling of the liner during the inversion of the liner, but the weighted devices on the right are only useful for nearly vertical travel of the liner. The weighted aids can be lowered, with a cord attachment, downward over the tether passing over the connection of the tether to the liner and over the liner to the EP for a liner already in place in a borehole. The weighted aids are more helpful for preventing buckling of the liner during inversion from an oversize hole or in long enlargements of the borehole. An example of that use would be the inversion of a liner from a 4″ borehole into a 10″ surface casing. However, in practice, the eversion aids are seldom used as other methods have avoided the problems addressed by the aids. These aids are only practical for blank liners since liner attachments do not pass easily through these devices.

Since eversion and inversion are essential features of many liner applications, numerous other eversion aids have been built and tested. Some were not as effective as desired. Liquid mercury is attractive as the ideal heavy mud but, of course, never allowed.

10 Theory and Application of FLUTe Liner Methods

Characteristics of liner mechanics and liner devices are described in Chapter 3. The use of those methods of eversion or installation for hydrologic measurements and other methods are called applications. A wide variety of these applications use much the same mechanisms to perform very different tasks. This chapter describes the applications of liner characteristics to those functions. The theory of the physical processes invoked is also described. Most of the applications, but not all, are related to hydrologic measurements. Several of the methods described serve as alternatives for some traditional hydrologic measurement methods such as straddle packers. The physical processes of the liner methods and the construction of the devices involve a wide range of processes from diffusion to hydraulic flows, turbulent and laminar, to thermal conduction, optics, radiation transport, electronics, mechanics, and chemical reactions. Other methods such as a variety of drilling methods are used in conjunction with liners to provide useful subsurface results. Some methods and hardware even borrow from climbing and sailing experience. Data recording and processing are other aspects of the applications described. Laboratory analytical methods are employed. Radio frequency welding is a major operation in building flexible liners for these applications. The science of textiles is yet another contribution included in these applications. Of course, mathematical models guided the designs and the data interpretations. The flexible liner applications are simply applied science as needed.

Each application is described hereafter as a separate method with the problem addressed, the theory of the mechanism, the method description, and the results. Finally, the advantages and limitations are described. There is some limited comparison of the methods with alterative measurements, but that is not a detailed account of the differences. In many cases, the liner measurements are complimentary to other methods such as geophysical measurements. There are circumstances where other methods are more useful. Some are noted.

10.1 BLANK SEALING LINERS

Blank liners are those flexible liners without any attachments. Blank liners as used for sealing of boreholes are addressed in detail in the chapter on liner mechanics (Chapter 3). The need to seal an open hole is addressed in Sterling et al (2005). In that paper is described the consequence of an open hole in a contaminated environment allowing contaminated water to connect with fractures deeper in the borehole. Shapiro (2002) also cautions against leaving boreholes open due to the high flow rates possible in open boreholes. The ability to seal the entire borehole quickly and well over long distances is unique as compared to packers, both wireline and straddle packers. It is not easy to install many packers simultaneously in boreholes. It is especially difficult to install packer devices in high angle holes or vertically upward. Clearly that capability of flexible liners is further enhanced with the ability to isolate measurements at different elevations in boreholes. The ability to install cure-in-place liners is not an option with packer systems for more than very short distances. As the experience with sealing liners increased, the use of sealing liners was expanded to the use of the everting liner to measurements of flow and head by application of the simple sealing liner and by the installation of everting liners. Those applications are addressed in subsequent chapters. Adding a variety of cover materials to everting liners, or using a flexible liner simply lowered into the boreholes, has led to further measurements of contaminant distributions and identifications of flowing fractures. Others have used the sealing liner with their own measurements of temperature distributions and pressure distributions. Those methods are described in their publications and only addressed briefly here.

10.1.1   Installation of a Blank Liner

The most common use of a blank liner is to quickly seal the borehole to prevent cross-connection of contaminants in the open borehole. As described in Chapter 3, the installation is relatively simple. However, certain precautions are advised:

- Maintain ~10 lb of tension on the liner and tether to assure a ΔP sufficient to seal the borehole during the eversion.
- One does not always need to evert the liner the entire borehole depth to achieve a sufficient seal.
- Fill the liner to at least 5 ft. above the highest head in the formation and preferably 10 ft. above the highest head to obtain a seal. Installations usually use a higher excess head in the liner.

Without sufficient tension on the liner during installation, the liner may not seal the borehole as it is being installed and the liner can pass the last flow zone in the borehole. Upon inverting the liner for removal, the ΔP can be very high and with no inflow below the EP, the liner cannot be inverted as described in Section 3.3. When only sealing the borehole, stop the installation when the liner installation rate is at ~20″/min and with good tension on the liner, the removal is much faster. Relatively little transmissivity remains below the liner when the installation is halted as suggested.

A common misunderstanding is that the liner must be everted the full depth of the hole for a sufficient seal. As the liner seals the flow zones and slows down, the remaining transmissivity decreases. Most blank liners are removed not long after installation. When inserting a blank liner, to the recommended rate of 20″/min (the liner or tether velocity, not the EP velocity), the transmissivity beneath a 4″ liner is 0.1 cm²/s. That remaining transmissivity assures an easy removal rate and normally has a total installation time of less than 2 h. The T profile is usually done to a 70-fold lower installation rate before stopping. There is little inflow and outflow (the cross-connecting concern) beneath the liner at such low transmissivities.

The liner fill during installation can be an excess head of 15–20 ft. When stopped, 10 ft. of excess head is sufficient to seal the hole.

10.1.2 Transparent Blank Liners

The ability of FLUTe liners to seal the borehole everywhere and still allow a view of the hole wall is unique to hydrologic methods. Transparent liners are described in liner materials (Chapter 5). This kind of liner has the advantage of sealing the borehole and still allowing visible, UV, infrared or other frequencies for viewing the borehole wall. Other kinds of views are also possible through liners such as acoustic tele-viewer images, acoustic coupling of seismic instruments, neutron and gamma emissions from the formations as well as thermal emissions. In this respect, there are many kinds of images "visible" through the flexible sealing liner.

10.1.3 Measurements by Others Using FLUTe Flexible Liners

Once a borehole is sealed with a FLUTe flexible liner, the flow in the open hole is stopped and the borehole is supported against slough. In that situation, the following measurements have been performed by other individuals:

Temperature measurements are done inside the liner to detect changes due to flow in the formation. The borehole water is first heated to be able to map temperature changes due to a cooling flow past the heated interior water of the liner.

The same principle is used to detect cooling of a heated cable emplaced in the open borehole before sealing with a liner. This method uses a fiberoptic cable instead of a traveling heat detector inside the liner.

Recording pressure transducers suspended on a thin cable have been isolated in an open borehole using a flexible liner. In time, the liner is removed and the transducers recovered for the download of the recorded pressure history.

Acoustic tele-viewer images have been obtained with the sondes normally used in open boreholes. The acoustic images are little affected by the liner (see Chapter 5).

Other geophysical measurements such as induction-coupled electric resistance measurements, gamma logs of several kinds, and natural and unnatural radiation measurements are performed in otherwise open holes without the concern of borehole slough entrapment of the sonde. Optical tele-viewer measurements can see the presence of fractures, vugs, and breakouts in the wall of the borehole due to the ability of the liner to conform closely with the hole wall (see Figure 3.9).

These kinds of measurements are not addressed here except to demonstrate the utility of the blank liner.

The FLUTe use of blank liners for other kinds of measurements is addressed next in this publication with more details available in some peer reviewed publications and the FLUTe website, www.flut.com. It is noteworthy that liner stretch is not important to the installation of sealing liners alone.

10.2 FLUTE BLANK LINERS WITH SPECIAL COVERINGS

10.2.1 The NAPL FLUTe

10.2.1.1 History of NAPL FLUTe Development

The first use of a special covering of a blank liner was in response to a request by Joe Rossabi when he was at the SRS (Savannah River Site). His need was for identification of DNAPL subsurface in cone penetrometer holes. Joe first used a covering which he and Brian Riha called a Ribbon NAPL Sampler (RNS) in open stable holes left after withdrawal of the cone penetrometer rods in the vadose zone. He covered a 2″ FLUTe liner with the RNS and everted the liner from an air canister built by FLUTe into the open vadose zone cone penetrometer hole. It worked very well because in the vadose zone the cone penetrometer holes stayed open after withdrawal of the rods. Later, in 1997, he requested that FLUTe invent a means of mapping NAPL using the same reactive covering in the unstable portion of the cone penetrometer holes below the water table. FLUTe invented the method now called a NAPL FLUTe. The system consisted of a slender liner covered with the RNS material. The RNS was a thin hydrophobic material dusted with Sudan IV, a toxic powder that turns red upon contact with NAPL.

Because the RNS used a compound called Sudan IV to produce a red stain upon contact with NAPL including any oils, FLUTe initially used the same reactive material. However, a big problem with the Sudan IV is that it became known to be a hazardous material that can only be applied in full Tyvek suits with at least a half mask respirator. Being a powder, it is difficult to control in a manufacturing setting and appears as red stains on door knobs and other oily surfaces. As a hazardous material, FLUTe abandoned its use and invented a different approach using a dye striping of the same hydrophobic liner cover material made of Tyvek. The original dye stripe was black. But the black mud of the peat bogs of Denmark made the black dye difficult to see. The black stripe was replaced with a red, a blue and a black dye stripe alternated every 1/4″ since few muds match those colors.

The dye-striped cover uses a different principle for NAPL detection. The solvent such as TCE or PCE upon contact with the dye stripes dissolves the dye and carries the dye through the cover to the white inside surface. The result is a bright stain on the white side of the cover where it has contacted NAPL. The dye is labeled nontoxic. The method to apply the stripes of different color required the design of a special machine to apply the dye with no bleed to the other side of the cover. NAPLs that are not solvents such as oils will not transport the dye, but produce a stain on the cover that is translucent, like oil on white paper, making the dye strips more visible. Some NAPLs have their own color readily apparent on the cover white side as they are absorbed in the cover.

This NAPL FLUTe method was used at several sites beyond the SRS such as Cape Canaveral, FL. The method was extended to installations in Geoprobe direct push rods of larger diameter (2.25″). Twelve such dye-striped liners were installed in Denmark for Anders Christensen of NIRAS in the early most slender Geoprobe rods (1.5″ OD). It was found after two failures that the rods needed to be clean and smooth, free of rust and old grout deposits, to avoid excessive friction on the liner. The rods were thoroughly cleaned with a brush. The subsequent installations were better. As the diameter of both the interior and exterior of direct push rods were increased by the manufacturers, the same method was used to install somewhat larger diameter NAPL FLUTe liner systems. The same technique is now used in sonic casing.

10.2.1.2   How the NAPL FLUTe Is Installed in Direct Push Rods

The NAPL FLUTe for the application in direct push rods has a central water addition tube and the liner and cover are compression wrapped with a weak thread to allow

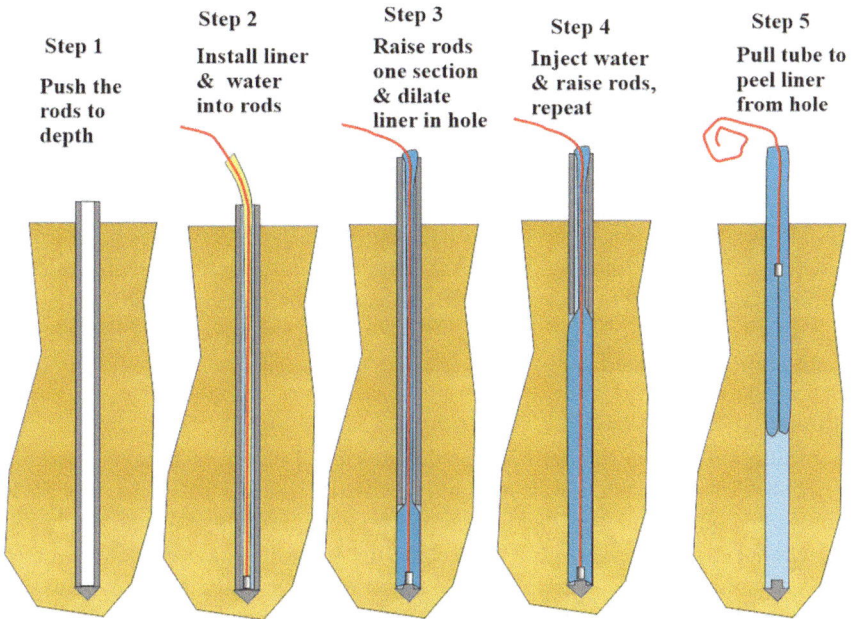

Step 1

Push the rods to depth

Step 2

Install liner & water into rods

Step 3

Raise rods one section & dilate liner in hole

Step 4

Inject water & raise rods, repeat

Step 5

Pull tube to peel liner from hole

FIGURE 10.1 Steps of the NAPL FLUTe installation through direct push rods.

the installation in the very small (~1″) interior hole of a cone penetrometer rod. The rod is driven to depth with a disposable tip (see Figure 10.1). The compressed NAPL FLUTe liner is installed through the central hole in the rod. Water or air is added to the rod interior between the compressed liner and the rod wall. The rod is withdrawn a short distance and the liner dilated immediately with the injection of a measured amount of water or air through the central tube to anchor the liner in the open hole beneath the rods. The dilation bursts the compression threads and forces the liner and cover against the hole wall. More water or air is added to the liner to further burst the compression wrap and to partially fill the liner still inside the rods. Then the rods are withdrawn one more section length. More water or air is added to the liner through the central tube and the water level, or air pressure, in the annular space between the liner and rod wall is kept filled to prevent significant drag of the liner on the rod as it is withdrawn. The sequence depicted in the drawings of Figure 10.1 is often used to emplace the RNS and NAPL FLUTe against the hole wall. After a relatively short time (15–20 min, longer sometimes) tension is applied to the central tube to invert the water-filled liner and cover from the hole.

The contact of the dye-stripped cover with NAPL underwater causes the NAPL to dissolve the dye and transport it to the other side of the cover that is normally white. The dye displacement results in a bright obvious stain on the white inside surface of the cover material. Using the same installation method as for the RNS, FLUTe was able to map the contact with NAPL at the hole wall in the United States and Denmark. The installations in Denmark showed the presence of coal oil and other NAPLs, which were not strong solvents like TCE and PCE.

The dye displacement is much reduced with common oils, but the contact produces an obvious wetting of the cover much like a drop of oil on paper. The dye stripes are more visible, but did not show the melding of the different stripes. The strong solvents produce a purple stain of the melded stripes. For less strong solvents like gasoline and diesel fuel, the dye strips are prominent on the backside of the cover, but not combined so much as with TCE or PCE.

The NAPL FLUTe cover will not react to an exposure to the dissolved phase of the NAPL. An exception is the weak diffusion of the dye if the cover is left in place subsurface for a long time (e.g., 2 weeks). In that case, the red dye strip can cause a faint pink discoloration of the cover in the presence of a probable high concentration of the dissolved phase of TCE or PCE. If the cover does not contact the NAPL, no obvious stain is produced.

10.2.1.2.1   Installations in Direct Push Rods with Water

The installation in such slender holes as the interior of direct push rods is dominated by friction. First is the friction of sliding a flexible liner down a 1″ ID rod for 50 ft. or more which requires that the liner be compression wrapped about a central tube. After being emplaced in the rod with a disposable tip, the rod is raised to expose a short portion of the formation. The previous addition of water to the annulus between the compressed liner and rod wall compresses the liner further and reduces the friction as the rod is withdrawn. Without the water in the annulus, the liner will rise with the rod withdrawal. Another advantage of the annular water is to resist water flow into the bottom of the rod with the associated "sand heave" which carries flowing sand into the annulus between the liner and rod further locking the rod to the liner assembly.

Once the rod is lifted, exposing the formation, the liner is immediately dilated in the open hole by water injection through the central tube to the bottom of the liner. The annular water head above the formation head prevents immediate collapse of the hole wall on the liner. The liner dilation forces the annular water into the formation. The dilation also forces some annular water upward in the rod. The liner dilation also bursts the compression wrap and presses the liner firmly against the formation, anchoring the liner in the hole. More water is injected into the liner at the bottom end further bursting the compression wrap of the liner higher in the rod. The flexible liner is now unconstrained in the rod with the liner interior head greater than that in the formation.

However, that lower formation head at the bottom of the rod allows some annular water to pass from the rod into the hole which dilates the liner against the rod wall for some distance up inside the rod. At that time, the full advantage of the annular water is evident. Above that dilated interval at the bottom of the rod, the annular water and the interior liner water are in equilibrium because the liner is flexible. The annular water minimizes the drag of the liner on the rod interior. Therefore, the liner is only pressed against the rod wall for a short distance at the bottom of the rod (determined experimentally to be about 8″ long).

Because of the large surface area of the interior of the rod, any pressure, ΔP, in the liner would produce such high rod drag on the liner as to tear off the liner with the next rod withdrawal. The surface area of 50 ft. of 1″ rod wall is about 1800 sq. in. A 1-psi ΔP and 0.5 friction coefficient would produce 900 lb of drag. By maintaining the annular water, and with the reduced liner friction against the rod wall, the rods can be withdrawn without damage to the liner or reactive cover. The large surface area of the hole wall below the rod provides an excellent anchor of the liner as the rods are withdrawn. The effect of the large surface area and associated drag with a small differential pressure is often an advantage, and also a disadvantage, in several liner applications. After each subsequent rod withdrawal, another measured volume is injected to dilate the compression wrap and to store more water inside the liner. As the rod is withdrawn, the water inside the liner at the head of the annular water dilates the liner immediately below the rod preventing the formation collapse on the liner. This procedure is continued until all the rods have been withdrawn or until the injected water dilates the liner above the top of the rod.

10.2.1.2.2   Installation through Direct Push Rods with Air

The water addition method of NAPL FLUTes is less attractive in the vadose zone due to concerns about the effect of the annular water in the rod, exterior to the liner, on the NAPL in the formation when that annular water moves into the formation as the rod is withdrawn. For that reason, the installation of the NAPL FLUTe with air as the driving fluid was perfected. The same principles were applied. The liner is dilated with air instead of water and the air pressure between the liner and the rod wall must be the same as the air pressure inside the liner to prevent the otherwise excessive friction of the liner on the rod wall as the rod is being withdrawn.

The air installation requires that the central tube be sealed to the top of the liner to be pressurized. Also, an air tight seal on the top of the rod is required to allow air to be injected into the annulus to reduce friction as the rod is withdrawn. The control of the air flow and pressure makes the installation more complicated. After the rod has been removed, the air pressure maintains the liner dilation in the vadose zone. Water is then added to the interior of the liner to displace the air to maintain the interior pressure and to allow the liner to be inverted in the same manner as for the water installation. As the water is added, the air is allowed to vent in a controlled manner via a small slit in the top of the liner. The method was employed for the first installation of a FACT (FLUTe Activated Carbon Technique) which was done in the first commercial installation of the FACT in Denmark. The method is most desirable when installing a NAPL FLUTe with the addition of the FACT in the vadose zone. For the detection of the dissolved phase, the device called a FACT was invented as described in Section 10.2.2.

10.2.1.2.3   Imperfect Installations in Direct Push Rods

While most direct push installations work well, the following situations, with possible remedies, can frustrate the method of installation:

1. If the disposable tip does not pull out of the rod upon lifting of the rod. (The remedy is, lubricate the o-ring of the tip prior to emplacement in the rod.)
2. When the rod is first raised, before the liner is dilated with the water addition, water inflow from the formation into the hole below the rod may carry a sand fill into the bottom of the hole before the liner is dilated. (Make sure the annular water is maintained to resist the inflow.)

3. As the rod is withdrawn, a layer of very plastic clay is exposed, which squeezes the hole shut despite the liner interior water pressure. (Difficult to fix. Not common. The plastic clay is loaded by the lithostatic load above and cannot be resisted by a water excess head in the liner.)

10.2.1.3 NAPL FLUTe Installations in an Open Stable Borehole

The current most common installation of the NAPL FLUTe is in open stable boreholes as a cover on an everting liner. Figure 10.2 shows a dye-striped cover on a 4″ diameter everting liner. The liner is installed as described for a blank liner.

Installations of NAPL FLUTe covers in open holes are usually done with a pump tube first installed in the borehole to remove the contaminated borehole water from beneath the liner as the liner is everted to the bottom of the borehole. Since the cover reacts very quickly to the contact with a NAPL, the pump tube can be left in place until the NAPL FLUTe is removed by inversion. The pump tube aids the liner inversion upon removal by allowing flow from fractures anywhere in the borehole to supply water to beneath the inverting liner by flow through the interstitial space next to the pump tube. In some situations, with DNAPL ponded in the bottom of the hole, the liner can force the DNAPL to rise adjacent to the pump tube. The resulting long stain associated with the large stain on the bottom end of the cover, makes that situation obvious.

FIGURE 10.2   Dye-striped exterior of a NAPL FLUTe cover as mounted on the exterior of an everting blank carrier liner.

10.2.1.4 NAPL FLUTe Covers over Core

As with many FLUTe inventions, the customers often wish to use it in different circumstances. The application to core ex-situ on the surface, versus in situ as designed, was desired as less expensive. The usual admonition regarding evaporation applies to wrapping or enclosing the core against the NAPL FLUTe cover. If evaporation can be limited as when extruding the core into a plastic covered NAPL FLUTe sleeve, the method can be useful. See the method of installation of a cover over the core barrel for sonic core in the photos of Figures 10.2.1 and 10.2.2 with the plastic cover emplacement over the NAPL FLUTe cover. In covering the sonic core, the dye stripes should be on the interior of the cover and the cover completely enclosed in the typical heavy clear plastic sleeve, usually used to enclose sonic core, as the core is extruded into the interior of the NAPL FLUTe cover.

After the core is in the cover, the plastic cover is knotted on each end to seal the plastic sheath reducing evaporation. The concern is that the core extracted from subsurface will have lost the normal pore fluids due to borehole water flushing the core during removal from the borehole. If the core contains clay layers to reduce the water flow through the core as it is brought to the surface, the pore water loss should be less. Laying the covered core on the ground allows DNAPL to drain to stain the cover as seen through the clear plastic. The core can often be flattened in the plastic covering to obtain better contact with any entrained DNAPL. Inserting a photo ionization detector tube through the plastic covering

FIGURE 10.2.1 Sliding NAPL cover over sonic core barrel with stripes on the inside of cover.

FIGURE 10.2.2 Sliding clear poly cover over NAPL FLUTe cover on core barrel. Core is then extruded into the concentric covers.

allows detection of NAPL vapors which can pass through the perforated hydrophobic cover material.

The use of simple application of the cover material to core extracted, such as with Geoprobe core, is not recommended because of rapid evaporation of the volatile NAPL before a substantial stain can develop. The application of drilled rock core to the NAPL FLUTe cover is also unlikely to produce a stain since evaporation of the NAPL in the core may be too fast. FLUTe has a video available showing the stain development under water and the rapid evaporation of TCE in air. The in situ measurement on a sealing liner is recommended rather than assessment of core ex situ.

10.2.1.5 NAPL FLUTe Sand Bags

Another use of the NAPL FLUTe cover material is with the NAPL sandbag. A sandbag smaller in diameter than the borehole is covered with the NAPL FLUTe cover material with the dye stripes on the outside. The sand bag can be lowered to the bottom of the well on a strong cord to contact any NAPL on the bottom of the borehole or screened casing. The DNAPL will be clearly evident on the cover material. This is useful for DNAPL layers too thin to be detected with an interface probe or too sparse to be pumped out of the well. The advantage of the sand bag is that it is much less expensive than a covered liner which is used to locate vertically the source of DNAPL. With extra cover material and with cleaning of the sandbag made of coated nylon, the method can be used in multiple wells.

FIGURE 10.3 Stained NAPL cover pulled out of direct push rod hole. The tape measure determines at what depth the NAPL is found in the formation.

10.2.1.6 Examples of NAPL FLUTe Stains

10.2.1.6.1 Direct Push Stains

The first application of the NAPL FLUTe was with the RNS. Later installations used the dye-striped version installed in direct push rods.

The direct push installations of NAPL FLUTe in an old quarry in Elton, MD provided some excellent examples of NAPL stains as seen in Figures 10.3 and 10.4.

The photo in Figure 10.5 shows how well the liner dilates below the rod in the borehole as the rod is withdrawn.

10.2.1.6.2 Open Borehole Stains

The NAPL cover shown in Figure 10.6 was installed by eversion into a 3″ diameter borehole in PA after the hole was cored. It was noted that the core did not show

FIGURE 10.4   The stain upon passing through a NAPL pool. The long streak may have been pushed down by the rod insertion.

FIGURE 10.5   The liner dilation as the rod is pulled from the ground. Note how well the water-filled liner dilates to more than the rod diameter to seal the open hole as the cover wicks the NAPL into the cover to produce the stains observed. The canister on the right side is the pressure injection device used for early installations to inject the water inside the liner as the rods are withdrawn expanding the liner cover to the hole wall.

FIGURE 10.6 Stains from TCE in fractures. The normally white inside surface of the NAPL cover shows where TCE has contacted the cover in a 3″ core-hole. The core from this borehole showed no NAPL in the fractures.

any TCE in the core fractures and it was decided to install the NAPL FLUTe only because it had already been purchased. The resulting massive stains showed DNAPL in the fractures, and also rivulets of TCE trickling down the borehole wall that were intercepted by the everting liner as the liner presses against the hole wall. It is often the case that the core does not show NAPL and yet it shows on the NAPL FLUTe. At another site in the UK, FLUTe, with the NAPL FLUTe liner, was waiting for the coring to be completed at a xylene spill site. Again, the core showed no NAPL. Yet the NAPL FLUTe after installation and removal did show stains well below the water table. The customer speculated that the liner eversion had simply displaced the xylene down the borehole to produce the stains because there was no evidence in the core. However, several days later it was observed that while baking in the sun in the core boxes, the porous sandstone core began to bleed xylene at the fracture interface elevations seen on the NAPL FLUTe. The generalization is that the core often does not reveal the presence of NAPL in the fractures. Only crushing and analysis of the core samples would indicate the earlier presence of NAPL in the fractures. However, a formation with small porosity may not have NAPL in the pore space because of the difficulty of NAPL displacement of water in small pores.

Figures 10.7 and 10.8 show other stains of TCE from a site in Denver and of coal tar in boreholes in Rye NY. The NAPL FLUTe has been used at numerous manufactured gas sites to locate coal tar. Other NAPLs have been tested against the NAPL FLUTe cover. The xylene stain in Figure 10.9 was from the core hole in the UK described above. The example of a gasoline stain, an LNAPL, is shown in Figure 10.10 as seen in a test of a small amount of gasoline on top of water in a

FIGURE 10.7 Stains on the inside surface of a NAPL FLUTe showing the location of the DNAPL. The tape measure on the right provides the depth BGS.

graduated cylinder. Because the gasoline is wicked upward and downward from the water interface, the NAPL FLUTe stain is not a good measure of the thickness of LNAPL, but rather only an indication of the presence of LNAPL.

The timing of the stain development on the cover material is shown in a video available from FLUTe, but the cover absorbs TCE in 10 s to 1 min. A 15 min residence in the borehole is sufficient unless the installation has displaced the NAPL and time is needed for the NAPL to return to the borehole wall for contact with the cover. Some installations are left in place overnight.

A more reliable measure of nearby NAPL that does not require contact is provided by the FACT (Section 10.2.2) which maps the dissolved phase of the NAPL. If NAPL is nearby, the relatively high concentrations in the FACT are indicative.

It is noteworthy that if one simply drips a drop of TCE on the dye-striped side of the cover material in air, the TCE instantly spreads and evaporates, leaving a very

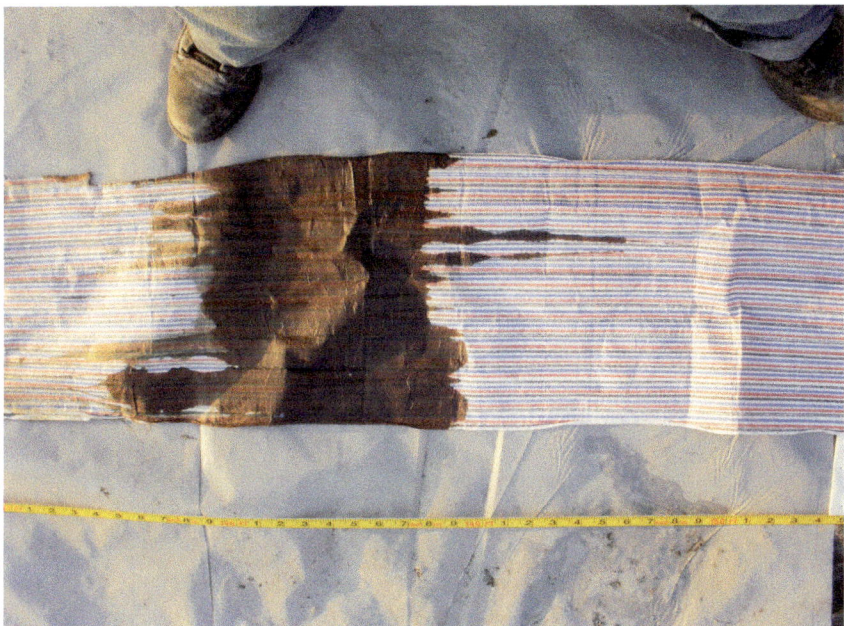

FIGURE 10.8 Coal tar stains at a fracture observed on the NAPL FLUTe cover in a 4″ core hole.

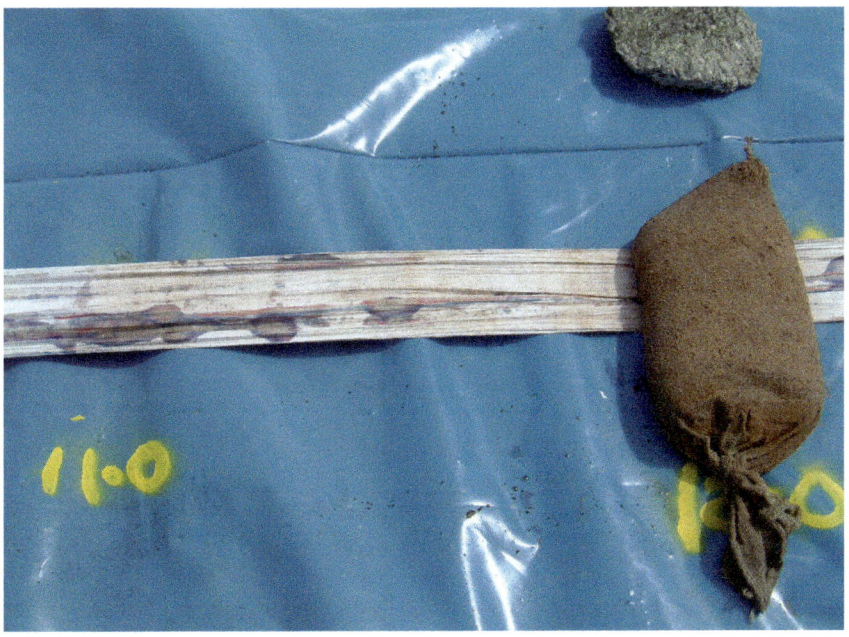

FIGURE 10.9 Xylene as mapped at UK site in porous sandstone.

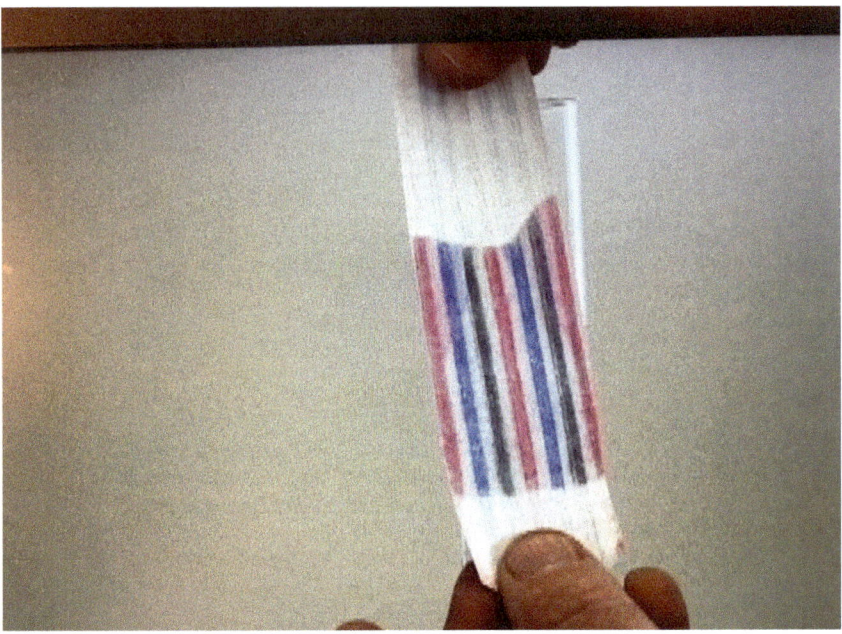

FIGURE 10.10   LNAPL as a thin layer on water. Note that the migration in the NAPL cover is not an indication of thickness of a thin layer on the water.

faint stain on the cover material, whereas if the contact is underwater, the typical prominent stain is produced. Therefore, the use or test of the core with the cover material in the open air is not recommended.

The stretch of the liner is important to the depth determination of NAPL stains. The method for estimates of stretch and the significance to NAPL FLUTe measurements is addressed in Section 10.6.

10.2.2   The FACT Application

10.2.2.1   History and Experience

FACT is the name given to the FLUTe Activated Carbon Technique. The FACT was developed to map the dissolved phase of contaminants which is not done by the NAPL FLUTe which only reacts to the pure product phase of NAPL. It was reasoned that a continuous map was more useful than water samples drawn from discreet intervals based on incomplete information on the contaminant distribution. Core assessment is another method of mapping the dissolved phase but core does not include the pure product or a fracture water measurement. Also, the core samples are also selected at discrete locations in the core and are not a continuous map of the contaminants.

The FACT method was first tested in the vadose zone in 2010 by Anders Christensen of NIRAS in Denmark, far from FLUTe's plant in New Mexico. It is surprising that the first test was so far from the United States. NIRAS was also one of the earlier users of the NAPL FLUTe in Denmark about 11 years earlier. Bill Davis of Triad Environmental, a U.S. company, did the first FACT carbon sample analysis in Denmark using his portable magnetic spectrometer at that same time. The first test, installed with air, was especially rigorous, as will be described, because of the comparison with two other kinds of measurements of the same contaminant only about 1 ft. from the FACT measurement. The FACT worked very well. Now the FACT method is used in many boreholes in other parts of the world for mapping the dissolved phase of contaminants often in combination with the NAPL FLUTe for mapping the NAPL phase. Not long after the first test in Denmark, the Danish Technical University (DTU, Broholm et al, 2016) used the method at a field site in the saturated zone in open stable boreholes. The DTU developed a master's thesis (Beyer, 2012) which investigated many of the questions arising for the use of the FACT method. The Broholm paper describes many other FLUTe measurements made at the same site as the FACT measurement, providing useful information on the situation of the FACT test.

The FACT measurement has become a common FLUTe measurement with over 800 FACT samples in 2021 analyzed for just one site. Many hundreds of FACT samples have been analyzed more recently for other sites in the United States and Sweden. The combination of the FACT contaminant distribution and the transmissivity profile method described in Section 10.3 is particularly useful in assessment of the contaminant transport in fractured media. Comparison of FACT results and other measurements are shown later.

10.2.2.2   The FACT Method

The FACT (FLUTe activated carbon technique) concept is simple. An activated carbon felt strip is added to the NAPL FLUTe cover (now called a NAPL/FACT). A diffusion barrier is added to isolate the carbon from direct contact with the liner to prevent diffusion from the liner material (e.g., toluene) or diffusion through the liner material from water inside the liner which may not be potable water. Figure 10.10.1 shows the FACT geometry in cross-section. The photo in Figure 10.10.2 shows the carbon between the dye-striped cover and the aluminized, silver-colored, diffusion barrier. On each side of the carbon, the diffusion barrier is stitched to the NAPL cover material encapsulating the carbon. The cover material is very thin (~1 mm) and perforated for easy diffusion through the cover to the carbon. The NAPL striping is not always used with the FACT, but more often than not. The cover material with the FACT, and often the NAPL striping, is a covering on the outside of a flexible liner, which is emplaced in the borehole with the inflated liner pressing the cover and carbon strip firmly against the hole wall while also sealing the borehole.

The purpose of the 1.5″ wide by ~1/8″ thick activated carbon felt along the entire length of the liner is to adsorb contaminants from the liquids in the formation. The carbon adsorbs by diffusion from both the pore space of the formation and from any fractures in the formation. The FACT works equally well in the vadose zone and in the saturated zone of the formation. Interestingly, the very first test of the FACT in Denmark was in the vadose zone. It was remarkable how well the FACT was tested the first time and how well it worked. The test comparison was with both Geoprobe MIP measurements and soil samples, both only ~1 ft. distant from the FACT hole.

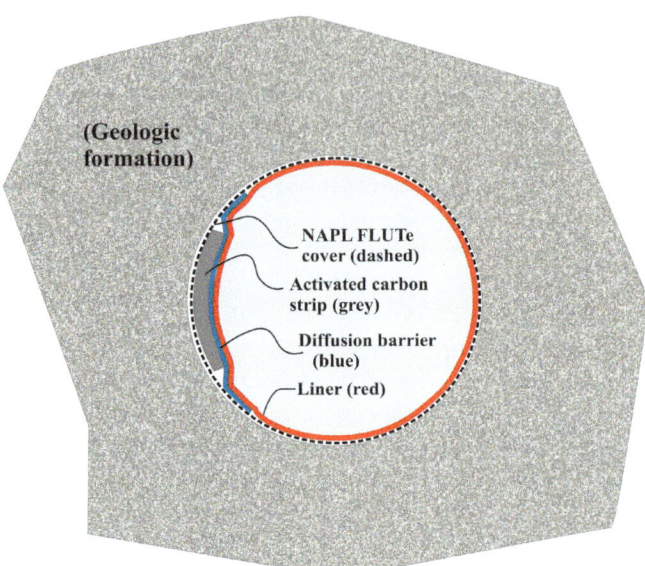

FIGURE 10.10.1   The cover for the liner (dashed curve) contains an interior-activated felt strip (gray) separated from the liner by a diffusion barrier (blue). The diffusion barrier is sewn to the cover and encapsulates the carbon. The cover assembly is exterior to the liner (red) and everted with the liner into position against the hole wall.

Each method showed essentially the same contaminant distribution. The first measurements were through drilled holes in the floor of a commercial dry-cleaning company in Denmark. The FACT was installed through the interior of direct push rods using a Geoprobe machine. Figure 10.10.3 shows the comparison of the gas sampled at many elevations during the MIP traverse and the FACT samples analyzed with a mass spectrometer using EPA method 8260B. The distribution of the contaminant is the same for the two methods and compared equally well with soil core samples. The following description in Section 10.2.2.4 of the FACT measurements includes results obtained at the Naval Air Warfare Center (NAWC) near Trenton, NJ. In that installation, there were many other kinds of measurements to compare to the FACT results to define the environment of the measurement.

When the activated carbon is installed against the borehole wall, the diffusion process is initiated. The activated carbon is composed of a carbonaceous felt material which is then heated to extreme temperatures to leave only the carbon material. The carbon has the extreme surface area typical of activated carbon. The activated carbon is useful for filtration of many kinds of fluids such as air and water to remove contaminants of many kinds. Absorbers were first used on a flexible liner in California in 1991 to collect tritiated water samples in the vadose zone. In the FACT application, the contaminants of interest are those VOCs typically extracted with EPA method 8260 which is suitable for solids and works well for activated carbon. In short, the process involves submerging the FACT carbon in methanol, shaking, extracting a small volume for analysis in a GC/MS (gas chromatograph mass spectrometer) to determine the compounds in the small methanol sample. That

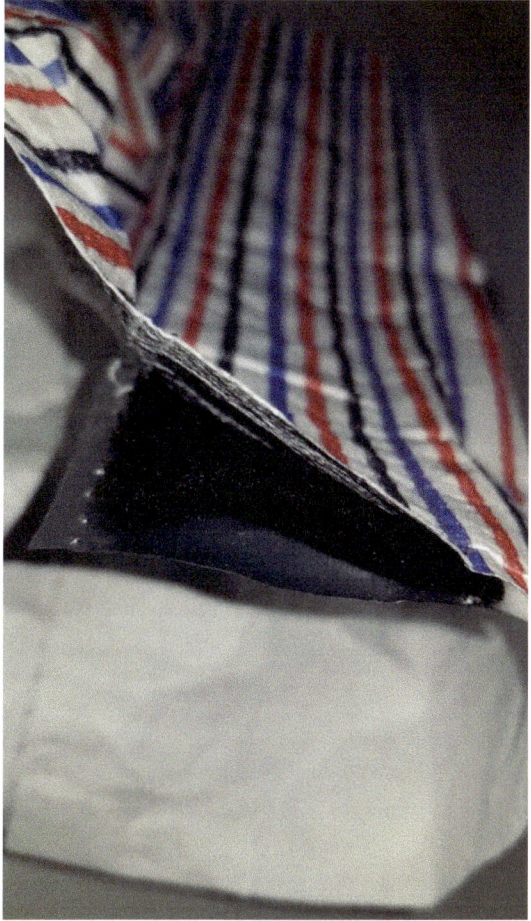

FIGURE 10.10.2 The cover is dye striped for NAPL detection. The carbon felt (black layer) is isolated from the liner by the aluminized diffusion barrier as shown in Figure 10.10.1

contaminant mass is used to calculate, with the carbon weight, the mass of contaminant per gram of carbon of that compound which is the data reported. The activated carbon strongly holds the compounds adsorbed and has not, in FLUTe applications, ever been saturated.

As a sink/adsorber of VOCs from both air and water, the carbon contaminant mass is only limited by the rate of diffusion from the surrounding medium to the carbon. The diffusion rate in air is nearly 10^4 times the diffusion rate in water. The water diffusion rate for TCE is 8.16×10^{-6} cm^2/s (Schaefer et al, 2012). Therefore, the TCE, for example, diffuses to the carbon in the vadose zone much more quickly than when under water. In the vadose zone in Denmark, it was found that the content of PCE was sufficient for analysis in two days. The master's thesis of Beyer addresses the rate of adsorption by the FACT material in water. A much longer exposure is used for the saturated zone.

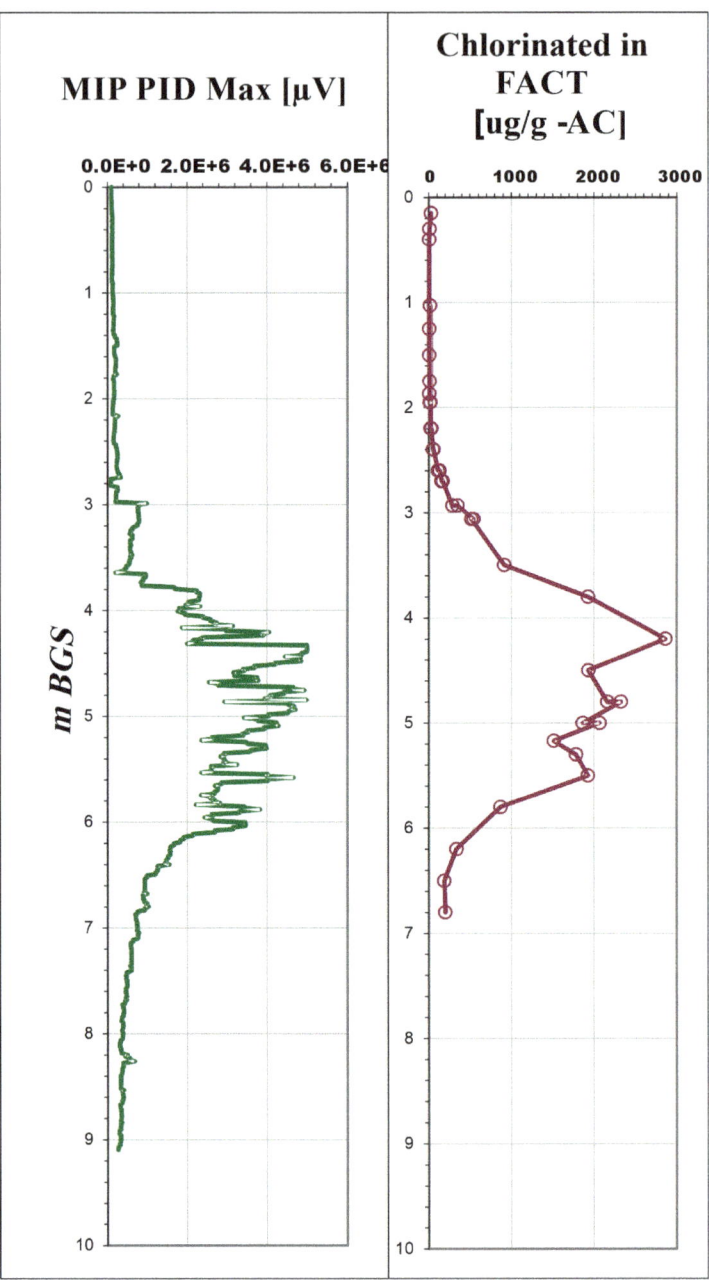

FIGURE 10.10.3   The first test of the FACT in Denmark shows how well the FACT distribution compares to an MIP measurement separated by 1 ft. from the FACT installation.

The rate of adsorption in the carbon, QD, is controlled by the concentration gradient $\delta C/\delta z$, and the diffusion coefficient FC and the surface area are exposed such that $QD = \delta C/\delta z$ FC A, where A is the surface area through which the compound is diffusing. Using that relationship, a finite difference model was written and tested against an analytical solution. The model was used to estimate the rate of transport through the formation matrix material with a porosity of ε filled with water. The particular formation of interest was the NAWC site. The paper by Schaefer et al (2012) provides the relationship of the effective diffusion rate, FC_{eff}, through the pore space of the NAWC formation as $FC_{eff} = FC\ \varepsilon^m A$, where m is 1.7 and ε is 7% porosity (ranging from 4 to 10%). For these values suggested by Schaefer, the normal diffusion coefficient in water is reduced by 99%. This suggests the diffusion rate from the pore space is ~1% of that from fracture water in contact with the FACT carbon. However, for the diffusion calculation provided for the NAWC site discussion in Section 10.2.2.4.2, another important factor is the gradient in the case of fracture water versus pore volume contribution to the FACT. As shown there, the concentration gradient from the pore space is much less and the much larger surface area of the matrix in contact with the FACT carbon may not offset the concentration gradient from the fracture water. The drawing in Figure 10.11 shows the

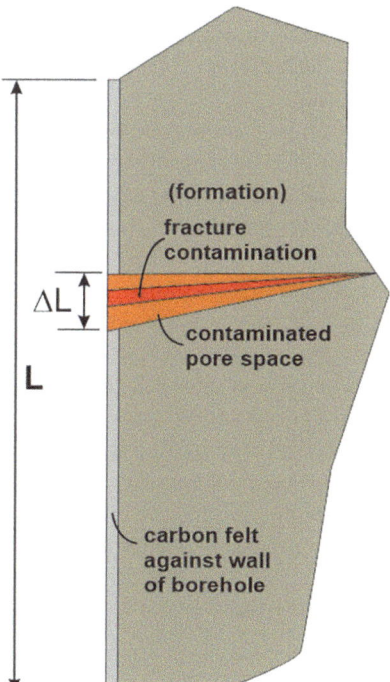

FIGURE 10.11 The FACT adsorption from the fracture and the adjacent pore volume. Only a short interval of the carbon may contain the fracture and pore water contamination. This relevant to the length of sections analyzed when comparing analytical results to the fractures identified.

typical geometry of the carbon against a fracture and which part of the carbon strip is exposed to the fracture water and to the matrix. Some contaminant is expected to diffuse from the fracture water into the pore space over time. Therefore, the carbon will adsorb contamination from the fracture and from the adjacent pore space. However, as discussed in Section 10.2.2, the fracture water, if flowing, is maintaining a very steep concentration gradient for diffusion to the carbon, whereas the concentration gradient in the pore space is decreasing in time as the carbon draws contaminants from the pore space. The expectation is that the fracture water flow will contribute contaminants to the carbon at a higher rate than the pore water. Those generalizations are reasonable unless the fracture water is at a much lower concentration than the pore water. That is possible if the pore space was contaminated by extremely contaminated fracture water and then later the fracture water is flushed by infiltration and the fracture water concentration is much lower. A later comparison of the carbon levels to the water samples obtained, primarily from the fractures, suggests the fracture contamination to the FACT is often dominant. This is supported by the comparison to crushed core analysis and water samples. It was observed at one fracture location that the pore water was higher than the fracture water as deduced from a comparison of core water with the FACT and a water sample.

The measurements along the borehole must be comparable in a relative sense, because the purpose of the FACT is to measure the contaminant distribution. If as shown in Figure 10.11, only a short interval of the carbon is exposed to the fracture contaminants, and the contaminant μg/g of carbon measured for that single fracture depends on the mass of carbon in a section whose mass depends on the length of the carbon; then, the distribution is most useful if all the carbon sections analyzed are of the same length. It is also very important that the entire carbon length be analyzed and the samples are not picked at different locations such as at fractures. The reason is that, as shown in the next chapter, the high flow zones often do not yield the most contaminated intervals. Another question is how long should be the carbon sections. If twice as long, the number of samples to be analyzed is half as many and half the expense. As will be shown, the sectioned lengths depend on the spatial resolution desired. The typical intervals are half a foot to 3 ft. The concentration in the methanol extract is the actual mass of contaminant in each section of carbon. The contaminant mass/gram of carbon is less for the longer carbon sample if the carbon section only intercepts a single fracture. All the carbon segments are of the same length, which does not change the relative contaminant distribution. It is worth remembering that the entire strip should be analyzed because the carbon does not allow contaminant to migrate along the length of the carbon; therefore, some contamination would be missed with an analysis less than of the full carbon length.

10.2.2.3 Assessment of the FACT Method

As a new method of possible use by the hydrologic community, the FACT method met the usual skepticism. It was suggested that it would not work as proposed and could be misleading. FLUTe recognized the concerns and has addressed them. An obvious question was how does the FACT data compare to the other accepted methods of measurement of contaminant distributions? Another question was what are the perturbations of the method that could be misleading? The question list is:

1. How does the FACT contaminant distribution compare to that obtained with:
   a. Crushed core analysis
   b. Water samples from multilevel sampling systems or straddle packers
   c. Measurements using direct push methods
   d. Soil samples

2. In open stable holes, what are the effects of:
   a. Drilling
   b. Open borehole contaminated water flow in the hole
        i. Contaminant diffusion from the borehole water into the formation
        ii. Borehole water flow into fractures before the FACT emplacement
   c. Exposure of the FACT to borehole water contamination during installation
   d. Trapped borehole water between the liner and a rough hole wall
3. What is the influence of handling and analysis methods on results?

Assessment of potential perturbations by comparison with traditional measurements is subject to the question of whether traditional measurements are comparable or even as reliable. The comparison with the MIP and soil samples described above should be acceptable. However, in open holes for the above questions, one can make the comparisons or estimate the significance with calculations of known processes to assess potential perturbations. Without direct comparisons or calculations, the opinions are unbounded. In some matters, the concerns are not easily quantified and the concern can be reduced by eliminating or reducing the possible effect.

10.2.2.3.1   The Drilling Effect on the FACT Measurement

Drilling is a potential perturbation of all measurements in boreholes. It is somewhat surprising how the effects of drilling have become accepted and ignored. An example is drilling with recirculation of the drilling fluid as is common in core drilling. The recirculation of the drill water is sure to redistribute the contamination when drilling through NAPL as compared to when not recirculating the drilling fluid. A concern about that effect has led to a removal of the outer layer of core that is to be analyzed to avoid the effect. However, the borehole wall also is affected by the recirculation. The effect may be avoided by using a drilling method that is more rapid and does not recirculate the drilling fluid.

10.2.2.3.2   The Open-Hole Flow of Contaminated Water

During the drilling process, especially the slow core drilling, it is not uncommon to leave the drill casing in the hole overnight which allows cross-connecting flow. After the hole is completed, until it is sealed, the open hole also allows cross-connection of aquifers and migration of contaminants to intervals of higher or lower contamination depending upon the hydrologic characteristics of head and transmissivity in the formation. The hole is often left open to allow for the traverse of geophysical instruments. It is recognized that the time the borehole is open is best limited when there is a need to map the natural contaminant distribution in the formation.

The open-hole effects on the FACT include both diffusion into the pore space of the formation and the flow of borehole water into the fractures. Those are addressed later, but it is prudent to develop the well to both remove the drilling cuttings from fractures and to somewhat restore the natural fluids in the fractures. In this chapter, both effects are addressed in a discussion of the data and the calculations done.

10.2.2.3.3   FACT Exposure during Installation

As shown in Figure 3.3 of Chapter 3, the liner carrying the carbon in the cover passes from the inverted state inside the liner to the everted state against the hole wall. During that transition, the cover is exposed to the borehole water. That time of exposure is addressed later. As the liner everts, the seal to the wall is generally good as seen in Figure 3.9. However, some borehole water can be trapped in the vugs, fractures, and breakouts seen in the photo. Obviously, a smooth hole wall is desirable, but some drilling methods especially in some formations produce a rough hole wall. That is also a problem for straddle packers. The effect of the trapped water on the FACT is addressed with a caliper log and a calculation described later.

10.2.2.3.4   Handling and Analytical Effects

The FACT uses an activated carbon felt for the adsorption of contaminants. The standard operating procedure involves some air exposure which can result in some contaminant loss. That is addressed by the Danish Technical University measurements (Beyer, 2012). The loss is small in the exposure times of handling (less than 50% loss in two days exposure in air). The use of DI water as the preferred shipping medium of samples, versus methanol or pentane for the extraction fluids, is addressed in the DTU thesis by Beyer and in tests done by the University of Guelph. The loss from the carbon is undetectable in the same two days of exposure in DI water. The EPA method 8260 is used for the carbon assessment. That method is described with precautions to minimize errors in the relative concentrations in samples.

Finally, it is important to recognize which potential perturbations can affect the distribution of contaminants from sample to sample. The purpose of the FACT is to assess the relative distribution of contaminants and not the actual concentrations per liter of water. However, the comparison of FACT levels with water samples suggests that an empirical correlation may be useful.

These issues are addressed quantitatively in a consideration of actual field data gained at a site with many relevant kinds of measurements. The site is in New Jersey at a site called the NAWC (Naval Air Warfare Center) near Trenton, NJ.

10.2.2.3.5   The Geometry of the FACT and Fracture Flow

The FACT activated carbon normally used is a single strip 1.5″ wide. Broholm et al (2016) addressed the question of carbon adsorption relative to the fracture flow direction. That is summarized here in a discussion of the daFACT (Section 10.2.2.5). The pore space contribution is not expected to be affected by the fracture flow direction. An alternative design using three FACT carbon strips is offered to negate any flow directional effects and theoretically can be used under the right circumstances to deduce the direction of fracture flow. That is described in Section 10.2.2.6 as the daFACT.

In a fractured rock emplacement, the FACT as a continuous strip will adsorb contaminants from both the pore space and fractures in a discrete manner. The contaminant cannot diffuse along the carbon, so the contaminant is adsorbed at the location of the source and the source is not continuous along the carbon length; therefore, each section of carbon may have one or more hot spots of high contamination. Figure 10.11 is a drawing of fracture in a porous matrix with contamination in the fracture and a halo of contaminants which have diffused into the adjacent pore space. The carbon will adsorb from both the fracture and the pore space in relatively short segments of the carbon. Therefore, a single section of carbon may have multiple "hot spots" which contribute to the total contaminant in the analysis of the strip. Since the analytical results are reported in micrograms of contaminant per gram of carbon (i.e., the mass of the entire section of carbon analyzed), in order for the relative distribution of contaminants to be obvious, the carbon sections should be of the same length. It is also advantageous for the carbon sections to be short, but that leads to many samples to be analyzed since the entire carbon strip should be assessed in order to not miss any contaminated intervals. The compromise due to cost is to determine what spatial resolution fits the budget and the needs of the investigation.

10.2.2.4   Quantitative FACT Assessment at the NAWC Site

The best FACT assessment is done with many other measurements of the hydrologic state and contaminant distribution as they can affect the FACT results. It is useful to remember that the primary purpose of the FACT is to map the contaminant distribution in the natural state. Therefore, it is necessary to determine what may cause deviations from that objective. It is very important to have data that is as nearly as possible that of the natural state of contaminant distribution. Additional data should aid in assessing perturbations of the FACT measurements and guide any corrections. At least there should be guidance on the expected resolution. A list of concerns is described above. The test at the NAWC site in borehole 94BR provides the best current data set to date for assessment of the FACT.

The test at the NAWC site was a project conducted by Prof. Beth Parker of the University of Guelph with help from Steve Chapman and Seth Pitkin. Funding was provided by SERDP (ESTCP no. 16-EB-ER1-035). The personnel involved were mainly Univ. of Guelph and FLUTe staff. This was an especially thorough test of the FACT method even beyond the scope of the very fruitful tests in Denmark. The advantages were the variety of measurements for comparison, the saturated fractured rock environment, and the time span and redundancy of the several measurements.

The data obtained were:

1. Drilling method with timing
2. Crushed core data
3. Geologic situation
4. FACT data on the 6″ scale

5. Water samples over 2 years after the FACT measurement
6. NAPL FLUTe DNAPL map
7. Head measurements of two kinds
8. Head histories at discrete elevations
9. Geophysics including ATV, OTV, caliper log

The NAWC site has been extensively investigated by the USGS as reported in numerous papers. A report by Dan Goode et al (2014) is relevant to this borehole. The geology is described as a formation of mudstone, shale, and sandstone in a sequence of gently dipping beds. Of particular interest is the TCE measured in nearby boreholes and the natural distribution of carbon in the formation. The strata are gently dipping to the NW.

Relevant measurements to the FACT assessment

Detailed measurements were:

1. Geophysics
   a. Strata with the OTV and ATV
   b. Caliper
   c. Fracture frequency and apparent aperture and dip
2. Core contaminant levels with depth
3. Button sampler samples of formation water with a sealing liner in place
4. Head distribution history with transducers isolated by the sealing blank liner
5. FACT analysis on the 6″ scale
6. Transmissivity profile using a blank liner
7. Head profile using the blank liner
8. NAPL stains on a NAPL FLUTe cover
9. Water samples over nearly two years at 10 sampling intervals in the sealed hole
10. Head measurements at the 10 ports of the MLS (multilevel sampling) system

Not all of these measurements are reported here, but in general, they were in agreement with the results reported here. (For example, the button sampler (a device developed by the Univ. of Guelph) results are not described.

The primary data useful to the FACT assessment:

1. Drilling procedure for cross-connection in the borehole
2. Core contamination levels and distribution with depth
3. Fracture geometries
4. Conductive intervals and conductivity
5. Head distribution
6. Open borehole flow rates deduced from T and head measurements
7. Water sample concentrations versus depth over two years in a sealed hole
8. Contaminant concentration in the open borehole from 4, 5, and 7.
9. Head history measurements at different times and elevations
10. NAPL stains vs. depth

11. Caliper log
12. FACT contaminant levels every 6″ of borehole

Using this information, the concerns described were addressed and comparisons were made. It is not expected that any of the concerns expressed were not possible. Rather, the objective of the FACT evaluation is to define which perturbations are significant and under what circumstances. In other words, which FACT results should be judged to be irrelevant to the natural state of the formation before the borehole was drilled? It is also useful to note where the FACT results may raise concerns about other kinds of measurements, some not performed (e.g., straddle packer tests).

10.2.2.4.1 *The Drilling Procedure and Geophysics*

The 94BR borehole was HQ core drilled to 150 ft. over a period of three days with the drill rods left in place overnight providing an imperfect seal. The drilling fluids were recirculated. No samples were collected of the drilling fluid (which would be useful). The core was sampled on somewhat irregular intervals near fractures during the drilling process. The samples were crushed and analyzed for contaminants, mainly TCE and DCE. The same compounds were assessed in the FACT samples collected later. Geophysical measurements were performed after the drilling to identify strata and fractures. The flow meter was not reported possibly due to over ranging of the heat pulse flow meter due to high vertical flow in the open borehole. Later measurements suggest the flow rate in the borehole was very high.

10.2.2.4.1.1 *The Drilling Process Concern*   There is little doubt that the drilling process perturbs the natural state both for hydrologic parameters and contaminant distribution. Not only does the clogging of fractures change fracture flows, but the open hole is also a major perturbation of the natural head distribution and associated flows. We live with the necessity of drilling effects. Sealing the borehole can partially restore the natural head distribution, and in time, it should also tend to restore the contaminant distribution. That is why there is often a delay from the well or MLS installation before water sampling is considered reliable.

The sealing liner of the FACT measurement is an advantage. Furthermore, there is no flow in the borehole past the measurement while the measurement is being performed as there is for some measurements in open contaminated boreholes. Examples of those open-hole measurements are straddle packers, passive diffusion bags, open-hole low-flow measurements, or in long screened wells. How quickly the sealing liner is installed is important to the extent of open-hole effects. Another precaution is that instantaneous measurement of the current conditions (e.g., immediately after the borehole is drilled) may not be so well related to the natural state. However, delaying measurements until the open hole is suffering a greater perturbation is not attractive either. It seems fair to generalize that some perturbations are accepted because of limited historical options and because of the expense. The FACT was not an option available earlier. However, the FACT is usually installed soon after the borehole is drilled and how much that affects the FACT result is addressed later in the comparison with water samples.

10.2.2.4.2   The Open-Hole Flow Effect on the FACT

10.2.2.4.2.1   Diffusion of Contaminants into the Hole Wall   The advantage of
the NAWC site measurements in 94BR is that there is a wealth of data that defines
the actual condition of the FACT measurement. It was learned from subsequent measurements that the open hole is flowing ~10 borehole volumes per day from above the
70 ft. depth into the formation below 70 ft., particularly into a large fracture at 130 ft.
where the natural fracture contamination was much higher than in the inflow portion of the borehole above 70. A simple obvious question is how does that affect the
FACT adsorption from the hole wall pore space and fractures. Based on later water
samples and head measurements, the source for the flow through the borehole was at
~5000 µg/l of TCE. If the natural pore space concentration is well below 5000 µg/l
of TCE, the borehole water contamination will diffuse into the pore space of the hole
wall material. The question of how far, how fast, and what may be adsorbed by the
FACT carbon is very amenable to calculation.

A diffusion model was defined with the diffusion coefficient and pore space
defined above in Section 10.2.2.2 for the NAWC site. The 5000 µg/l boundary condition for borehole water was set at the wall surface. By using a unit boundary source,
the results can be scaled for any borehole water concentration. The calculation results
in Figure 10.12 show the concentration of the diffused contaminant at different times.

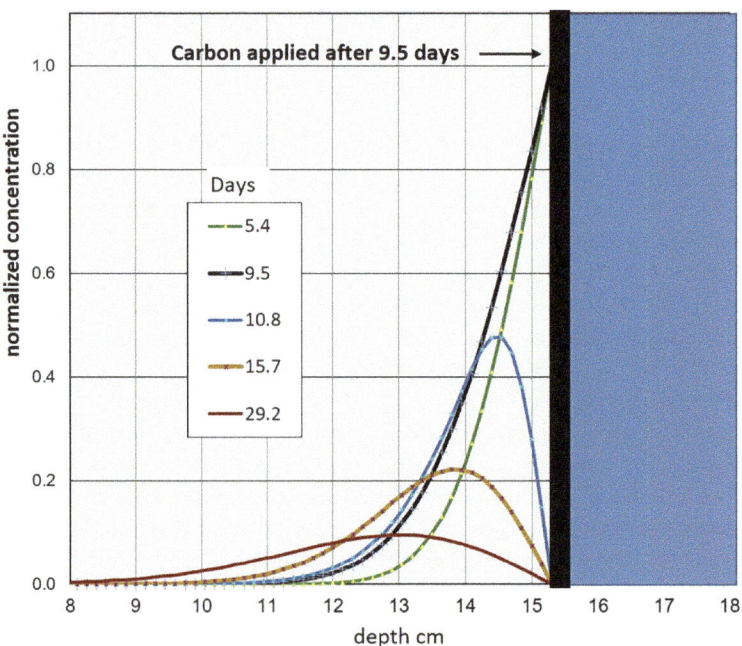

FIGURE 10.12   Concentration of contaminant from the borehole water at the time of the
carbon application, 9.5 days. Level builds at 5.4 days and after the carbon is applied the contaminant diffuses to the carbon and further into the borehole wall. The concentration in the
borehole is defined as 1.0 and can be scaled as needed as needed in units of borehole water
concentration. The diffusion coefficient is that of TCE in water with porosity of 7%.

By five days substantial contaminant had migrated into the hole wall. After 9.5 days, the concentration of TCE in the wall material was the black curve. It looks like a typical diffusion curve with the hole wall value at the full source strength. That was then defined as the state of contaminant distribution when the FACT carbon was applied to the hole wall by the liner eversion. The time of 9.5 days was picked as the time the borehole was open before the FACT was emplaced. That included the drilling time and the open-hole time during various activities in the borehole.

The carbon contact changed the boundary condition to a sink, instead of a source, as the aggressive carbon extracted contaminant from the hole wall. Because the carbon levels are normally far from saturation of the carbon, it is effectively an infinite sink. The carbon adsorption changed the contaminant distribution in the hole wall at later times as shown by the several plots (10.8–29.2 days) in Figure 10.12. While the carbon extracts contaminant, some of the distribution at 9.5 days does continue to migrate into the wall material.

The carbon does not extract all of the contaminant diffusing from the borehole water into the wall material, but it extracts most of it. In the calculation, the concentration per unit volume is known, so the carbon adsorption is the difference between the total contaminant when the carbon was first applied and the total contaminant still in the wall pore space when the carbon is removed, 14 days later. After two more weeks (23.5 days), the carbon content extracted in the calculation can be compared to the actual carbon levels measured in the FACT carbon to determine if the diffusion from the contaminated borehole water is a significant perturbation of the FACT carbon content. The important term is "significant". There is no doubt that contaminated borehole water will contribute to the matrix contamination and that some of that will be removed by the carbon. In this calculation, the carbon gain is ~180 µg/g of carbon. The plot of the actual carbon levels measured in the FACT is shown in Figure 10.13 on a log scale. Any contribution from the borehole wall due to the borehole water should be a background level in the carbon analysis. It is obvious from the carbon results in Figure 10.13 that there are very few TCE samples below 100 µg/g and many lower values of ~200 µg/g. This in general agrees with a background of ~180 µg/g. There were no non-detects in the carbon. The background from the open borehole water will vary with the material in the formation, and the borehole water variations. Based on the water samples measured as discussed later, the borehole water was apparently flowing from low concentration intervals in the borehole downward to high concentration intervals at depth. That may be unusual but implies a relatively low open borehole water contaminant level as determined from the water samples.

If the FACT is emplaced after only two days of open hole, the total estimated carbon contaminant content is about half that of a 9.5 day delay. Part of the explanation of why a fourfold longer exposure does not give a fourfold increase is that the longer exposure leaves more in the hole wall which is not extracted in 2 weeks. Another explanation is that the concentration gradient into the hole wall decreases in time after the carbon is emplaced. However, the amount of contaminant in the carbon scales directly with the borehole water source concentration. It is also noteworthy that the known variation of porosity from 4 to 10 percent will change the diffusion rate by a factor of about 5 in the term $\varepsilon^{1.7}$ where ε is the porosity.

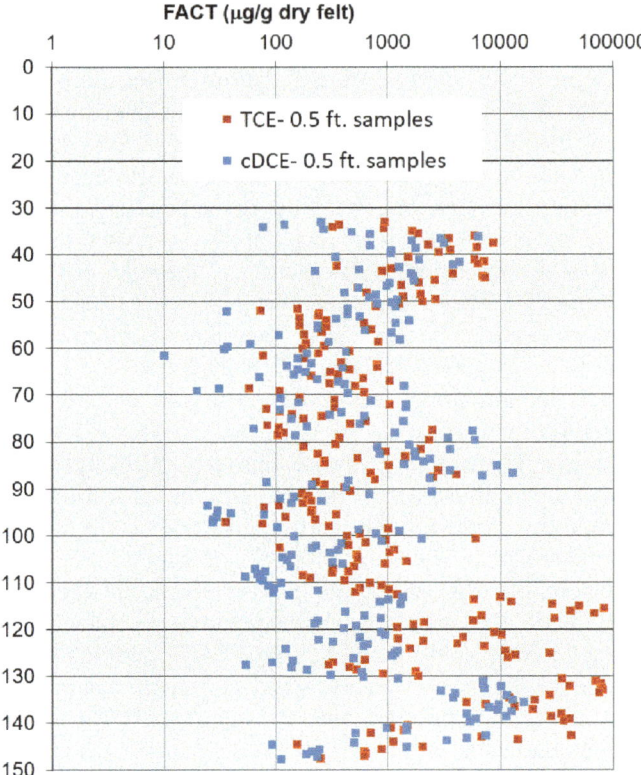

FIGURE 10.13 FACT analysis results in 94BR with 6″ section lengths. There are no NDs suggesting a background level from the open borehole water. The background is 3 orders of magnitude less than the peaks. Borehole water levels were probably near 5000 µg/l flowing downward.

Using the calculation to estimate the effect of the borehole water on the background of the FACT values, it suggests that most of the contaminant in the carbon is far above the "background" due to the borehole water. The remaining questions are whether the other concerns suggested were significant.

10.2.2.4.2.2   The Migration of DNAPL   The NAPL FLUTe obvious stain observed was only a single small dime size stain at ~43 ft. depth. Other discolorations were observed but were not the typical stains seen for TCE or strong solvents like PCE. There were no DNAPL stains at the bottom of the borehole. The conclusion was that the high FACT levels in the lower half of the hole are probably not related to DNAPL in the formation or descending in the open hole, but similar to high levels measured at depth in nearby holes (Goode et al, 2014). The stains in Figure 10.6 show what is normally seen when TCE is moving down the open borehole.

10.2.2.4.2.3   The Trapped Water Effect   As the liner everts, some borehole water is expected to be trapped between the hole wall and the liner where the borehole

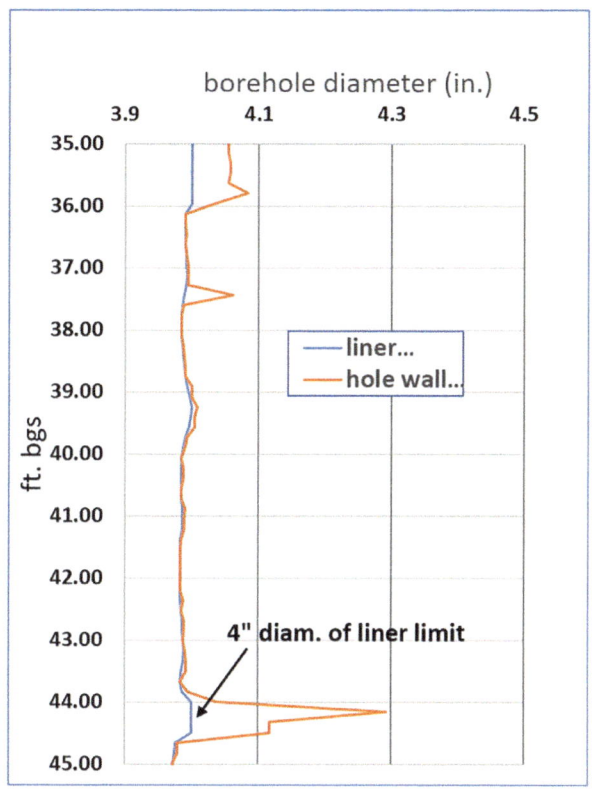

FIGURE 10.14 Liner drape over enlargements in interval of 35–45 ft. with one of the larger breakouts. Liner dilation into an enlargement is limited to its diameter.

is enlarged. In the photo of Figure 3.9, it is expected that borehole water would be isolated in the vugs and in other borehole enlargements, especially if the enlargement is of a diameter greater than the liner. An estimate of the volume of the NAWC borehole water between the liner and the hole wall was developed as follows. The borehole caliper plot with depth was smoothed in a manner to approximate the liner drape over the variations in diameter. Figure 10.14 shows an enlargement of the caliper and liner contour between 35 and 45 ft., with a substantial enlargement. The volume of trapped water is the volume between the caliper curve and the liner contour draped across the enlargements. The further assumption is that all of the contamination in the trapped volume of borehole water will migrate to the FACT carbon. A calculation of that mass of contaminant was then assumed to be adsorbed in the carbon providing a mass of contaminant per mass of carbon estimate. A plot of the trapped water contribution is shown in Figure 10.15. The plot is smoothed over 6″ to approximate what would be seen in a 6″ carbon segment.

The contaminant content of the enlargement is assumed to be that of the borehole water. Because of the water samples obtained and the calculated flow in the open hole from the measurement of transmissivity and head, the borehole water concentration

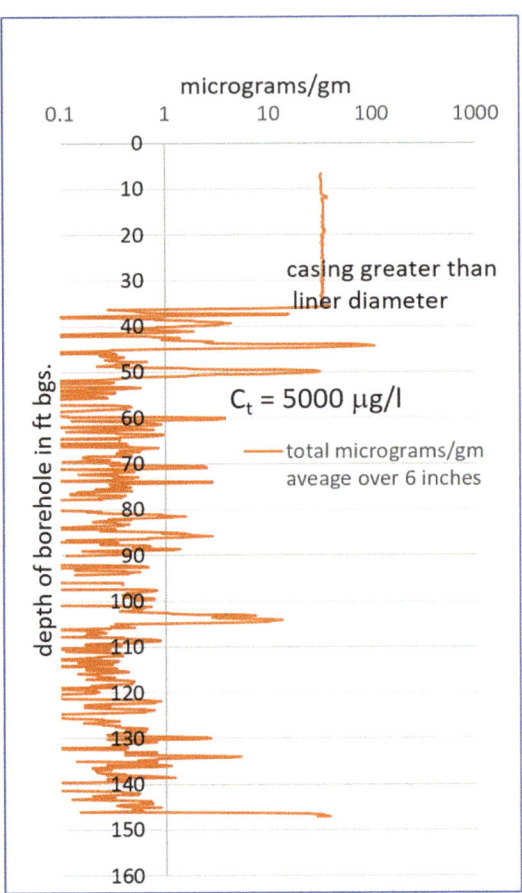

FIGURE 10.15 Trapped water contribution to carbon in μg/g. Contribution averaged over 6″ as in 6″ chapters of carbon. Longer carbon sections would reduce peak values but still provide the correct distribution of contaminant.

is reasonably estimated. However, without an open borehole water sample, that open hole number is not known. In order to assess the effect of the open borehole on measured contaminant distributions, it is important to know the borehole flow and the contaminant distribution. That sample measurement would best be collected from the borehole at several depths. It was not done for this borehole.

It is further assumed that the contaminants in the enlargement will not migrate from the far side of the liner but in the 4″ hole will only contribute from an annular enlargement from one fourth of the circumference (twice the carbon width). If for some reason the contaminants diffuse from the entire annulus to the carbon, the estimate would be increased by a factor of four. However, it is also probable that the enlargements occur at fractures intersecting the borehole and that fracture flow may be major contributors to the carbon. That is discussed more with the daFACT design later in this chapter. With those caveats, the calculated trapped water contributions

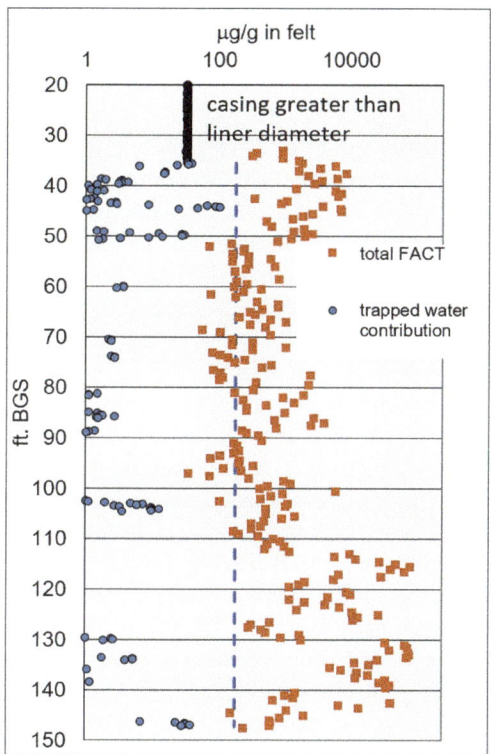

FIGURE 10.16 Comparison of trapped contribution to carbon (blue points) with smoothing over carbon length. Carbon contribution of trapped water over the same 6″ length as for TCE levels (in red). For this borehole, the trapped water contribution is far less than the measured values of TCE and well below the apparent background of 100 µg/g.

are shown in Figure 10.16 on a log plot of the FACT results from the 94BR hole. The trapped water contribution is far less than the values measured in the carbon and less than the apparent background of about 100 µg/g. Because the carbon was cut into 6″ sections, the trapped water contributions were averaged over 6″ for the comparison. That means that the integral of the averaged values is about the total contribution, but that the value in the 6″ section would not be higher than any point on the averaged values. Averaged is defined as summed over 6″ and divided by the 3 caliper intervals included in the sum. That means that a single high value would be less because the total absorbed is divided by the total carbon weight in the strip. Figure 10.11 addresses that effect.

Since the liner does not dilate more than its diameter, the volume of trapped water in major enlargements can be high. A plot like Figure 10.14 would be helpful in identifying trapped water effects. It is noteworthy that if the potential constant background of the open borehole water (Section 10.2.2.4.2.1) of ~100 µg/g from the wall pore water, were added to the trapped water contribution, the influence of the trapped water is quite insignificant.

While not the same as trapped water, the open hole does allow major flow into fractures of lower head. That water flow can affect the FACT results and is discussed in the comparison with water samples in Section 10.2.2.4.4.

10.2.2.4.3   Summary of the Potential Perturbations on the FACT Results

The drilling and open-hole effects are the main perturbations on the FACT results as a background. They cannot be separated without samples of the drilling fluid and the open borehole water. However, water samples and measurements of head and transmissivity do allow a reasonable estimate of the open borehole water contamination. Calculations of diffusion of the open-hole contaminants suggest that the diffusion of contaminants from the open hole may have provided the nominal background of the carbon measurements. The major effect on the FACT is that of fracture flows past the FACT based on later comparisons with the core and transmissivity information gained from the same borehole. The other potential concerns only add to the background of the FACT distribution. It is best to sample the drill water and the open borehole water to allow a more relevant assessment of the open-hole effects and the time the borehole is left open should be as little as possible.

In the comparison of the actual FACT levels and the water sample in Section 10.2.2.4.5, the effect of the fracture flows on the FACT are dominant. The water sample is drawn primarily from the fractures. The contribution of the pore space contamination is a second-order effect and should relate well to the crushed core analysis.

10.2.2.4.4   Comparison of the FACT to the Core Values

Figures 10.17 and 10.18 are a comparison of the FACT to the crushed core levels of contaminants. The comparison shows that the high core values compare well with some of the high FACT values. An exception is the high core value at ~70 ft. However, the high FACT values do not compare well with the core values. The primary reason is probably because the core measurements are only of the pore space near fractures where core is recovered. Core samples are generally only selected near fractures. The FACT measurement is continuous and relevant to both pore and fracture flows and is dependent on how well the core pore space is connected to fracture concentrations. Also, if core is not recovered, there is no measurement of that interval, whereas the FACT does span intervals of those fractures that can reduce core recovery. In general, one would expect that the distribution would be related, but if the fracture flows are more recent in time and the pump and treat system is changing the fracture flow concentrations, they will not be so closely related to the core. The core is relevant to the historical flow before the hole is drilled and the core is collected. The FACT measurement is more relevant to the state after the hole is drilled. The correlation of the FACT with the water samples in the next chapter suggests that the FACT measurement relates to the fracture flows since fracture water is collected by the water sampling procedure.

10.2.2.4.5   Comparison of the FACT with the Water Samples and Flow Zones

Figure 10.20 compares the FACT samples with the water samples collected at 10 months and at 23 months after the FACT measurement. In general, the correlation is good. Adjacent to the FACT data is the FLUTe transmissivity profile (Figure 10.19) integrated

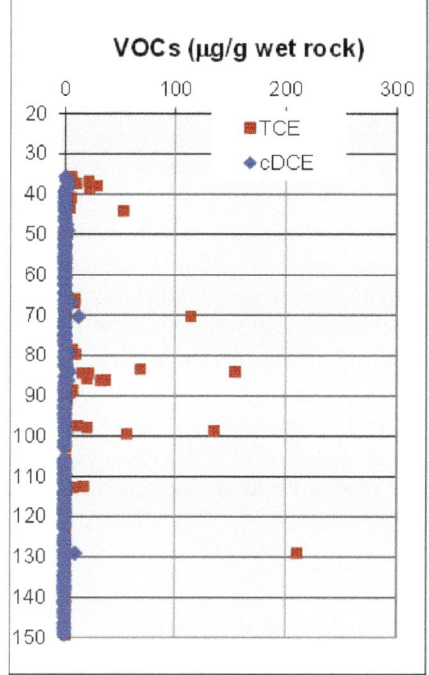

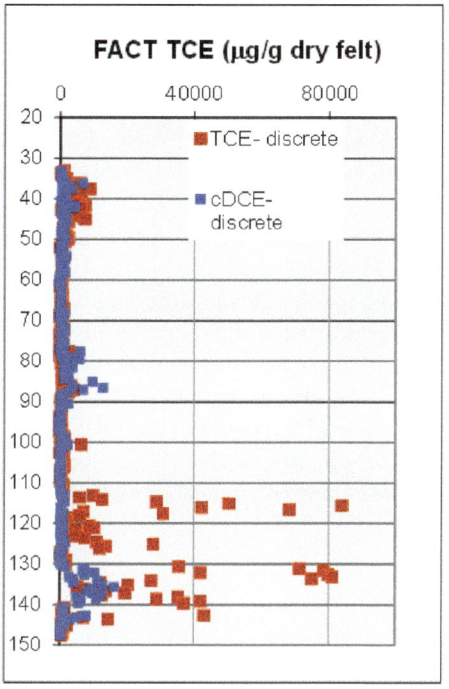

FIGURE 10.17 Core contaminant levels in same 94BR borehole as FACT samples. Core peaks compare well to FACT peaks, but distribution is not the same.

FIGURE 10.18 FACT peak at 115 ft. matches only small core peak at 115-ft. FACT is a different measurement including both pore and fracture contaminants.

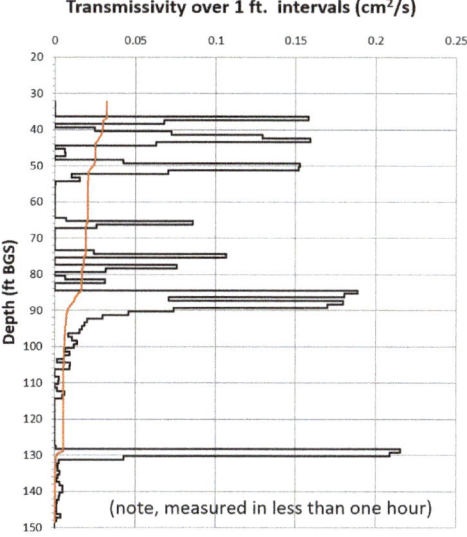

FIGURE 10.19 Transmissivity in one foot increments of the borehole with FACT measurements of Figures 10.18 and 10.20.

on the 1-ft. scale (a method explained in Section 10.3). It is interesting that the highest water samples and the highest FACT levels are not well correlated with the highest flow zones of the T profile. In some intervals such as 87 and 130 ft., they agree with high flow zones, but at 115 ft., the small closely spaced fractures visible in the OTV (optical tele-viewer), with relatively low conductivity, show extremely high FACT concentrations. The conclusion is that there may be DNAPL nearby in the small fractures below the interval of higher conductivity from 85 to 93 ft.

Another interesting comparison is the high flow zone at 43 ft., which shows both high FACT and high core values, but lower water sample levels. One must conclude that both the FACT and the core are influenced by high pore concentrations, perhaps developed at an earlier time by high fracture levels, but that perhaps infiltration of recent precipitation at this shallow depth has flushed the shallow high flow interval yielding lowered water sample levels.

It is useful to note how well the FACT compares to the water samples long after the borehole was drilled. Since the borehole was sealed with a flexible liner with sampling intervals indicated by the black bars in Figure 10.20, the drilling influence

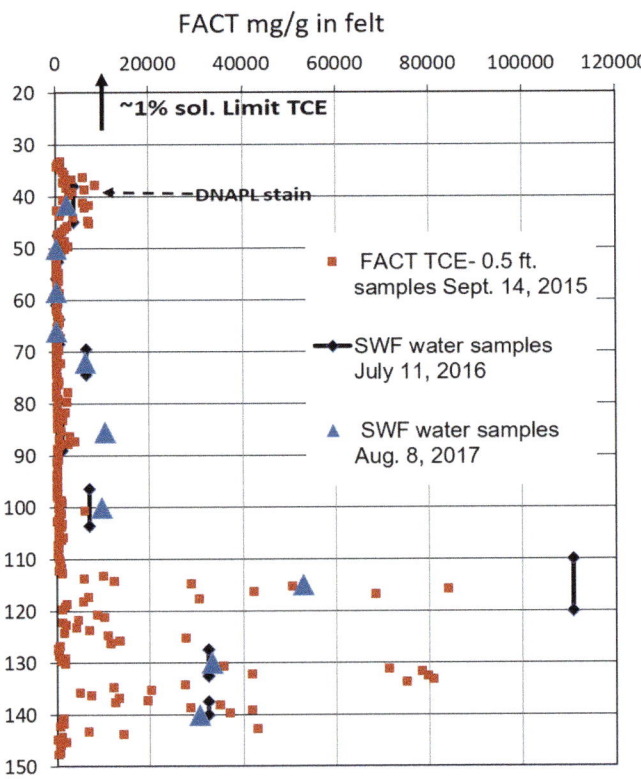

FIGURE 10.20 The FACT results compared to the water samples obtained nearly one and two years later.

should have diminished greatly and the FACT and water samples are indicative of the current state with the pump and treat system operating in that time.

The FACT measurement includes both current fracture flows and the residual of earlier times in the pore space. The core samples are more indicative of the earlier state of fracture water near the core pore space. It is interesting to also consider the total information available for this site from the many USGS measurements, but those were not performed in this borehole for comparison. However, many of the USGS contaminant measurements were done with passive diffusion bags, low-flow sampling (LFS), and straddle packers in open holes which may provide much different values than measurements in a sealed borehole.

10.2.2.4.6   *Assessing Active Flow Zones Using Toluene in the FACT*

It has been shown in the paper by Keller et al (2017) that the concentration of contaminant from the FACT method is not always related to the high flow zones, but the contaminant peaks of DNAPL are often seen in less conductive intervals near and below the high flow zones near the source zones. For boreholes more distant from the source region, there is a better correlation of high flow zones with contaminant since the high flow zones are the major conduits for contaminant transport.

Another interesting observation of FACT results is the presence of toluene in the carbon. It has been noticed that the toluene peaks often are related to contaminant peaks, but not always or directly. Since the diffusion barrier is to prevent the liner direct contact with the carbon felt, one must assume that the toluene reaches the FACT carbon by transport in the water flowing past the FACT which has passed the liner and gained some toluene from the liner. The low level of toluene in FLUTe liners is a result of the manufacturing process. The typical levels of toluene seen in water samples from FLUTe MLS systems are well below the drinking water limit. The toluene has been shown to normally decay to non-detect levels in ~3 months, or less, in intervals with high lateral water flow. The toluene can persist for longer periods when there is little lateral flow past the liner. However, when the FACT is emplaced, the toluene level in the liner is at its peak value with no time to have decayed.

The fact that the toluene is leached from the liner was an early concern for water samples collected if toluene is a contaminant of interest. However, since liner material without toluene is of inferior mechanical quality, FLUTe has elected to use the liner material with traces of toluene since it is not long lived. The large recommended FLUTe purge volumes are also shown to reduce the effect of toluene from the liners (as noted in tests done with less than FLUTe prescribed purge volumes at the SSFL site in California).

However, the toluene is an effective tracer leached from the liner and the contaminant is a tracer in the water flowing past the liner, and it was noted that high levels of contamination often correlated with peaks in the toluene as both compounds were collected in the FACT activated carbon. FLUTe includes a diffusion barrier between the FACT carbon and the liner in order to eliminate the direct adsorption of toluene from the liner.

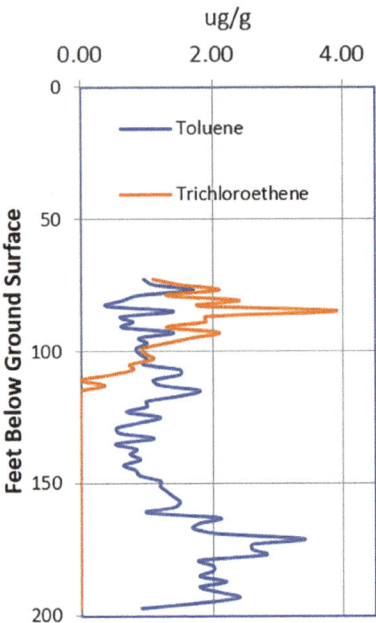

FIGURE 10.21 Comparison of toluene with contaminant distribution. FACT depths adjusted for apparent stretch for comparison with Transmissivity profile.

Therefore, the toluene in the carbon can be expected to have diffused from the water traveling past the liner in the formation. That toluene may be one of the contaminants of concern, but if not, it is a very useful measure of the flow past the liner and therefore as an indicator of the lateral flow in the formation.

Figure 10.21 is an example of an excellent correlation of the toluene peaks with the contaminant peaks above 100 ft. Figure 10.22 shows a similar correlation with the flow paths measured with a FLUTe transmissivity profile. Although the T peak at 85–89 ft. is barely above the resolution limit, it is noteworthy that the high flow zone near the bottom of the borehole shows high toluene values but no contaminant. For that reason, it is generally assumed that toluene leached from the liner is a useful measure of the existence of lateral flow past the borehole in the formation since the borehole is sealed with the continuous liner for the FACT assessment. However, that generalization is qualified in the discussion of the daFACT below in Section 10.2.2.5.

The description of the daFACT concept uses the leached toluene for both an assessment of active flow zones and the possibility of assessing the direction of flow if three equally spaced carbon strips are used, each isolated from the liner by a diffusion barrier. That concept and the FACT are both patented or patent pending FLUTe methods. The daFACT has not been used yet (2021). A significant disadvantage of the concept is that it triples the number of samples to be analyzed and adds to the fabrication cost. An intermediate less costly method is the combined analysis

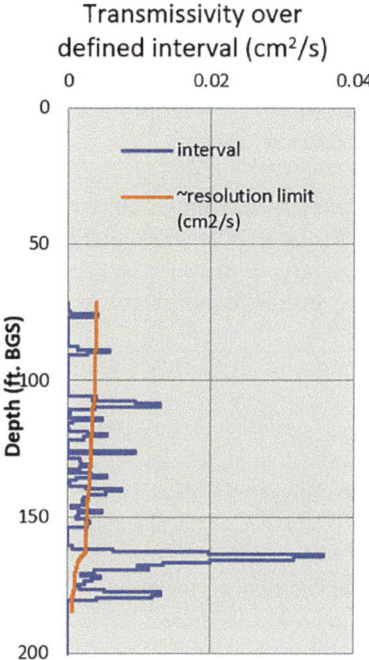

FIGURE 10.22 Transmissivity over 1-ft. intervals from FLUTe T profile. Same borehole as Figure 10.21 (borehole in NJ).

of the three carbon segments over equal intervals to avoid any dependence of the FACT result on the direction of flow, as is described in the daFACT description.

Note that the flow past the FACT depends upon not only the local transmissivity but also the head driving the flow and the velocity of the flow which can affect the rate of leaching of the trace of toluene from the liner. A very slow flow rate for a finite toluene leach rate will have a higher concentration on arrival at the FACT than a fast flow rate. A larger exposure area of the liner will also result in a higher total contaminant adsorbed in a FACT section. Therefore, the toluene peaks are considered an indication of flow but not a measure of the relative flow rate. That is a similar limitation of several other active fracture indicator methods. Since there is no additional cost of the toluene peak assessment of active flow after the cost of mapping the contaminant distribution, the observed correlation seems especially practical and economical.

It should be noted that the control of the depth of the FACT samples is not as good as for the T profile depths. This is because the FACT cover material is often stretched somewhat by the process of removal from the inverted liner. We have noted that the FACT depths may best be adjusted upward by 1–2% of the actual depths or more in very long liners with higher tension on the FACT cover material during removal from inside the inverted liner. For that reason, it is useful to mark a depth scale on the cover during fabrication. The use of a blower to inflate the liner during the cover removal reduces the drag, and therefore, the potential stretch of the FACT cover during removal from the inverted liner.

10.2.2.5 Comparisons of the FACT with Other Methods

A comparison of the FACT results with some other traditional measurements would not be especially useful as a test of the FACT results because the FACT is a very different measurement. The following list is of measurements that are not expected to be comparable for the limitations noted:

1. Comparison with packer results since packers draw from fractures that may be connected to the open hole above and below the straddled interval. Packer samples can be largely of water entering the fractures due to open borehole flow unless a sufficient purge removes all the water and there are no residual effects.
2. Low-flow measurements in open holes are subject to extensive cross-connection.
3. Passive diffusion bags are subject to extreme cross-connection in open boreholes.
4. Core is an interesting comparison, but it only assesses the pore water, not fracture flows.
5. Active fracture flow assessments with temperature measurements are not quantitative and while interesting for comparison to the daFACT provide no data on contaminant concentrations.

However, comparisons with the above measurements may define the limitations of those methods.

10.2.2.5.1 A Special Problem with Packer Testing as Revealed in This NAWC Borehole

It is considered standard practice to purge three times the straddled interval to obtain a representative formation water sample (Holloway and Waddell 2008). In this ~4″ diameter, NAWC borehole described above with a downward flow rate of nearly 10 borehole volumes per day, the cross-connection volume (~600 gal./day) does not compare well to a three 10-foot straddled interval purged volume of 15 gal. A packer sample in that hole would not be representative of other than the open borehole flow concentrations unless collected in the source interval of the borehole flow. However, that borehole water concentration would compare well to passive diffusion bag and grab sample data at the same interval as is reported in one packer development article (Holloway and Waddell, 2008) as proof offered that the packers were obtaining representative samples. One would expect all three sampling methods to compare well with a borehole water grab sample. Are they not all open borehole water samples? See Section 10.5.1 for information on LFS in open boreholes as opposed to formation water measurements in sealed boreholes.

10.2.2.6 The daFACT

In Broholm et al (2016), a very thorough paper, they addressed the question of whether the FACT position on the liner relative to the direction of flow could be

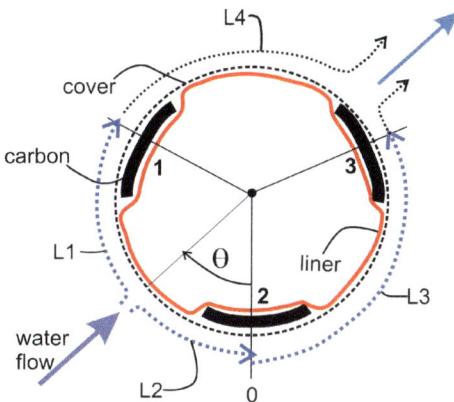

FIGURE 10.23 daFACT three carbon strips in cross section of liner. Flow direction shown as q relative to strip no. 2. Flow distances adjacent to the liner are shown as L1–L3. Not shown is the diffusion barrier isolating each carbon strip from direct contact with the liner.

significant to the level measured by the FACT. The conclusion was that it will have some influence but mainly in highly conductive formations (>10^{-2} cm/s). FLUTe has suggested a somewhat different design of the FACT called the daFACT ("day fact") that would be more sensitive to the flow direction and yet provide a reasonable relative concentration with elevation. The cross-section drawing in Figure 10.23 shows the important features of the design with indicated flow around the lined hole for a hypothetical flow direction. The hypothesis is that for any flow direction, the several daFACT carbon strips if combined would reflect the same contaminant level regardless of the flow direction. An additional advantage of the daFACT is dependent on the observation that the toluene level in FACT samples correlates sometimes with high levels of concentration of toluene, but if there are high levels of toluene and low contamination, the toluene may be associated with the known leaching of toluene from new liner installations. If one assumes a relatively uniform leaching with length of exposure of toluene from the passing flows in fractures, the active fractures would carry the toluene to the FACT. But, since the FACT carbon is isolated from the liner with a diffusion barrier, the leached toluene would be that toluene carried from the liner via fracture flow to the FACT carbon exposed to the fracture flow, but not exposed to direct contact with the liner. Furthermore, it is assumed that the amount of toluene leached to the ground water depends upon the distance the fracture flow is exposed to the liner. As shown in Figure 10.23 that depends on the direction of the flow in the plane of the fracture. In a simple concept, the amount of toluene in each carbon strip would depend on the direction of the fracture flow as illustrated by the length of the flow paths past the liner to each carbon strip.

Theoretically, one could analyze the amount of toluene in each strip and deduce the direction of the fracture flow, assuming of course that the orientation in the borehole of the strips is known. In light of the expense of analyzing three carbon sections

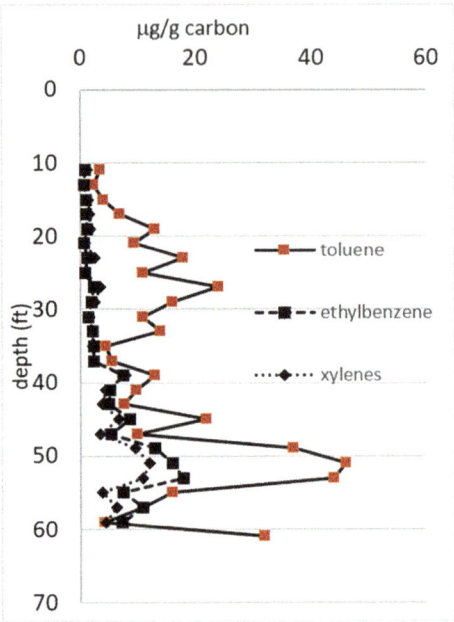

FIGURE 10.24 Comparison of toluene peaks to contaminant peaks. Toluene flow appears to be related to flow past the liner, that is to say, the active fracture flow.

for the daFACT instead of one for the FACT, that is not a popular concept. And, the flow past a borehole may not be well related to the overall flow field direction in the formation. To avoid the expense of analyzing three strips, it is proposed that the three strips be combined to avoid a sensitivity to the direction of flow and then used to distinguish the active fracture flow relative to the toluene measured in each sample. With no fractures, the toluene levels would be low because of the diffusion barrier. Since the carbon integrates flow contamination passing the carbon and diffusing into the carbon, flowing fractures should contribute more toluene to the carbon than non-flowing fractures or no fractures.

A test of the concept is to examine toluene levels in the carbon compared to contaminant levels. If the toluene level is relevant to fracture flow, it should be associated with the conductivity deduced from the T profile. It would also correlate with high levels of contaminant, but only if contaminants are occurring in flow zones. Such a comparison has been done. One example is shown in Figure 10.24 where the high contaminant levels are associated with some known (measured) flow zones, and high toluene is associated with other known high flow zones. This is a useful consideration only if toluene is not a known contaminant in the ground water. In a simple sense, the question is: Are known flow zones related to high toluene levels and also do high contaminants in high flow zones also relate to high toluene levels? The answer to both questions may seem to be yes. However, with a single FACT strip, the flow past the carbon can depend on the direction past the FACT carbon as suggested by Broholm, et al (2016). If the flow direction is directly at the carbon, the toluene uptake would be less than at perhaps 90 degrees from the carbon.

This use of the toluene level is better with the daFACT geometry if all three strips in a sampling interval are not combined for one analysis. The downside is the additional cost of building three strips into the cover material. That fabrication is being tested. The gathering of three strips into one sample container is not significant additional labor.

The FLUTe T profile measurement includes all flowing fractures driven by the everting liner (Section 10.3), and it does not differentiate active from inactive fractures under natural gradients. Contrary to some individual's erroneous claims that the T profile also maps dead end fractures, dead end fractures are not mapped by the T profile, since the measurement is of flow changes, and dead end fractures by definition do not flow relatively incompressible water.

The simple relation of toluene peaks to active fractures should not be misinterpreted as a relation of the toluene peak magnitude to the amount of flow past the FACT. The data set shown in Figure 10.21 suggests the toluene peaks seem to be directly related to the flow capacity of the fractures when compared with Figure 10.22, the T profile data. However, another data set is shown in Figures 10.24.1 and 10.24.2 of toluene peaks versus the T profile for the same hole. There is not a good correlation of toluene with the flow capacity. This demonstrates that there is more to the process of toluene leached from the liner and the adsorption in the FACT. The FACT adsorbs in proportion to the concentration gradient and the concentration in the water. Since the flow maintains a high gradient, the concentration of toluene in the water is most relevant to the adsorption in the carbon. However, the flow path can affect the concentration of the toluene in the water as described above, but the flow rate can also have a very large effect on the toluene concentration in the water passing the carbon. Given a constant leach rate of toluene from the liner, the high velocity flow will have a lower toluene concentration since the exposure to the liner is for a shorter time. Conversely, very low-flow rates can have very high levels of toluene as they pass the carbon. Given the same exposure area/length of the carbon, the fast flow will produce lower carbon toluene levels over the time that the carbon is in the borehole. Therefore, small slow-flowing fractures which are probably associated with low T values can produce high toluene levels in the carbon if the carbon does not extract all toluene from the passing flow, which it does not. Many fractures intersecting a single carbon section can also provide higher levels per gram of carbon.

This means that toluene peaks are still indicative of active fractures, but the relative toluene peak levels may not be related to the flow rate in active fractures. A very close look at the low but measured conductive fractures above the detection limit of Figure 10.24.2, below the depth of 150 ft., does correlate with the high toluene peaks at those levels. The high toluene peak is probably not related to the flow rate past the liner and FACT. Why then (Figure 10.24.1) are high toluene levels also seen in the high T intervals? High T intervals are not necessarily flowing faster under the natural lateral gradients in the formation, and they may also expose a longer section of the carbon to the flow as suggested in the drawing of Figure 10.11. The larger aperture of high T fractures also increases the exposure to the liner and can

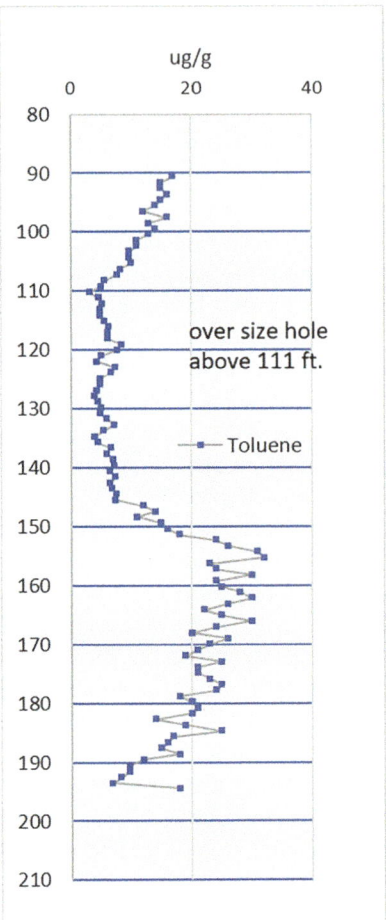

FIGURE 10.24.1 Toluene in 1-ft. FACT sections. Compared to T profile in Figure 10.24.2, toluene peaks match very low T peaks.

add to the carbon content of toluene. The carbon content is the sum of all flow zones along the full length of the carbon section. A large breakout with no flow can lead to a large carbon exposure to the liner toluene, just as a large breakout can provide a higher cooling rate in a temperature measurement used to assess active fractures. An example is in Figure 10.24.1 above 111 ft. where the hole is larger than the liner as in an extreme breakout. For a single fracture at a single flow rate and the daFACT geometry, the toluene can still be an indicator of flow direction at that borehole. Therefore, flow direction, flow rate and length of flow past the liner, plus the leach rate from the liner, can all affect the carbon level of toluene. Fortunately, the contaminant concentration is the primary parameter relevant to the carbon contaminant levels and not the flow rate past the carbon.

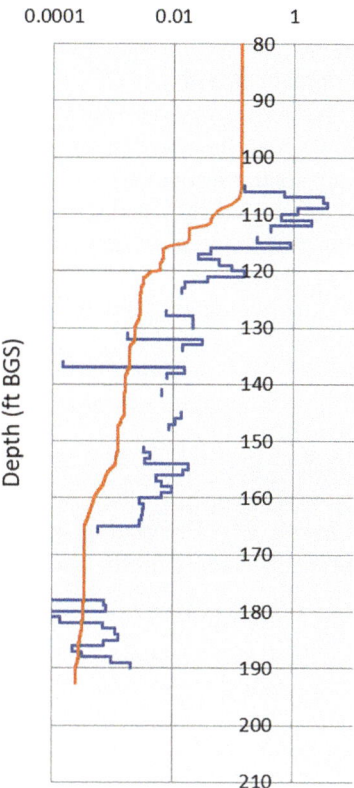

FIGURE 10.24.2 T profile in same hole as 10.24.1 over 1-ft. intervals. Red curve is the resolution limit of the T profile.

None of these considerations affect the contaminant distribution in the carbon, since the concentration in the formation flow is the primary factor in the carbon levels.

It is noteworthy that the use of toluene to recognize flowing fractures depends on the activated carbon felt being free of toluene before the installation. Unfortunately, some samples of new lots of activated carbon have shown contamination of BTEX. Therefore, it is important that the presence of such compounds be recognized in the interpretation of results. Fortunately, frequent tests show that the other compounds measured in the method 8260 are not in the carbon and are reliable as a relative distribution within the uncertainties discussed above. There are a limited number of sources of activated carbon felt and most seem to have originated in China.

The depth of FACT samples in the borehole is affected by the stretch of the liner during installation and during the removal as well as when the cover is removed from the inverted liner on the surface. The estimate of stretch for different liner materials and the adjustments for assignments of more precise depths to FACT samples are addressed in Section 10.6.

10.2.2.7   Advantages and Limitations of the FACT Measurement

Advantages of the FACT measurement:

- The addition of the FACT carbon strip to the NAPL FLUTe cover allows a map of both the NAPL and dissolved phase distribution in the formation with exceptionally high spatial resolution.
- The measurement is performed in a sealed borehole and relatively independent of the contaminant in the open flowing borehole unlike open-hole measurements.
- The FACT maps both the pore space contamination and the fracture fluid contamination.
- The FACT is very complimentary to the NAPL mapping capability of the NAPL FLUTe in the same borehole at the same time. Extremely high FACT values suggest nearby NAPL.
- The FACT is an excellent guide of where to locate MLS system sampling intervals.
- The same liner used for the FACT measurement is also used for other subsequent measurements such as the T profile and RHP.

Limitations of the FACT measurement:

- The cost of analysis of the FACT carbon as compared to collection of a few water samples alone, over much larger intervals.
- The FACT does not provide a µg/l measurement as desired by some regulatory agencies. However, it is an excellent guide of where to best collect water samples with µg/l assessments.
- While a strong indicator of nearby NAPL, the FACT, like the NAPL FLUTe, does not prove the presence of NAPL nearby. That proof requires the stain on the NAPL FLUTe.
- While indicative of flowing fractures, the toluene level derived from the liner is not a reliable indication of flow rate past the liner in that high flow rates can lead to lower toluene levels in the FACT. But toluene peaks still correlate well with the location of permeable features. The preexisting presence of toluene as a contaminant in the carbon can be misleading in the active fracture assessment. Until the carbon is reliably free of BTEX, it cannot be used to assess BTEX in the ground water.

10.2.3   Absorbers of Other Kinds on Blank Liners

10.2.3.1   Pore Water Collection in the Vadose Zone

In 1991, the first installation of absorbers on a flexible liner was done at LLNL to collect tritiated water samples. This was the first commercial use of a flexible liner.

In 1993, a paper given at the NGWA Outdoor Action Conference by Keller and Travis (1993) included the measurements of the capillary tension versus saturation of a wide range of different absorbent materials. Bryan Travis, in that paper, also provided the calculations of the rate of absorption of pore water in the absorbers for a wide range of formation characteristics.

A later application of absorbers was by Stan Martin at LLNL (~1997) who used a FLUTe method of monitoring relative absorption in absorbers on the outside of a flexible liner using a measurement of the conductivity/resistance of the absorber between two contacts on the exterior of the liner about an inch apart. The contacts were connected to wire pairs to the surface which were used to monitor the resistance in time of each absorber as it gained water by wicking water from the formation. In that use, Stan was able to monitor the passage of a wetting front generated by infiltration of rainwater in the vadose zone. The resistance measurement of absorbers required the use of an alternating voltage source to avoid the charge accumulation of a DC source.

10.2.3.2   Radioactive Contamination Absorbers

Bill Lowry, of SEA (where he earlier worked with Carl Keller to perfect flexible liner methods), developed several methods for using a variety of devices on the outside of everting liners. One approach was to carry photographic film into piping which upon exposure could be developed to locate radioactive deposits on the wall of the pipe. This was used by the DOE in several nuclear facilities, but I am not aware if the reports are available.

In about 1997, FLUTe developed an everting liner called a caterpillar which had inflatable extensions in all directions (~30 of them, and hence the caterpillar appearance). As the liner everted, the inflatable arms extended from the liner surface about 6″ to contact the wall of a pipe as the liner was everted into the pipe. To the end of each arm was attached an absorbent swipe to collect a sample of radioactive dust from the wall of the pipe as the arm/leg swept into position as it emerged from the EP. The arms extended radially from the liner surface when inflated. The liner did not actually touch the pipe wall but "walked" on the legs. Upon inversion the arms collapsed as the liner was rolled back onto the reel in the air canister. It worked well at a demonstration in Denver, CO and was purchased for use at the DOE Hanford site. However, further assessment of the probable intensity of the radiation from samples collected on the liner system, when recovered into the canister, was expected to be too high for the safety of the operator of the canister system and the use was abandoned. It was an intense radioactive environment and hence the need for a remote mapping technique. The construction of the system with only a hot air gun and urethane-coated nylon was a major achievement of a FLUTe employee named Ann Nguyen at the FLUTe plant in Houston, TX.

10.3   THE TRANSMISSIVITY MEASUREMENT METHOD

10.3.1   HISTORY OF THE TRANSMISSIVITY PROFILE METHOD

The earliest liner installations were in the vadose zone using air driven installations from canisters as described in Section 3.5.1. The use of a water-filled blank liner to quickly seal a borehole was recognized to be a potential advantaged of the everting liner. A demonstration of a blank sealing liner was arranged at the USGS facility in Storrs, CT in ~2002. When the everting liner was installed to below the water table to seal the borehole, using water as the driving fluid, it was obvious that the rate of installation decreased with depth. The reason was that the blank liner sealed the conductive intervals in the borehole. Since the sealing of the flow paths intersecting the borehole was slowing the liner eversion rate, it seemed reasonable that measuring the installation velocity versus depth would locate the conductive paths where the velocity of the liner changed. FLUTe also deduced that if one considered the everting liner as a flow meter that it was possible not only to determine the location of the flow zone but also the flow rate into each flow zone sealed by the everting liner. The rest of the transmissivity profile (T profile) formulation was developed for the flow into the formation using the well-known Thiem equation as used for straddle packer testing in order to assess the conductivity/transmissivity of the straddled interval. One journal article reviewer recently objected to the use of the term transmissivity for other than a full aquifer. In this document, the term refers to the product of the average conductivity multiplied by the length of the borehole being considered. It is not limited to the thickness of an aquifer and the length being considered is often a few inches, the distance traveled by the everting liner in a time step. The mathematical concept as described here is relatively simple. The major challenge was to perfect the apparatus needed to provide the control needed for the measurement and the spreadsheet to reduce the data. The first of several tests of the T profile method was at a borehole in Guelph, Ontario in a borehole provided by the University of Guelph. Packer tests done by Pat Quinn of the Univ. of Guelph were compared to the T profile results. Later papers by Pat Quinn (Quinn et al, 2011) have compared T profiles to Packer transmissivity results at other sites. The T profile calculational model is described, then the equipment design, the results, the comparison to straddle packers, and the limitations. In theory, the flow in the very short intervals measured by the T profile can be used to estimate an effective fracture aperture. That has also been done.

The transmissivity profile has not been used in the vadose zone, although it may be possible with an inverting liner with a downhole pressure transducer and the linear capstan. FLUTe has patents on the T profiling method and apparatus in the United States, Canada, and Europe.

T profiles have been used as of 2020 at over 210 sites in over 650 boreholes since 2004. The design has steadily evolved in the equipment design and the data processing. T profiles are done by FLUTe personnel or personnel trained by FLUTe.

10.3.2   THE TRANSMISSIVITY MEASUREMENT METHOD

10.3.2.1   The Liner Behavior

The procedure requires that the excess head driving the liner and the tension on the liner be maintained relatively constant. Under those conditions, the liner descends at a rate controlled entirely by the flow rate out of the borehole beneath the liner. The liner will suffer a drop in its descent velocity each time a flowing fracture is sealed by the descending everting liner reducing the transmissivity beneath the liner. That velocity change multiplied by the cross-sectional area of the borehole is the flow rate that existed in the fracture before it was sealed. Since

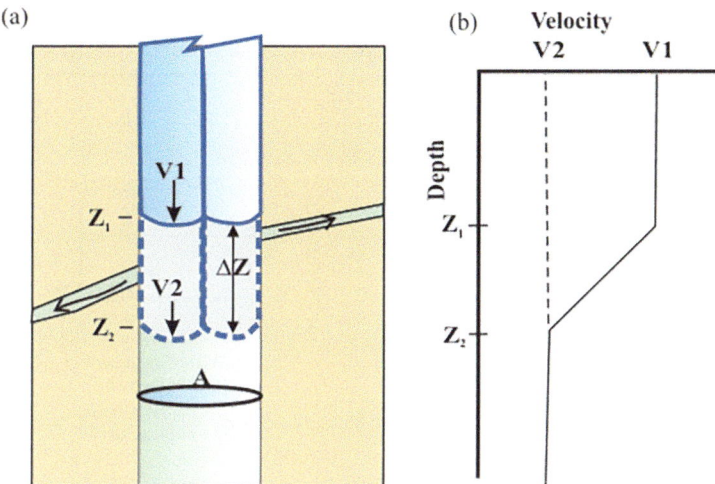

FIGURE 10.25 The liner velocity change upon sealing a flowing fracture. In (a), the liner traveling from z_1 to z_2 seals the fracture. The liner velocity plot in (b) shows the velocity drops from the v_1 at z_1 to v_2 at z_2 as the liner everts from z_1 to z_2 in time step Δt.

the liner displacement of water must equal the flow into the borehole wall beneath the liner, a change in T, ΔT, must be matched by a change in the liner velocity. Figure 10.25 depicts the liner position and velocity at depth z_1 and after passing a fracture to depth z_2 in one time step. The velocity change, v_1-v_2, can be used to determine the flow rate into the fracture before it was sealed by the liner. That detailed calculation is described hereafter. However, there are other changes in the liner velocity that are not related to flow into the borehole wall. The drawing of Figure 10.26 shows what occurs when the liner propagates through an enlargement of the borehole. Because the effective cross-section of the borehole is larger in the enlargement, the product of velocity and the liner cross-section (i.e., the flow rate out of the hole below the liner) must be unchanged, unless there is flow into the formation in the enlargement. In other words, if the liner cross-section is larger, the velocity must be less. Figure 10.26a shows the liner above the enlargement, z1, as it enters the enlargement, z2, midway through the enlargement, z3, as it exits, z4, and into the original hole diameter at z5. As the liner propagates out of the enlargement, the velocity will increase due to the smaller cross-section of the liner. In the enlargement at z3, the liner can only dilate to the extent of its fabricated diameter.

Fortunately, the liner tends to evert in a relatively straight line for a short distance. That allows it to enter the borehole below the enlargement. A plot of an actual liner velocity with depth in the borehole (Figure 10.27) shows numerous drops in velocity at enlargements followed by an increase in velocity as the liner enters the smaller diameter borehole beyond the enlargements. This is a relatively irregular borehole with many enlargements. The temporary drop in velocity caused by the enlargement

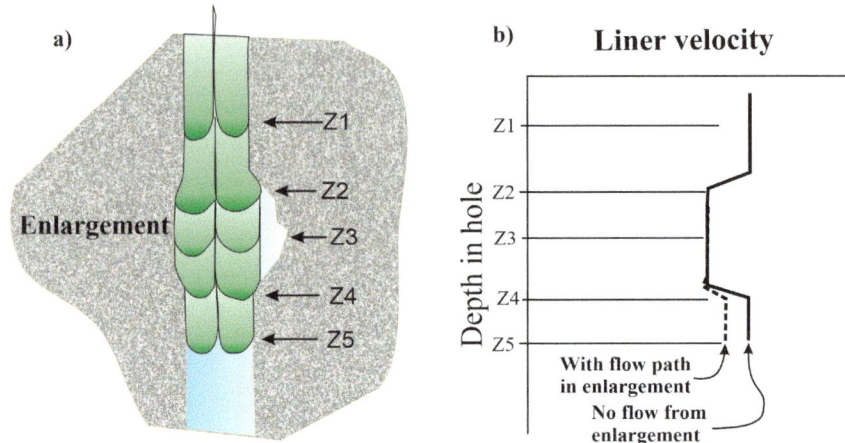

FIGURE 10.26   Effect of an enlargement on (a) liner shape and (b) liner velocity. The liner velocity is reduced in proportion to the enlargement of the liner cross section. If a flow path exits in the enlargement, the liner will exit the enlargement with a lower velocity than the entrance.

is not necessarily relevant to a change in the transmissivity beneath the liner. In order to ignore such temporary drops in velocity, a monotonically decreasing curve is fit to the data of Figure 10.27 as shown in Figure 10.28. When one sees such a drop in velocity, there is probably flow in the enlarged portion of the hole only if the monotonic fit does not have the same value before and after the drop in velocity. If there is a flow path in the enlarged interval, the liner will exit the enlargement with a lower velocity than the liner entered the enlargement (dashed line in Figure 10.26b). The monotonic fit will assign any velocity change (transmissivity) due to a slower exit from the enlargement (the dashed line) to the entrance of the enlargement. The transmissivity is not lost, only displaced upward to the entrance of the enlargement. The transmissivity is calculated from the monotonic fit. Since the liner of fixed diameter does not always dilate to the full diameter of an enlargement, the relative effect of an enlargement may depend upon the size of the liner relative to the borehole diameter. In Figure 10.28, the drops in velocity at 88 and 100 ft. are less than would occur if the enlargement exceeded the diameter of the liner. The change of velocity in an enlargement depends on the square of the change in liner diameter. In the method description above, the borehole diameter is assumed constant, and therefore, the calculated flow rate plotted is in error in the enlargements and results in the intervals with velocity drops below the monotonic fit. If the hole diameter was measured with a caliper log of the hole, the actual hole diameter can be used for the borehole diameter, but is still limited to the diameter of the liner. That correction can reduce the size of the flow curve dips below the monotonic fit, but it does not always change the deviations. Another cause for the temporary differences is there may have been an abrupt, but brief change in the drag term described in Section 3.2.2. That is usually recognized in the measured head beneath the liner. Dividing by the reduced driving pressure for the T calculation also offsets such dips in the flow curve.

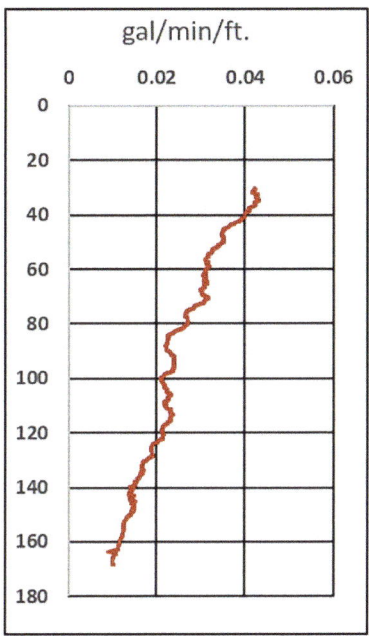

FIGURE 10.27 Flow rate per unit driving pressure with depth. This borehole had many enlargements causing the irregular velocity.

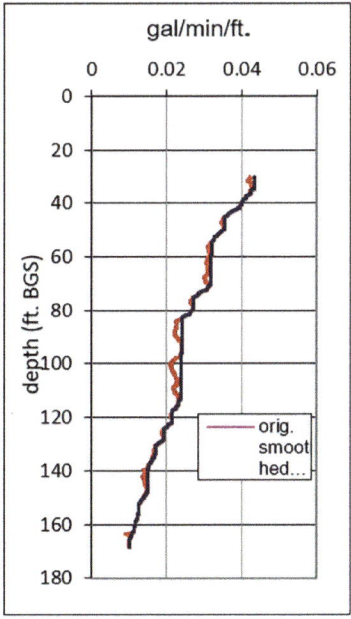

FIGURE 10.28 Monotonic curve (black line) fit to flow curve of Figure 10.27. Monotonic fit ignores liner velocity drops due to enlargements.

Since the driving pressure can affect the velocity, and the actual driving pressure is not easily kept constant, the liner velocity in Figure 10.27 has been divided by the driving pressure/head in the borehole which removes some of the variation due to head changes in the borehole beneath the liner. Multiplying the monotonic velocity per unit driving pressure by the liner cross-section provides the rate of flow per unit driving pressure (Figure 10.27) out of the borehole beneath the descending liner as a function of the liner depth. The flow rate changes are the actual flow rates into the fractures before each fracture is sealed by the liner.

The same logic applies to any permeable feature intersecting the borehole such as a bedding plane or permeable bed. For a uniform permeable bed, the velocity change is distributed over the thickness of the bed. The velocity of the liner descent is calculated from the measured depth of the liner at each time step. The typical time step is half of a second; therefore, velocity changes are calculated for each interval traversed by the liner in half of a second. Hence this allows the very high spatial resolution which can be achieved for the transmissivity profile.

10.3.2.2 The Calculational Model

The transmissivity of the interval traversed by the liner in one recording time step is calculated using the Thiem equation:

$$Q = T\Delta H \ 2\pi/\ln(R), \text{ or } \Delta Q = \Delta T\Delta H \ 2\pi/\ln(R)$$

Q is the flow rate into the interval of transmissivity, T, where T is the conductivity, C, times the interval ΔZ. That is, $\Delta T = C \ \Delta Z$ for a short interval or a long interval. Many traditional borehole measurements are limited to the screen interval as ΔZ or an aquifer bounded by aquitards or aquicludes. Some T measurements are of the entire saturated borehole. In straddle packer measurements, Δz is the separation of the packers.

ΔT is the transmissivity of the interval; R is the ratio of r_a/r_h (see Figure 10.29 for the terminology and geometry). ΔH is the difference between the head, Ha, at the range of the ambient head, ra, and the head, Hh, in the borehole of radius r_h. Or, $\Delta H = Ha - Hh$. Although transmissivity is only applied to an aquifer, as I was told by one reviewer, in this model, ΔT is the transmissivity of relatively short intervals, ΔZ, in the borehole. Writing Q as ΔQ, and T as ΔT, the flow rate into the short interval, ΔZ, one can solve for ΔT.

$$\Delta T = \Delta Q/\Delta H \ \ln(r_a/r_h)/2\pi = (v_i - v_{i+1}) \ A \ \ln(r_a/r_h)/(2\pi(H_h - H_a)) \quad (10.1)$$

In this equation, the flow rate into the borehole wall of area, $2\pi r_h \ \Delta Z$, is that flow which is interrupted by the sealing liner in everting a distance ΔZ. That flow rate

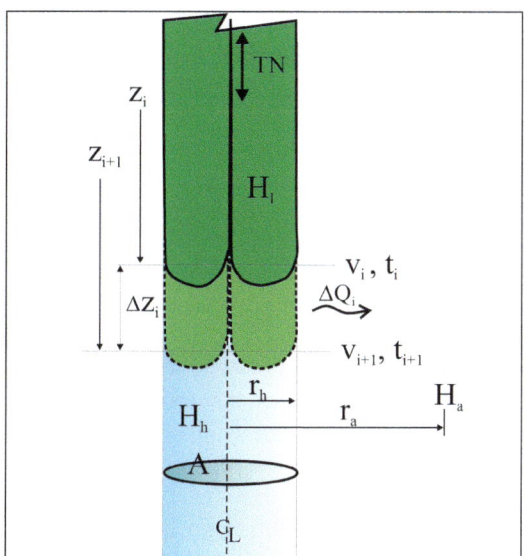

FIGURE 10.29 Drawing of spatial terms used for the calculation of the transmissivity of the ith interval traversed in the ith time step. H_h is the borehole pressure/head and H_a is the formation ambient pressure/head at radius r_a.

change in the liner displacement is $\Delta Q = \Delta v \, A$, where A is the cross-section of the borehole or liner, Δv is the liner velocity change, Δvi, between the depth Z_i and $Zi+_{1\,i+1}$. The subscript, i, is related to the time step between the liner EP depth at time t_i and when at the greater depth at t_{i+1}. In other words, the flow rate of the sealed intervals is the displacement volume per unit time difference between the liner at Z_i and Z_{i+1} per time step $\Delta t_i = t_{i+1} - t_i$.

Since r_a is not well known, R is often not well known, it is estimated. Because R occurs in the natural log term, an error in the ratio is not a major error in the calculation. It is true that ra is probably not the same range for flow from the borehole everywhere. The Thiem equation is the common relationship used to deduce T from straddle packer tests. Note the Thiem equation is based upon a steady-state flow from the borehole. That is addressed later in Section 10.4.3. Throughout most of the boreholes, the flow into the formation is near steady state as the liner everts.

The above relationship is the basis for the T profile method that has served very well for mapping the conductive features in the borehole. The time step Δt is typically one half second, and therefore, ΔZ is the everting liner distance traveled in each half second. As one approaches the bottom of the borehole with little transmissivity remaining below the liner, the distances traveled in a half second approaches $0.01''$. Fortunately, the calculated ΔT values can be summed over longer more realistic distances allowing fracture measurements on less than the half foot scale. Near the top of the borehole, the flow rates out of the borehole beneath the liner can be a thousand-fold higher and the ΔZ intervals nearer 10″. As the liner slows with depth, the resolution drops to better than half a foot. The net result is that the resolution is far better than packer typical test intervals of 5 ft. or more.

As described earlier, H_h is the head in the borehole during the liner descent, H_a is the head in the formation before the borehole pressure was increased by the descending liner. Because the head, H_a, in the formation is not known, the initial assumption is that it is the blended head in the open hole. If the head distribution in the formation is known, it can be used in the $T(z)$ calculation as a refinement of the ΔH.

10.3.2.3   When to Terminate the T Profile

As the liner everts and seals the flow zones, the descent rate drops. If the descent rate drops below 0.001 ft./s, (3.6 ft./h) and there is much more borehole left, the T profile is usually terminated. At that velocity, the remaining transmissivity in a 6″ hole is about 0.02 cm²/s which is usually an insignificant transmissivity. Occasionally, the profile is continued if there are observed fractures near that depth (e.g., observed in an earlier OTV).

The velocity at which the liner eversion is halted can be much higher for the liner if the purpose is to only provide a temporary seal of the borehole. An earlier halt of the liner eversion allows a more rapid inversion when removing the liner from the borehole because the remaining transmissivity is higher but still not significant to the temporary seal of the borehole.

10.3.3   The T Profile Results

Figure 10.28 is an actual flow rate curve (red) as a function of borehole depth. The small drops below the monotonic curve (blue) at 88 and 100 ft. are due to borehole enlargements. The transmissivity calculated from the flow rate for each time step is shown in Figure 10.30. The spatial interval is dependent upon the recording time step (typically 0.5 s) and the velocity of the liner. In other words, the faster the liner descent, the longer the interval traversed in one time step. In order to make the result more comparable throughout the borehole, the short interval transmissivity was integrated (summed) over one foot intervals to produce the result of Figure 10.31 which one should get from a series of one foot long straddle packer tests with no leakage. The red curve of the graph of Figure 10.31 is an estimate of the transmissivity resolution threshold estimated as 1% of Figure 10.32.

Figure 10.32 is the integral from the bottom of the borehole to the top of the discrete transmissivity results shown in Figure 10.30. Not surprising, it has the same shape as the monotonic curve of Figure 10.28.

The graph of Figure 10.32 is especially useful in determining the transmissivity of any interval in the borehole. The difference in the graph values in Figure 10.32 at an upper elevation and a lower elevation is the transmissivity of that interval. For example, the transmissivity at 60 ft. is 0.06439 cm²/s and 0.0506 cm²/s at 100 ft. The

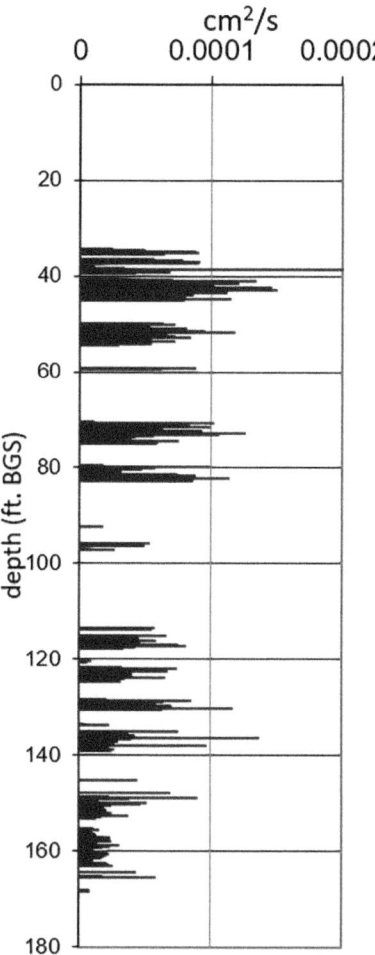

FIGURE 10.30 The transmissivity as calculated for each time step over the travel distance of Δz_i using the monotonic fit of Figure 10.28. At lower portions of the borehole the data points are much more closely spaced which may divide the flow of a high angle fracture into numerous transmissivity steps.

40-ft. interval has a transmissivity of 0.00933 cm²/s. This ability to determine the transmissivity of any interval from the T profile will be used more in Section 10.4 for the head profile calculation.

The advantage of the liner transmissivity measurement is twofold. First, an entire borehole of 100–600 ft. depth may be measured with high spatial resolution in 1–2 h. Second, there is no concern about leakage past the liner as can occur with straddle packer tests. All other transmissivity profiling methods require more time and provide lower spatial resolution in that time. Because the liner is usually moving faster

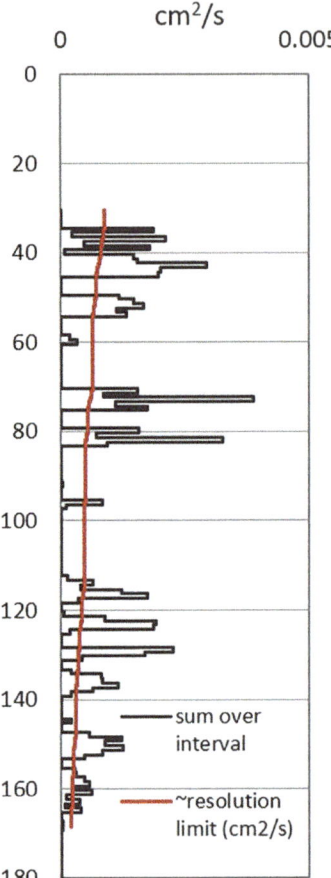

FIGURE 10.31   The integration of Figure 10.30 over 1-ft. intervals. This is equivalent to 140 1-ft. packer tests. The red curve is an estimate of the level of reliability. Peaks below that level may not be reliable.

at the upper end of the borehole, the spatial resolution is not as fine as near the bottom of the profile. The journal article by Keller et al (2014) describes in greater detail how various factors can affect the transmissivity results. The transmissivity profile is often used to aid in the selection of sampling intervals to be included in the multi-level system (MLS) design described in Section 10.5. For example, in Figure 10.32, sampling intervals centered on 43, 52, 73, 82, 117, 123, 129, 137, and 151 ft. might be considered for monitoring of water quality and head with the MLS described in Section 10.5.1.

The calculation of ΔH requires knowing the formation head at the time of the measurement. This is true of other measurements such as straddle packer tests. With no other information on the head distribution in the formation, it is assumed to be equal to the blended head in the borehole. If the more detailed information on the

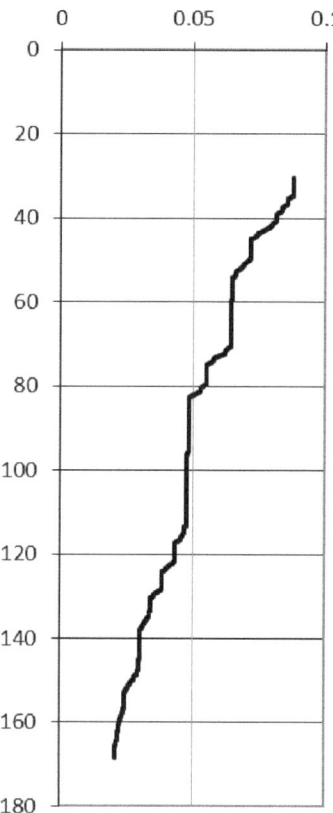

FIGURE 10.32 Integral of Figure 10.30 from the bottom of the borehole to the top. Note this has the same shape as the monotonic curve in Figure 10.28. This graph allows the determination of the transmissivity of any interval throughout the depth of the profile.

formation vertical head distribution is known, those values can be interpolated and used in the refinement of the T profile. This is often done when the RHP (reverse head profile) has been done as described in Section 10.4.

10.3.4 EXAMPLES OF OTHER T PROFILES

Figures in Section 10.3.3 are the actual results from a typical borehole transmissivity measurement. Another profile with a 90-fold higher flow rate than in Figure 10.27 is shown in Figure 10.33. With such a high flow rate, the resolution of velocity changes (on the order of 1% of the velocity) is much more limited. However, below 185 ft., the resolution is much better. Another profile such as in Figure 10.34 shows the difference that one can see in a profile of a different hole. The liner has a slight rise in velocity as it exits the casing into a smaller hole, there are some enlargements at 67 and 79 ft. (obvious, even without the monotonic fit) and a very large flow zone at 90 ft. Thereafter, the transmissivity is much more distributed in the formation until

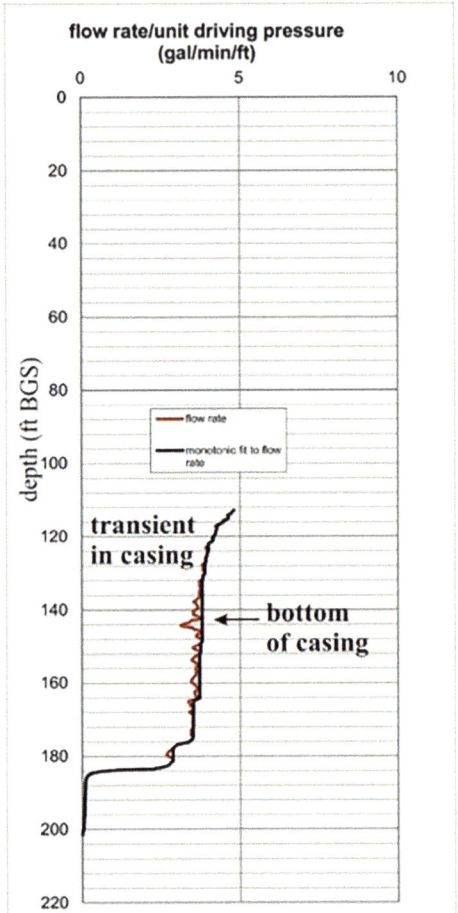

FIGURE 10.33   An example of fast flow near the bottom of the hole at 185 ft. The transient is seen decaying in the casing and did not need correction. The transmissivity of this borehole is 90-fold higher than that of Figure 10.27. Therefore, the limit of transmissivity resolution is ~90 times greater. Small flow zones like those of Figure 10.31 would not be reliably detected until below 185 ft.

210 ft. Every profile shows such differences. It is not easy to deal with flow rates over 150 gal./min. The profile of 10.33 with an excess head of 20 ft. requires a water addition to the liner of over 100 gal./min.

Note that in Figure 10.33 that the liner velocity is dropping in the casing above 145 ft., yet there can be no flow into the casing as occurs for the open borehole. That decay in the casing is due to the transient to steady-state flow discussed in the next Section 10.3.6.

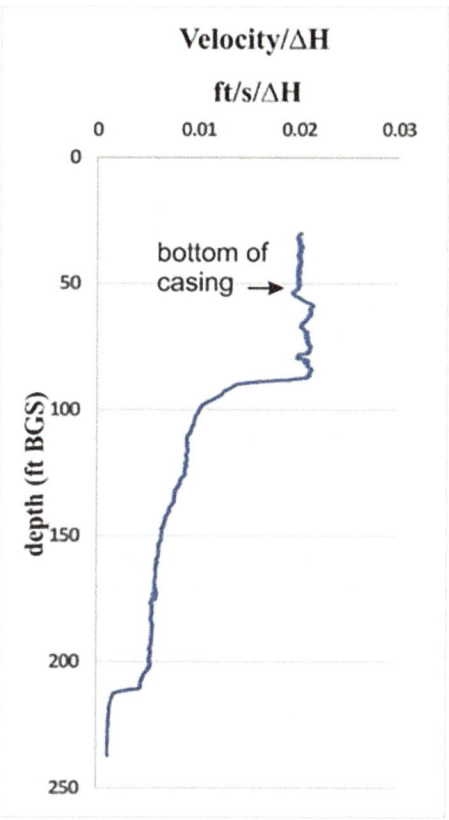

FIGURE 10.34 Another flow velocity curve showing a very different profile. There are two major flow zones (90 and 211 ft.). No monotonic fit is shown for this curve. Such a fit would show apparent enlargements at 67 and 79 ft. The borehole has a slight enlargement at the bottom of the casing after which the liner velocity increases due to the borehole being smaller than the casing.

10.3.5 CALCULATION OF THE EFFECTIVE FRACTURE APERTURE USING THE T PROFILE RESULTS

The T profile provides the flow rate and driving pressure for the flow into the borehole wall for each interval traversed by the liner in the short time step of the measurement. That data can be summed/integrated over any interval desired or at least within the spatial resolution limit available. From either the OTV or simply assuming a number of fractures in a calculated interval such as the 1-ft. interval of Figure 10.31, it is possible to estimate a fracture aperture based on the model in the Bird, Stewart and Lightfoot text, in which they derive a radial flow model for flow between two plates for a viscous fluid under laminar flow conditions. That aperture is denoted as 2b, the distance between the two plates. If one makes the leap of faith that the flow into the hole wall into a fracture is approximated by two parallel plates, the T profile data allows a calculation of that effective aperture. This is a common assumption. The details of the aperture calculation and examples of the result are described in Section 16.4.

Whereas such an aperture is only useful in modeling fracture flows, it also has a practical application within the assumptions. For a multilevel water sampling system as described later, the sampling interval may intersect relatively few fractures or a single large fracture. When that interval is purged of a prescribed volume of water, it is possible with an effective aperture based on the same model assumptions for the fracture volume to estimate the range from which the water sample is drawn after the purge volume is extracted. This is most useful in limiting the speculation on the effective "reach" of a water sampling event in fractured rock.

The main effect is that for large fracture apertures, the range of extraction is less than for fractures of small aperture for equal purge volumes.

It is interesting that this kind of quantitative assessment shows that some landfill monitoring boreholes do not monitor for more than a short range from the borehole in fractured rock and based on another flow model and measured porosity, the reach of common monitoring wells with a sand pack annulus also do not monitor a large volume adjacent to the well. The assumption of a regular geometry is not good, but the assumption that a monitoring well is drawing from a long range between monitoring wells is a common error in the reliability assessment of an effective monitoring well spacing.

10.3.6 Corrections to the Simple T Profile Calculational Model

10.3.6.1 Transient Correction

10.3.6.1.1 The Method of Correction

A basic assumption of the Thiem equation is that the flow into the formation is at a steady state or very near to it. After the liner has traveled some distance with a constant driving head and tension, the flow out of the borehole does approach a steady state. However, when the T profile is started, the head in the borehole goes from the blended head to a higher head nearly instantly. The initial instantaneous flow out of the borehole is very fast with the abrupt increase in the gradient out of the borehole and slows in time to a steady state as illustrated in Figure 10.35 (or seen in Figure 10.33 as the liner is near the bottom of the casing). The result is that the liner starts down the borehole very fast due to that initial high flow and then slows as the steady state develops throughout the formation intersected by the borehole. The lower the conductivity, the longer the time to achieve the steady state. (Fortunately, the liner is also traveling more slowly and travels a shorter distance before the steady state is achieved.). The problem is that the measurement method uses the drop in velocity to determine the flow rate out of the hole as each transmissive feature is sealed. If the liner velocity is decaying due to the transition to steady state that decay in velocity is not related to the flow out of an individual flow interval being sealed. The typical transient decay occurs over 10–20 ft. of travel and more quickly in fast flowing holes. In that interval, it is difficult to calculate the flow rates in flow features being sealed using the standard method.

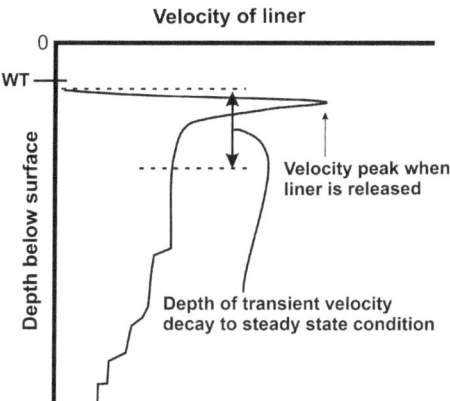

FIGURE 10.35 Transient in liner velocity upon release and during decay to steady-state outflow from the borehole.

The solution is to correct for the transient by subtracting the flow during that decay to a steady state. The question is what is the flow due to the transient? That was solved by doing a separate calculation in a simple 1D radial flow geometry of the flow decay to steady state. However, as the liner is started down the borehole, many layers of different characteristics are flowing at different rates toward that steady state. However, if one does a calculation of the convergence to steady state for many different conductivities, and then adds them, the curve of the sum still looks like a transient decay to steady state. It was reasoned that by picking an effective conductivity and storativity, one may be able to match the apparent transient of the initial flow seen in the T profile. Subtracting that simulated transient flow from the T profile flow rate in time would be a first-order correction of the transient to steady state and would leave more obvious the features in the flow due to sealing of conductive features in the interval of the transient. That worked well.

However, how does one pick the appropriate parameters for that calculation? The later portion of the T profile is measuring the flow out of most of the hole. One can pick a value to try, from the T profile data after the transient has decayed, as the conductivity in the calculation of the transient correction. The driving pressure in the borehole is also known from the measurement. The storativity is not known but tends to be relatively low in fractured rock sites of perhaps 10e–07. With those values, the first calculation is done and the transient is subtracted from the flow measured during the transient.

Fortunately, there are several hydrologic constraints. One is the subtraction cannot cause an increasing flow rate, a sign of over correction. If it does, the conductivity can be reduced. If the correction extends further in time (i.e., depth in the hole) than the apparent transient, the storativity is too large. When those two parameters are adjusted to where the corrected flow rate has some flat intervals (i.e., no flow into the hole wall), and is not increasing with depth, and does not extend beyond the apparent end of a transient in the measured curve, the transient correction is probably defensible. With those conditions, it is often clear in the corrected curve where flow features have been sealed without the confusion of the transient effect and the monotonic fit is effective in the calculation of the flow rates. If this did not actually work so well as it does, it may be considered an inappropriate manipulation of the data.

If the surface casing extends below the water table, it is best to start the T profile in the casing. Then the transient is obvious and may have decayed before exiting the casing. That is a good time to try the transient correction, since in the casing, only the transient is causing the eversion velocity to decay. However, many surface casings do not extend into the water table and the T profile can only be started in the water at the blended head level below the casing. With practice, the transient corrections require few iterations.

The transient correction was not used in some of the earliest profiles and when the T profile was compared to other measurements, the T profile conductivity was too large in the top part of the hole. Hence, there is a need for the correction.

10.3.6.1.2   Result of the Transient Correction

In Figure 10.36 is shown the effect of a transient correction on an actual measurement. The initial flow rate with depth is shown as the red curve. Picking a flow rate from the red curve of 1.0 gal./min/ft. as a median value, the conductivity of 6.6e–04 cm/s is calculated for the open saturated length of the borehole. The driving head was 8 psi (18.5 ft.). The storativity of 1.0e–07 was used, and the resulting transient flow was subtracted from the measured flow. The correction was judged insufficient. Raising the conductivity to 5.0e–03 cm/s provided a better correction (the green curve). This suggests that the similar very conductive flow at 112–114 ft. is dominating the transient. The result of the corrected flow rate is shown as the green curve. The monotonic fit of the green curve is shown as the black curve. In this corrected flow rate, the transmissive zones at 82 and 94 ft. are more obvious. The low-flow

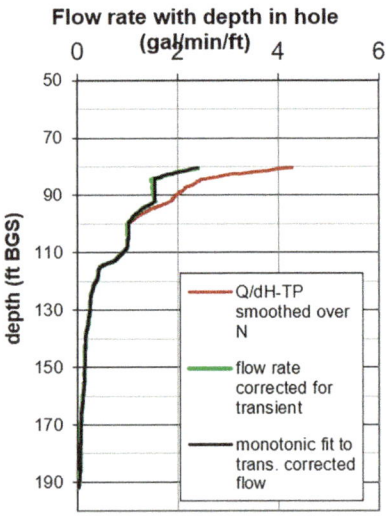

FIGURE 10.36   Actual correction of a transient. An excessive correction would lead to an unrealistic increasing velocity at 84–93 ft.

zone from 85 to 92 ft. is also more apparent. This correction method is used where it seems appropriate from the inspection of the initial flow rate. Were the correction not made, the data reduction would show very high flow into the upper portion of the borehole based on the rapid flow rate decay with depth. Once the transient is no longer significant, the borehole has essentially steady-state flow along its full length. However, in fact, theoretically, the flow never achieves a full steady state. Most hydrologic measurements don't.

As said above, the first transient correction did not seem sufficient. It is easier to judge if the transient correction is excessive. In the case of the decay in a casing, the overcorrection leads to an obvious increase in velocity in the casing which is no more realistic than a drop in velocity due to flow through the casing. In an uncased hole, it is also unrealistic to have a velocity increase apparent in the over corrected flow rate. In Figure 10.36, an overcorrection would cause the green curve between 82 and 92 ft. to rise to the black curve. In the plot, it (the green curve) does increase very slightly with depth to 90.8 ft. A slight reduction in the conductivity would be better, but as an example of a slight overcorrection, it is useful.

10.3.6.2   The Borehole Diameter Correction

As described above, variations in the hole diameter can produce drops in the liner velocity that are best ignored by using a monotonic fit of the velocity data. However, if a caliper log is available as after a suite of geophysical measurements, that better hole diameter can be incorporated in the calculation of the correct liner cross-section as described above. The corrected diameter allows the better flow rate to be calculated for enlargements (even if greater than the liner diameter). That caliper data is interpolated to provide an effective hole diameter at each time step and corresponding depth in the data reduction.

10.3.6.3   The Vertical Head Correction

In the first T profiles, it was assumed that the head in the formation was approximately the "blended head" or equilibrium head measured in the open hole, the water table. However, in many boreholes that blended head is not the head in the formation, but only somewhere between the lowest and highest water tables in the formation. In many holes, the head in the formation may vary significantly with depth and when the hole is drilled, connecting the several conductive aquifers, the blended head is developed. If those normal formation head variations are small, using the blended head for the formation head is not a grievous error if the driving head in the liner is much higher. That is, the error in the head difference ΔH is not significant. However, if the head variations are large, the calculation can be corrected if the head distribution in the formation is known. That is not known usually at the time of the T profile, but a new method called the "reverse head profile" (RHP, see Section 10.4) does allow the measurement of the head distribution as the liner is withdrawn after a T profile in a step wise manner, because at that time, the T profile has already been measured. After performing the RHP measurement, the head obtained can be included in the T profile calculation of the T profile as a first-order correction. That refined T profile is then used to obtain a better head profile as described in Keller (2016). The T profile and formation head calculated converge quickly after one iteration. If the head distribution is known from some other measurements as with packers or using MLS systems in nearby holes, that better head distribution is included in the T profile data reduction. Fortunately, using the blended head assumption still allows relatively easy identification of the high flow zones. It is always better to use a large excess head in the liner, but for very fast flowing holes that requires a very high capacity pump to keep up with the liner descent rate which can be as high as 200 gal./min.

10.3.7   THE TRANSMISSIVITY PROFILING EQUIPMENT

The transmissivity measurement includes the measurement and control of the liner tension, and recording of the liner depth in the borehole, the water level inside the liner, and the head in the borehole below the liner for each time step. The drawing of Figure 10.37 shows the basic features of the current profiling system which measures or controls the important parameters. The depth meter (the encoders) and tension control (the brake drum) are added to the normal blank liner installation with a recording transducer at the bottom of the borehole. The photo in Figure 10.38 shows a typical hardware arrangement for the actual transmissivity profile measurement. No crane truck or drill rig is required. The liner is fed from the shipping reel though the profiler system into the borehole. In some situations of slow flowing boreholes, the head beneath the liner can be reasonably calculated from the liner tension and the water level inside the liner according to the equation in Section 3.2. The water level in the liner is monitored with the bubbler tube and associated transducer. However,

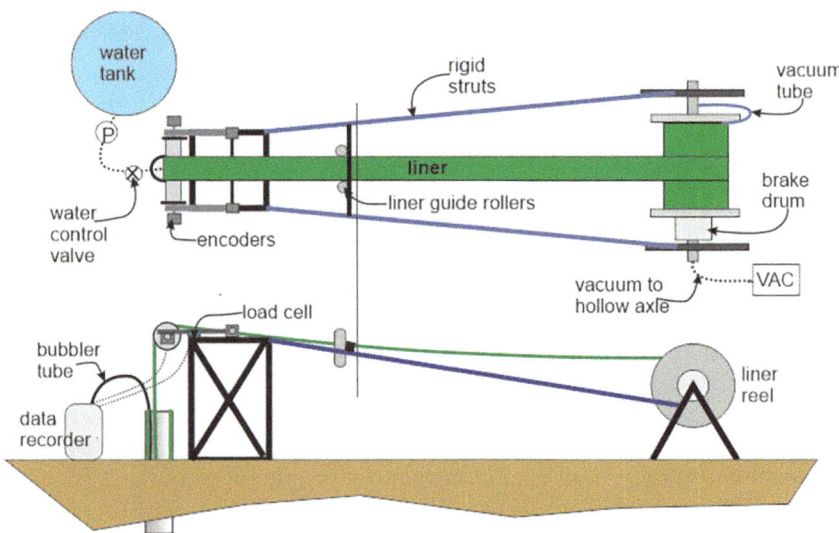

FIGURE 10.37 The components of the T profiling procedure. The water addition is controlled by the valve and supplied by the pump and water tank. The data is recorded form the bubbler, load calls, and encoders in laptop computer connected to the data collection system. As the liner velocity slows and the reel diameter decreases with the liner deployment, the brake drum torque is adjusted.

FIGURE 10.38 Photo of profiling equipment in the field. The data is recorded to a laptop computer during the installation. Water is usually pumped from a tank into the liner to maintain a nearly constant driving pressure in the borehole. The main features of the drawing in Figure 10.37 are seen in this photo.

for most T profiles a transducer is located at the bottom of the borehole (Figure 10.37) to obtain the pressure/head used to calculate the results of Figure 10.30. The struts prevent the differential motion of the shipping reel relative to the profiler which can otherwise cause tension variations leading to velocity variations. A vacuum tube to the hollow reel axle is connected to the rotating liner on the reel to remove the air trapped in the liner to avoid the formation of an air balloon in the liner during the T profile (see Section 3.2.4). The load cells on the arms supporting the main roller on the profiler allow the calculation of the tension on the liner during the installation.

10.3.7.1 Maintaining a Constant Tension on the Liner

The first T profiles used a simple hand tension on the liner for the measurement. That was a poor method, since the liner speed and the operator's skill were major factors in the measurement. With a pressure transducer in the bottom of the borehole, it was not so critical that the tension be constant, but it is still best if the liner tension is constant during the profile. The tension relationship to the pressure in the borehole and also the effect of the water level in the liner are both important to the head beneath the everting liner as explained in Section 3.2. However, large changes in either parameter are undesirable since the basic assumption in the Thiem equation is that the borehole pressure has developed a steady-state flow in the formation. Variations in the borehole pressure violate that assumption. However, early tensioning by hand still yielded useful results. But several methods were devised to avoid the tension variations. It was also learned that subtle variations such as the shipping reel diameter caused variations in the tension even with a braking system on the liner to control the tension. The first braking systems were elaborate. A disk brake was used on a roller with a large contact area over which the liner was draped as described for the linear capstan in Section 9.2. A tension monitoring device that controlled the disk brake caliper helped, but the variations in the disk surface and alignment were determined to be the cause of variations in the tension. A more simple system was developed which provides a relatively constant torque on a drum attached to the shipping reel. It worked better, but variation in the shipping reel drum diameter caused variations in the tension. The remedy was to use a reel for storage of the liner with a more constant radius. Another factor causing variation of the tension is the diameter of the liner rolled on the reel. As the liner is deployed the effective diameter of the reel decreases. However, the brake on the shipping reel is adjustable in order to compensate for the decreasing diameter of the liner on the reel. That is the current design on the system shown in Figure 10.37. The brake on the shipping reel is a spring-loaded wrap on a steel drum concentric with the shipping reel axle using the same principle as for the tension on the linear capstan. It is also necessary that the liner be fed uniformly with no variation in any drag on the liner and that it be reasonably centered on the roller with the encoders. Under very windy conditions, that is a challenge. The end result is that the current system has a relatively constant tension on the liner to maintain a minimal variation in the borehole head/pressure beneath the liner.

10.3.7.2   Maintaining the Constant Driving Head

The head in the borehole is still sensitive to variations in the excess head driving the liner eversion. Several methods were designed to maintain that constant head. The challenge is the changing velocity of the everting liner. As each flow zone is sealed, the rate of eversion drops. If a water supply to the liner, as it propagates, is set to a constant flow rate to match the liner eversion rate, at each drop in liner velocity, the head in the liner starts to rise. For large velocity changes (e.g., when passing large flow zones), the flow rate into the liner needs to be changed abruptly to keep the constant excess head in the borehole. In the more simple case, the liner is filled to the top of the casing and any excess flow drains over the top of the casing, but in many boreholes, the water table is too deep to fill the liner to the top of the casing. In that case, the water level in the liner is monitored with a bubbler system. A bubbler system is a simple tube submerged below the desired water level in the liner (usually the desired level of the bubbler is the depth of the blended head in the borehole). Given a constant flow rate of air through the tube, the air will bubble out the bottom of the tube at a pressure slightly greater than the depth of submergence. By maintaining a constant excess head, the submergence is constant, and the bubbler pressure is constant. But that requirement means monitoring of the bubbler pressure and controlling a valve system to keep the bubbler pressure (the excess head) constant. That is currently done by the operation of a valve downstream from a pump at the water tank while monitoring the bubbler pressure. Several systems were designed to do that automatically but were awkwardly heavy and bulky and not sufficiently responsive. The challenge besides the abrupt changes in the flow rate is that the water flow rate which is needed varies by three orders of magnitude during a typical T profile. We are working on that better automatic control system. The manual controller is not a problem, but watching the bubbler pressure for an hour or two is tedious and requires an additional operator. During the T profile, the actual liner depth in the hole is measured to compare the liner depth calculated from the encoder to that measured to assure a relatively precise match of the T profile with borehole features. The effective diameter of the encoder roller changes slightly with the liner versus when the tether is passing over the roller. The roller diameter is adjusted in the calculation of the depth for that effect in the comparison with the tagged depths of the liner during the T profile.

10.3.8 Effect of Well Development on the T Profile

Different drilling methods clog the fractures to a different degree. From our observations, core drilling tends to clog the fractures more than air drilling. The simple reason is that driving the cuttings to the surface with injected water past the drill pipe in a slender annulus takes more pressure. Air drilling often has a net inflow to the borehole and hence can produce much more water from the formation which should tend to clear the drill cuttings in fractures. Figure 10.39 shows the difference between a core hole undeveloped and the same hole reamed with subsequent development. The difference is great and may be related to the lack of development of the core hole, but the reaming with development clearly made a large difference.

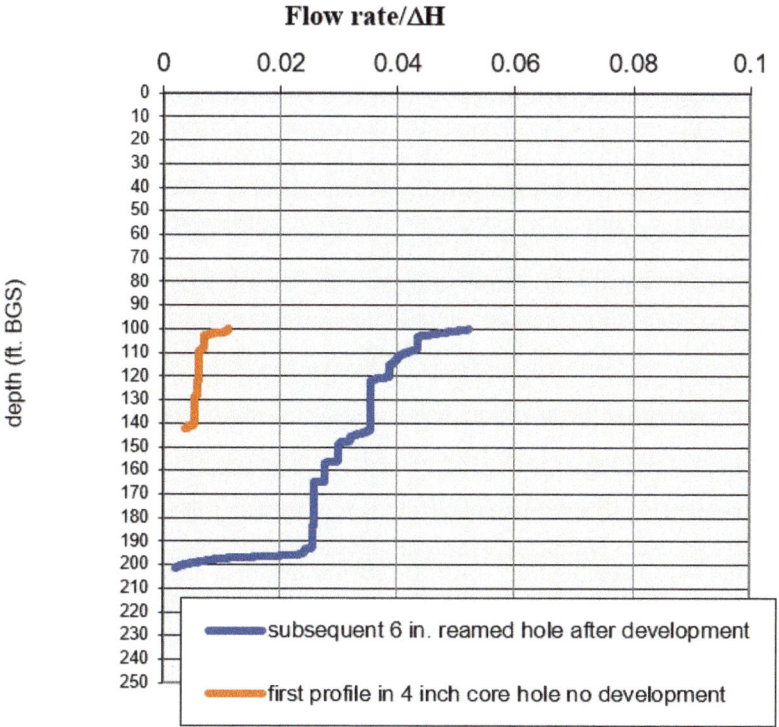

FIGURE 10.39 The orange curve is the T profile for a cored HQ hole without development. The blue curve is the same hole reamed to 6″ after being developed.

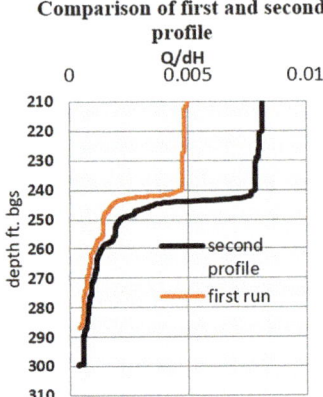

FIGURE 10.40 The comparison of T after one liner removal. The liner removal increased the flow at 243 ft. by ~65%.

Another example (Figure 10.40) is a T profile done after the first liner was removed and a second T profile was performed. Because the tension on the liner during removal causes a large drop in the head beneath the liner, it is a very effective well development tool. That is especially the case for the lower end of the hole when the tension would be high with a slow removal. Higher in the hole it depends on how hard the operator is tensioning the liner.

It is surprising that some well development is done with the pump located at several levels in the borehole "to enhance the development". However, since the vertical flow impedance in a vertical hole is minimal (Shapiro et al, 2002) (a simple laminar flow calculation provides an effective flow conductivity of 10^8 cm/s for an open hole), the location of the pump in the well is probably not significant. A surge block development with pumping is much more effective.

In the NAWC test of the FACT described in Section 10.2.2.4, after many liner removals the total transmissivity increased by 25%. In yet another cased hole with three-screened intervals, the first installation of a Water FLUTe (WF) encountered sediment above the deepest screen. After inverting the liner from the cased borehole, which increased the sediment level, and doing an air lift pumping removal of the sediment, the descent rate during the WF re-installation went three times faster. Well development should be done with explicit requirements to the driller. It is also useful to note the well development advantages of a blank liner removal. The remove would best be done with the linear capstan shown in Section 9.2 which can maintain a substantial tension on the liner inversion even at the higher speed and lower tension of the upper portion of the borehole. The utility of manual removal for development with the green machine depends on the operator and the formation conductivity.

10.3.9 A SPECIAL DESIGN FOR T PROFILES OF BOREHOLES WITH VERY HIGH ARTESIAN HEADS

When there was a need to perform a T profile on a well in New Brunswick, Canada, with 30 ft. of artesian head and on another artesian well with 150 gal./min. flow over

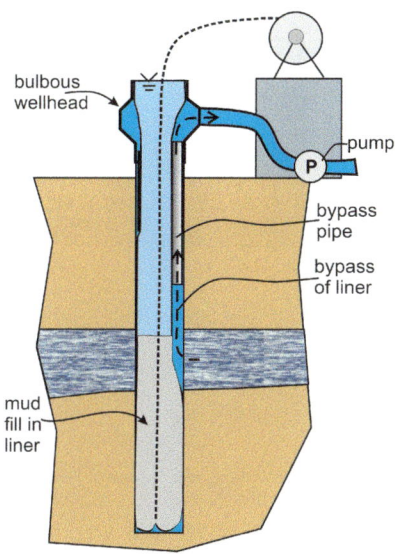

bulbous
wellhead

pump

P

bypass
pipe

bypass
of liner

mud
fill in
liner

FIGURE 10.41 Bulbous wellhead design. After the mud slug has passed the bypass pipe, water is added to the liner to perform the T profile.

the top of the casing, a design was developed to meet the need. The design called a bulbous wellhead allowed the T profile to be performed against the upward flow in the well. Figure 10.41 shows the geometry of the components. The major components were a bulbous wellhead to allow flow past the liner to be redirected to a side pipe connected to a vent hose or to a high-volume pump and a bypass pipe. The bypass pipe was lowered into the well for a distance greater than the artesian head before the liner installation. The bypass pipe allowed the water beneath the everting liner to flow past the descending liner. In order to obtain a sufficient head to allow the liner to propagate into the bulbous wellhead, a scaffold was erected to extend the casing above the bulbous wellhead.

The liner was driven with a heavy mud column to a depth greater than the length of the bypass pipe. As the liner descended with the heavy mud fill, it passed the bypass pipe with sufficient excess head to overcome the artesian head. Beyond the bypass pipe, the mud-filled liner sealed the borehole. As the liner everted, water was added to the liner to augment the mud pressure.

In this manner, the T profile was performed after the liner passed the bypass pipe. The artesian head was due to the presence of a brine-filled pit on the surface near a potash mine. A photo of a T profile using the bulbous wellhead is shown in Figure 10.42. In this installation, a hose is connected to the outlet of the bulbous wellhead to redirect the flow until the liner passes the bypass pipe and seals the borehole.

10.3.10   T PROFILE COMPARISON WITH STRADDLE PACKER RESULTS

Figure 10.43 shows the difference in the flow field expected for a descending liner and the flow for a straddle packer. The liner flow field beneath the liner is essentially

FIGURE 10.42   The bulbous wellhead used to perform a T Profile. Hose to the pump removes water bypassing the liner above the mud.

a one dimensional radial outward flow ignoring the actual 3D nature of fracture flow. In contrast, the straddle packer flow field diverges with distance from the straddled interval. Another difference is shown in the drawing in Figure 10.44. This illustrates potential leak paths for a straddle packer system for by-pass of a packer at either the contact of the packer with the borehole wall or through the formation. That leakage into the open borehole above or below the straddle packer system can lead to an error in the transmissivity calculated from the flow rate measured in the packer test. The liner flow field does not have such potential bypass to an open borehole from the pressurized borehole beneath the liner.

There are situations where the liner does not produce superior results to a straddle packer test. One is the situation where most of the borehole transmissivity for a highly transmissive borehole is at the bottom of the hole. In that case, most of the borehole is traversed in a very short time until the liner seals the highly transmissive zone. Figure 10.33 shows such a profile. For the curve in Figure 10.33, the resolution is relatively poor above 185 ft. In that case, the straddle packer measurements above 185 ft. should provide higher resolution unless bypassed.

If the straddle packers are not leaking, straddle packers can generally measure lower transmissivity intervals better than a liner measurement. In general, if the quest is for the transmissive zones and those are not located primarily at the bottom

a) Packer   b) Everting Liner

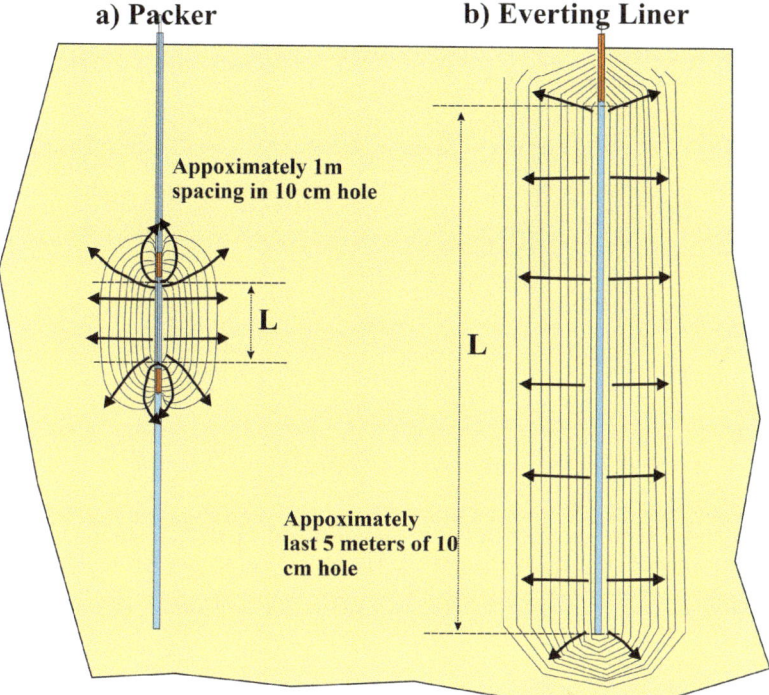

Appoximately 1m spacing in 10 cm hole

L

L

Appoximately last 5 meters of 10 cm hole

FIGURE 10.43 Flow field of straddle packer versus that of a liner measurement. The packer flow field during injection, even with a 1 m spacing, diverges to a more spherical flow field intersecting the borehole above and below the packer. The liner flow field after the transient is relatively 1D cylindrical, despite the random fracture orientations, and there is no bypass of the liner to an open hole.

of the borehole, the liner measurement can be more cost effective. Because of the very low velocity in a low transmissivity borehole such as one might expect in a rad-waste site assessment, the liner is not a practical transmissivity measurement.

Another situation where a straddle packer may be more useful is in the upper 10–20 ft. of the saturated borehole. In that interval, the liner goes through a transient velocity peak until the nominal steady-state flow condition is established throughout the borehole. In that transient region, the liner measurement has poorer resolution. However, a transient correction is commonly applied to that interval for the liner measurement. Those details are addressed in the paper by Keller et al (2014). Another practical limit for the liner method is if the open borehole has a vertical downward flow rate exceeding 150–200 gal./min. If that flow rate is concentrated in one flow feature, the resolution is very poor above that feature.

The paper by Keller et al (2014) compares the transmissivity results to packer test results. The comparison is qualitatively very good. However, for high flow zones, the assumption of laminar flow for the liner measurement can be violated resulting in an under estimate of the transmissivity. Quinn et al (2011) addresses the differences in packer measurements of transmissivity with varying driving pressures.

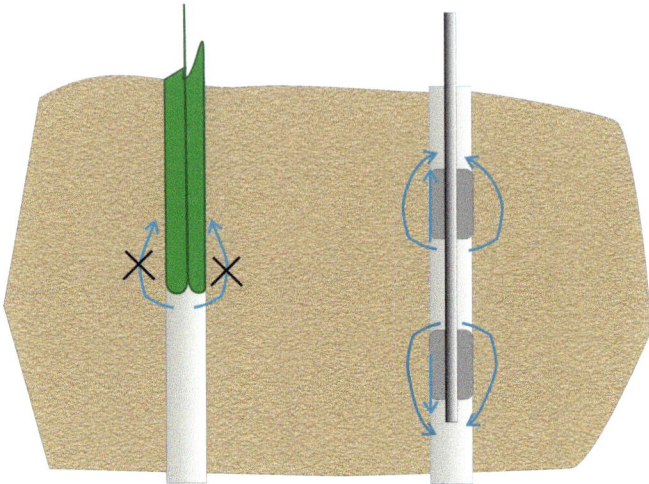

FIGURE 10.44   Showing the potential bypass of straddle packers. The liner advantage is that there is no open hole to allow connection. The packer bypass can occur both in the formation and at the contact with the borehole wall. For rough borehole walls and highly fractured media, the liner can still provide a useful seal.

The comparison in Keller (2014) with Quinn's measurements, which were done at relatively low flows, shows the error in the highest flow zones to be about a 50% under estimate of T with the liner measurement. However, the high flow zones are obvious in the liner measurement. The packer tests can be performed at different flow rates to define the laminar flow regime. The T profile unless redone uses one driving pressure and one flow rate in each permeable interval during the measurement. The liner inability to identify very low-flow zones is also apparent in the Quinn comparison. That limitation does not negate the general utility of the T profile.

A reasonable question is how does the observed stretch of a liner during emplacement affect the T profile results. The velocity calculation is not affected significantly by liner stretch over the short time or interval of the recording. However, the correlation of the flow zones with discrete depths could be affected by the stretch of the liner in that the depth of the liner measured at the encoder may not be the depth of the EP after any stretch of the liner. The EP marks at the surface are recorded in time as the liner moves into the borehole for comparison to the depths of the liner inferred from the encoder measurement. The definitive assurance of the correct depth association with the measurement is the tagging of the liner depth frequently during the T profile. Those measured depths are compared to the encoder-derived values and the encoder diameter is adjusted if needed to match the measured depths. Those adjustments are very small and the comparison of T profile results with the observed large fractures seen in the optical tele-viewer show the T profile depths to be within a few inches of the observed flow zones. When making such comparisons, it is essential that the two data sets have the same reference whether the GS or the top of the casing.

10.3.11   ADVANTAGES AND LIMITATIONS OF THE T PROFILE

10.3.11.1   Advantages

The first advantage is the speed with which one can map the major flow zones in a borehole of 300 ft. depth. While it depends on the transmissivity distribution, 1–2 h for the actual profile is typical, not counting the time to set up and take down the equipment. Packer tests take hours, a traditional alternative, or often take days, highly dependent on the hole depth and number of tests done. The T profile measurement time is more dependent on the transmissivity of the borehole than on the depth. Low-flow holes take longer to reach the cut off velocity 0.001 ft./s.

The second advantage is the high spatial resolution of the T profile with a resolution of better than a foot. The resolution limit of transmissivity seems to be about a velocity change of 1% which means that a fast profile has lower T resolution.

Another advantage is the relatively light weight and mobility of the equipment used, if not an artesian condition. A water tank is required but can be filled from the borehole if allowed. No crane truck is necessary as used for straddle packer installations.

There is no significant bypass of the sealing liner. Bypass calculations in the formation have been done and will be addressed in the CSC (continuously screened casing) design (Chapter 14).

A third major advantage is that the T profile provides a continuous profile which is used in other FLUTe methods.

10.3.11.2   Limitations

The main limitation is how low in transmissivity the measurement can go. If all flow zones are slow flowing, the T profile takes longer to perform but is usually much faster than packer testing since packers need time to achieve a steady-state flow at each interval as required by the Thiem equation used. The T profile is usually terminated when the descent rate drops to 0.001 ft./s. The remaining transmissivity beneath the liner is about 0.01 cm²/s. That is 3.6 ft./h and not worth the extra time to measure further in most situations. That slow rate typically occurs abruptly after passing the last significant flow feature.

If there is a high flow zone in the bottom of the hole, the high velocity above that zone reduces the transmissivity resolution of individual flow zones or features above the high flow zone.

The typical T profile is done once at a given driving pressure. Packer tests can be done again and again at different flow rates to assure laminar flow and to measure very low T values if there is no packer leakage.

Packer tests are available from nearly every driller. Only FLUTe, or FLUTe-trained personnel, do T profiles. That may change in time as the use of the method increases.

10.4   RHP (REVERSE HEAD PROFILE) MEASUREMENT OF A HEAD PROFILE

10.4.1   THE HISTORY OF THE RHP METHOD

Unlike several of the FLUTe methods, there was not a request from anyone to measure the RHP. It arrived as a simple consideration of the advantages of measuring the head beneath the liner, usually with a measurement of the tension on a liner, as can be deduced from the description of liner mechanics in Chapter 3. In many cases, the head beneath the liner, when the eversion is paused, is equal to the water table in the borehole or near it. If there is no gradient in the hole, the tension on the liner is 0.5 ΔH A. If the head beneath the liner is not near the blended head in the borehole, the head beneath the liner is also a blended head but different from the open-hole blended head. The head beneath the liner does control the tension on the liner. Therefore, a measurement of the tension on the liner should indicate the head beneath the liner, but that head beneath the liner is dependent on the transmissivity distribution and the head in each different aquifer intersecting the borehole below the liner when the liner has been stopped. Fortunately, the transducer at the bottom of the hole will always measure the new blended head when equilibrium has occurred independent of the water fill in the liner or the tension on the liner. At the stop of the liner for a T profile, the head beneath the liner is near that in the formation. Inverting the liner a short distance can produce a different equilibrium head. What is that head? From that line of reasoning the RHP concept was born. Only the continuous transmissivity profile makes the RHP possible because it defines the transmissivity of any interval in the borehole. The term RHP seemed a good name since it is done with the inversion, reverse, of the everting T profile liner. Both the T profile and the RHP are possible with the continuous seal of the liner above the EP. While the same kind of measurement can be done using a packer, the bypass of the packer and the inability to obtain the continuous transmissivity profile in a reasonable time make it impractical.

10.4.2   THE PURPOSE OF THE FORMATION HEAD MEASUREMENT

This head profile method is not a new description of the mathematical modeling of ground water flow. It is an application of relatively common concepts and mathematical models using a new procedure with a flexible liner and the continuous transmissivity measurement (Section 10.3). The liner head measurement method has been named the RHP, because it is a reverse/inversion of the T profiling measurement and usually done at the same time as the T profile. This method is only practical when using a sealing liner and with the relatively new continuous T profile. Because it is done in conjunction with the T profiling method while removing the liner used for the T profile, it is much less costly than many other methods such as straddle packers, separate wells, or using MLSs.

The definition of aquitards is particularly important to the understanding of the ground water flow (Maxey, 1964). Maxey pointed out long ago that aquifers are not necessarily associated with geologic units but rather on extensive lateral permeability which can vary substantially within a geologic unit. Therefore, other hydrologic measurements must be made to define aquifers, aquitards, or aquicludes. A head measurement can help define the presence of aquitards and assist in the definition of those intervals which may then be best monitored with a multilevel sampling system (such as described in Section 10.5) for water quality and long-term vertical gradient definition. Those same aquitards can be barriers to vertical migration of contaminants.

The use of a blank liner to measure the formation head profile is a relatively new addition to the flexible liner methods (Keller, 2016). The FLUTe transmissivity profiling technique (Section 10.3) has been used in over 300 boreholes since 2007. That method often assumes a constant head distribution in the formation. That is usually assumed to be the blended head in the open borehole if no other information is available on the head distribution in the formation. But the transmissivity calculation from the liner velocity data depends on the head in the formation (Section 10.3.2.2, Equation 10.1). Therefore, a head measurement technique is helpful to refine the transmissivity profile calculated from the liner measurement. Also, the vertical transport of contaminants and the recharge of subsurface aquifers depend upon the definition of the vertical head distribution.

As described above (Section 10.3.10), there are serious difficulties associated with the attempt to isolate an interval in the borehole with straddle packers due to the fact that the open borehole above and below the straddled interval can influence the measurements performed in the straddled interval. Only in a situation where the packers are located in aquitards can the straddled interval be well isolated, and yet the straddle packer measurement is still subject to the effects of the open hole before the packers are sealed against the borehole wall (Quinn et al, 2015). The method described hereafter is not open to flow from above the measurement interval because the entire upper hole is sealed by the liner.

The primary benefit of the RHP head profile method described (Keller, 2016) is that it is done in a very short time compared to some other traditional head measurements in boreholes, and it does not require a multilevel or clustered well installation. However, this technique only measures the head distribution at the time of the measurement, so an MLS still has merit for head measurements over time in well isolated intervals.

10.4.3   THE RHP CALCULATION

An initial assumption is that there is little flow into or out of the borehole beneath the stationary liner after a transmissivity profile has been completed and the liner is anchored at the surface and is still in the borehole. Figure 10.45 shows the geometry of the transducer at the bottom of the hole and the stationary liner when the transmissivity profile of Section 10.3 was completed. A simple assumption is that the remaining unsealed interval does not straddle an important aquitard. Since the remaining T is less than approximately 0.01 cm²/s, it is a very low-flow interval.

The only flow into or out of the borehole is that needed to achieve an equilibrium pressure in the unsealed interval beneath the liner. The unsealed portion of the hole after the liner is anchored at the end of the T profile is defined as the first interval, ΔZ_1. The equilibrium borehole pressure in that interval is defined as BH_1. The transmissivity for that interval is read from the integral transmissivity curve as T_1, which is calculated from the velocity of the liner at the time it is halted. The flow rate into the first interval would be:

$$Q_1 = T_1 \Delta H_1 2\pi/\ln(R), \qquad (10.2)$$

where $\Delta H_1 = BH_1 - H_1$, and H_1 is the formation head. But BH_1 equals H_1 when equilibrium has been achieved, so Q1 = 0. R is the ratio of ra/rh as defined in Section 10.3.

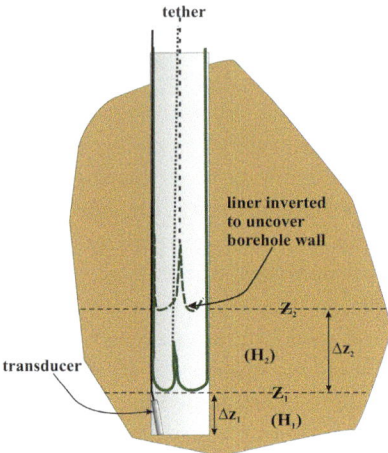

FIGURE 10.45   Liner geometry at the end of the transmissivity profile and after the first inversion for the RHP.

The liner is then inverted to a higher elevation, Z_2. The liner is then halted allowing the borehole beneath the liner to achieve a new equilibrium pressure, BH_2, measured by the transducer at the bottom of the hole, which may be different than in the first interval measured. The newly uncovered interval is defined as the second interval, ΔZ_2. A pair of flow equations is defined according to Equation (10.2) for the first and the second interval, respectively. The flow into or out of the first interval with the liner at Z_2 ft. is now:

$$Q_1 = T_1 \Delta H_1 \, 2\pi/\ln(R), \tag{10.3}$$

where ΔH_1 is the new difference between the equilibrium pressure measured, BH_2, and the pressure H_1 in the formation, now $\Delta H_1 = BH_2 - H_1$.

The equation of the second newly uncovered interval is:

$$Q_2 = T_2 \Delta H_2 \, 2\pi/\ln(R), \tag{10.4}$$

where ΔH_2 is $HB_2 - H_2$. T_2 is the transmissivity of the interval ΔZ_2 obtained from the T profile result such as Figure 10.32. Since the inflow is equal to the out flow in the total interval $Z_1 + Z_2$,

$$Q_1 + Q_2 = 0 \tag{10.5}$$

Using the expressions for Q_1 and Q_2 (Equations 10.3 and 10.4),

$$T_1 \Delta H_1 + T_2 \Delta H_2 = 0. \text{ In detail that is:}$$

$$T_1 (BH_2 - H_1) + T_2 (BH_2 - H_2) = 0,$$

where H_2 is the head in the newly uncovered portion of the formation and BH_2 is the equilibrium pressure measured in the borehole after the second interval was uncovered. Since T_1, T_2, BH_2, and H_1 are known, the equation can be solved for H_2 as a function of the known values. The result is:

$$H_2 = T_1 (BH_2 - H_1)/T_2 + BH_2 \qquad (10.6)$$

If another portion of the borehole is uncovered and allowed to equilibrate, the sum of the flows from all three intervals is equal to zero:

$$T_1 (BH_3 - H_1) + T_2 (BH_3 - H_2) + T_3 (BH_3 - H_3) = 0, \text{ which leads to}$$

$$H_3 = \left(T_1 (BH_3 - H_1) + T_2 (BH_3 - H_2) \right)/T_3 + BH_3$$

Or in general,

$$H_i = \left(T_1 (BH_i - H_1) + T_2 (BH_i - H_2) + \dots \right)/T_i + BH_i \qquad (10.7)$$

This procedure allows one to determine the head in the formation for each interval uncovered by inverting the liner in a stepwise manner from the bottom to the top of the borehole and measuring the equilibrium pressure after each new interval is unsealed. This is not possible without the knowledge of the transmissivity of each interval, T_i, as measured according to Section 10.3 in the T profile.

It should be noted that in Equation (10.7) that the division of the parameters of the lower intervals by T_i will produce an erroneous result if T_i is relatively very low and therefore not well measured as is usually the case with the T profile. That is one reason to pick stopping points in low T intervals such that each uncovered interval is of relatively higher T. In other words, the high T intervals are expected to control the head distribution, a well understood expectation.

10.4.3.1   The Times to Equilibration for Each Step of the RHP

The time to equilibration at each stop is in theory forever. In some boreholes, the time to effective equilibrium may be as short as 15 min. In most boreholes, it is much longer and often more than an hour. This can greatly prolong the measurement with numerous stopping points. Just as in straddle packer measurements, there is a wait time after each measurement has begun at each level for the head beneath the liner to equilibrate. Unlike a straddle packer measurement, which usually has a free surface in a stand pipe that equilibrates with flow into the straddled interval and a rise in the stand pipe, the RHP change in the liner elevation when inverting to the next stopping point causes a major inflow to the borehole. The liner is inverted until stopped and during that inversion, water must flow into the borehole beneath the liner. The inversion to a new stopping point is typically done in minutes rather than hours. Then the volume beneath the liner is essentially a closed volume, but the stretch of the liner when stopped at the new elevation with a relatively high differential pressure across the EP, will cause the liner to contract slightly as the head beneath the liner rises to equilibrium. As the borehole beneath the liner equilibrates, the differential pressure

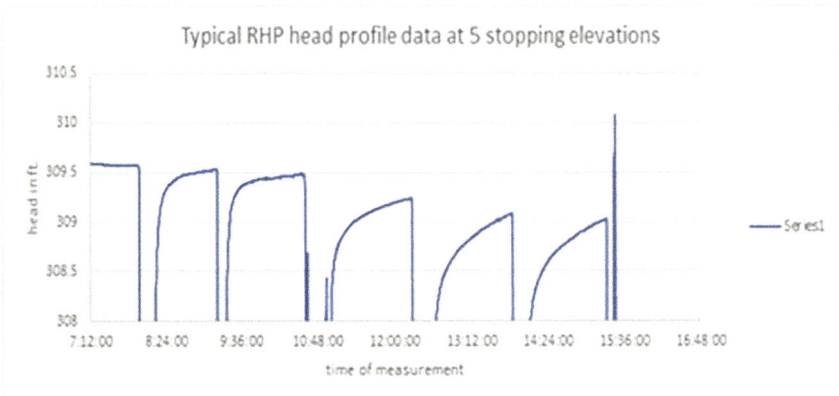

FIGURE 10.46 Typical convergence to equilibrium for RHP measurements at five stopping points. Each curve is allowed to converge for about 1 h.

across the end of the liner approaches a constant value dependent on the excess head in the liner. By that time the liner is no longer contracting. As with common approaches to equilibrium, the head beneath the liner in time looks like an exponential decay to a constant asymptote. In theory, that equilibrium never completely occurs. For the RHP approach to equilibrium, the final head is the value of interest. Figure 10.46 shows the measured head, H(t), beneath a liner after stopping at several different elevations. The actual equilibrium level is not obvious for these several stops. A method has been devised to determine the coefficient of the decay constant, c, in the expression $H = Lo - Ho\ e^{-ct}$, where Lo is the asymptote, $\Delta H = Lo - Ho$, when t = 0. The head change ΔH is the head rising to Lo over some long time after having been dropped by an initial amount Ho. The approach is to difference the value of H(t) at three times separated by equal time increments, Δt, to calculate the approximate first derivative of the level change. Using those values picked from the decay curve of head beneath the liner, one can deduce a final asymptote for Lo without waiting to achieve a nominal equilibrium. Using this approach, the calculated asymptotes for several data sets which had decayed to near the asymptotic values compared well with the actual apparent asymptotes. The data for the calculation should not be picked from near the asymptote since that pressure data is relatively noisy compared to earlier in the record which shows more curvature. The details of the calculational model for that extrapolation to the equilibrium value are explained in Section 16.3.

In the decay curves for several stopping points shown in Figure 10.46, it is interesting to note that the series from the deepest to the shallower stopping points has a very different rate of convergence to equilibrium. The main difference is that the borehole which is open below each stop is longer for later times and higher stopping points. That means that more flow paths and higher transmissivity are associated with each successive stop. As the stopping points approach the surface, the storativity should also increase. Regardless of the cause for the slower convergence to an asymptote, calculating the coefficient for the time constant allows the inversion to proceed to the next stopping point for the intervals higher in the hole in Figure 10.46.

An interesting possibility is that the rate of decay/rise to equilibrium probably contains useful information on the hydrologic properties of the formation such as storativity and perhaps more. For the RHP method, the head at equilibrium is the important parameter from the measurement. The diameter of the hole and volume to fill are probably important to the rate of convergence on the asymptote. The volume to fill with flow is also larger for the higher stopping points and may explain the different rates of the curves in Figure 10.46. A detailed comparison of the rate of convergence may provide some useful insight. The point is that the rise to equilibrium is very different in different boreholes and due most likely to hydrologic characteristics beyond the borehole diameter.

10.4.3.2   The Use of the RHP to Refine the Transmissivity Profile

The transmissivity measurement described in Section 10.3 is often done with no initial knowledge of the actual head distribution in the formation. That limitation is overcome with the use of a relatively high head inside the liner (when possible) to make the individual head variations less important in the transmissivity calculation. The head in the formation without any foreknowledge is assumed to be equal to the equilibrium head (the water table, in the open hole). However, with the result of the RHP measurement, the use of the open borehole water table assumption is not necessary.

The RHP results can be used in the original transmissivity profile calculation to obtain a better transmissivity profile and then a better integral T profile from which to obtain better interval transmissivities, T_i, for use in the RHP calculation. It is noteworthy that neither the original flow rates measured for the T profile nor the equilibrium heads measured for the RHP are changed in the iteration.

This refinement has been used on several RHP measurements. The changes have only been significant when the formation vertical head distribution was significantly different from the blended head in the open hole or when the T profile was done with a small excess head as may be the case with a shallow water table where a higher excess head is not available. The iteration was done for the following example in Section 10.4.4, but because the water table was relatively deep and the vertical gradient relatively small, the iteration did not produce a significant change.

10.4.3.3   Selection of the RHP Intervals to Be Measured

The first interval to be measured is that below the liner when the T profile is halted. The T profile is usually halted when the descent rate of the liner drops below 0.001 ft./s. With the transducer in the bottom of the borehole, it is useful to note whether the transducer returns to the head measured when the transducer was first lowered to the bottom of the hole. If there is little vertical gradient, the transducer equilibrates to near the open-hole reading after the borehole above the transducer is sealed with the liner. In some boreholes, when the T profile is halted, the transducer will decay to a head well below the original open-hole measurement indicating a downward gradient. In rare occasions, the transducer will equilibrate above the open-hole head measured. Because that lowest interval is measured directly, it is not susceptible to the limit of low T uncertainties.

For the RHP, it is best to use the T profile to select the intervals in which to measure the head. The intervals to be measured with the RHP are selected from the T profile in a manner that reduces the time/cost of the measurement and is based upon some simple hydrologic assumptions. It is assumed that an impermeable or low permeability interval may be an aquitard and that the more permeable intervals will control the head above or below the potential aquitards. Therefore, the stopping points are picked to be in the low T intervals between high T intervals.

The elevations of the top of each RHP interval to be measured are called the stopping points of the measurement. The stopping points are selected in the following manner for the following reasons:

If there is a significant vertical gradient, it must be due to aquitards or aquicludes or it may be just a low conductivity medium with a uniform gradient. However, the T profile indicates whether there are conductive layers interspersed with less conductive layers. If there are intervals of high lateral flow, they may dominate the head between layers of low transmissivity. Therefore, there is less interest in the head in low T intervals and more interest in the head in the more conductive layers. It seems rational that the head in the low conductive layers lies between the head in the more conductive layers above and below an interval of low T. This is the logic that leads to the selection of stopping points in the RHP that lie between the high flow intervals measured. It is possible that the low-flow interval is an aquitard and measuring the head above and below the less conductive layers is logical. Therefore, upon examination of the T profile data, especially the integral over one foot intervals, one can select the stopping points for the best efficiency of the method. If using the integral T curve, the stopping points would be in the flat portions of the curve, the low-flow zones.

The procedure for an RHP, once the T profile is done, is to examine the T profile for the low transmissive intervals, and pick the stopping points. Then reprogram the transducer to the longer time steps for the RHP. Then invert the liner to the first stopping point above the end of the T profile. The last stopping point is best inside the casing because the result should be the blended head water table of the open hole. However, it has been noted that if the RHP is much later (e.g., weeks or months) after the T profile that the blended head can change substantially in that time depending on the weather or pump and treat schedules or municipal pumping schedules of supply wells. It is noteworthy that head measurements are snapshots of the current state, whereas T profiles are relevant for longer times and not likely to change. As will be seen in the next section, the test of the RHP is best done in comparison with head values measured in a sealed hole the next day. Therefore, it is useful to monitor head variations in time to identify the variability (as described in Section 9.5, the ACT description). Differences in time are seen in RHP results in the same borehole.

10.4.4   A Result of the RHP Method

A test in 2015 produced the head profile in Figure 10.47. The transmissivity profile was done 3 months earlier and the liner was left in place to seal the hole. The RHP was performed as the liner was being removed to install a multilevel sampling

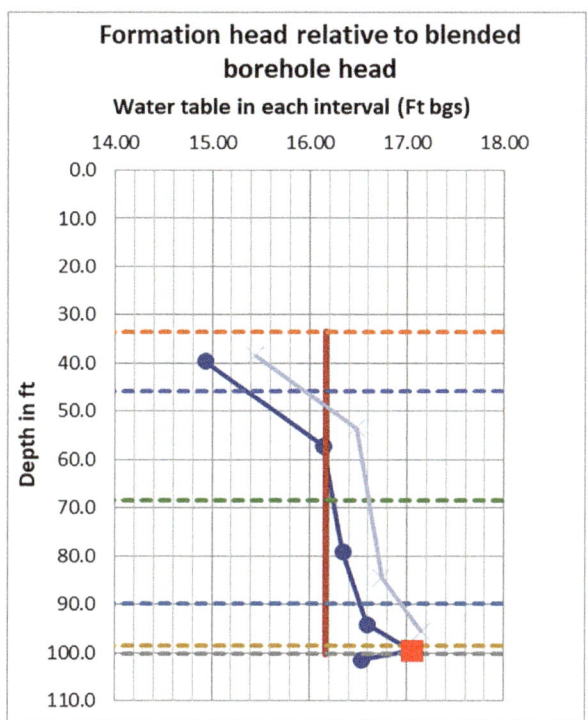

FIGURE 10.47   RHP result is dark blue curve compared to the measured water tables in the sealed MLS (gray curve) one day later. Red point is RHP with low uncertain T in the interval of measurement. The vertical red line is the blended head. Dashed lines are stopping points of RHP.

system. The RHP was done in 3 h. A FLUTe MLS which sealed the entire bore-hole was installed the day after the RHP profile was performed and the measured head profile in the MLS was obtained immediately. The MLS manual water table measurements in the MLS (gray curve) are compared to the RHP in Figure 10.47 (the blue curve). The blended head level is also plotted as 16.16 ft. BGS (red vertical line). The agreement is excellent with the possible exception of the red square. The red data point is where T_i in Equation (10.7) is a very low value and therefore not well measured. A test in the RHP data reduction procedure identifies the intervals of very low transmissivity and plots them as a red square. The uncertainty of the red data point is large since it is a low T_2 value. The probable head value in such a low conductivity layer should lie between the head in the more conductive layer above and the more conductive layer below the layer of low conductivity. This is so probable that the red square can logically be moved to lie on a line connecting the head above and below. The nearly constant small offset between the RHP values and the manually tagged levels is within the uncertainty of the depth and calibra-tion of the transducer in the borehole beneath the liner after 3 months in place. The water table depth at the time of the T profile is assumed to be the water table depth at the time of the RHP. That offset will be translated to the RHP head calculated. The offset is trivial for most RHP measurements since they are performed immediately after the T profile. The installation of an MLS system for comparison seldom occurs quickly after an RHP since the liner must be fabricated according to the T profile data.

Figure 10.48 (left graph, A) is the transmissivity distribution used for the head profile in Figure 10.47. It is noteworthy that the blended head is near the head in the interval of highest transmissivity (50–60 ft.) as it should be, since the blended head is dominated by the head in the most transmissive interval.

10.4.5   CALCULATION OF THE SYNTHETIC FLOW LOG

It is easy to use Equation (10.4) to calculate the flow into and out of the borehole for each measurement interval when the borehole is open (i.e., not sealed with the liner). This is because the formation head is known for each interval and the equilibrium head at the highest elevation in the borehole is essentially the blended head. Figure 10.48 (center graph) is the calculated flow into and out of the open borehole in each interval Δz_i of the RHP measurement. Integrating the flow of Figure 10.48 (center, B) from the bottom to the top produces a synthetic equivalent of a flow meter log as shown in Figure 10.48 (right graph, C). Comparison of the "synthetic flow log" from the RHP with actual flow logs measured as part of the geophysical measurements in the same boreholes as the RHP has shown a very satisfying agreement.

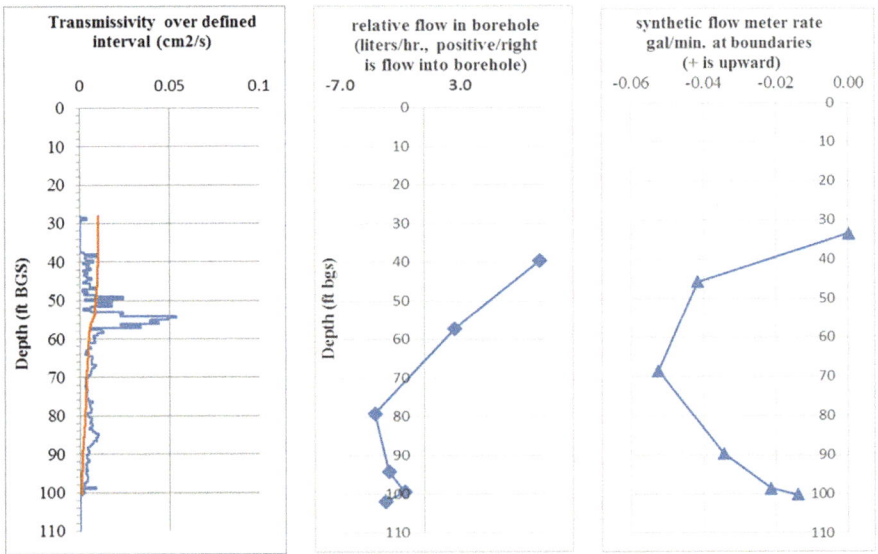

FIGURE 10.48   The left most graph (A) is the transmissivity for the RHP in the above example. The center graph (B) is the flow into and out of the open borehole using the RHP and the transmissivity values for each interval. The right graph (C) is the synthetic flowmeter log developed by integrating the center graph values from the bottom to the top.

The time to achieve near equilibrium in this borehole was less than 15 min for each halt of the liner. Subsequent measurements have shown that the time to obtain a reasonable approximation of the equilibrium pressure can be very long (>1 h) as seen in Figure 10.46. However, the lower the transmissivity, the smaller the storativity, and the larger the hole diameter, the longer is the time to approach equilibrium at each stopping point in the measurement. The equilibration time in Figure 10.46 also increases for stopping points higher in the borehole. This makes the extrapolation calculation for the asymptote especially useful.

There is not a very strong gradient in this result, and the head driving the liner for the transmissivity profile was well above the blended head in the borehole. Therefore, when the measured head profile was entered into the transmissivity profile calculation for the borehole in order to refine the transmissivity profile, and those refined results were then entered into the head profile calculation, there was little change in the head profile calculated. However, in some boreholes the use of the RHP to refine the initial transmissivity result, which is based on an assumed constant head in the formation, can lead to a significant improvement in the transmissivity profile and also in the subsequent head profile. It is noteworthy that the iteration does not change the flow measurement of the T profile nor the equilibrium heads of the RHP. They simply converge to a more consistent result. It is useful to make the adjustment of the red square values (as shown in Fig. 10.47) as suggested before entering the RHP result into the transmissivity calculation iteration. Otherwise, the poorly measured low T values can destabilize the iteration. The iteration of the transmissivity profile is most useful for shallow water table measurements where the liner is not so strongly driven and when the RHP shows a strong gradient in the formation.

10.4.6   RHP Profile Summary

RHP measurements are most convenient if made immediately after the transmissivity profile and not long after the borehole is drilled. Then the transmissivity profile and head profile are most useful in the development of the site conceptual model and in the selection of the sampling intervals for the MLS described in Section 10.5 for long-term monitoring. The results described above had the advantage of the head profile confirmation immediately after the RHP. It has been noted that the head profile can change with nearby pumping and infiltration events. Combined with the transmissivity profile, the RHP can generate a synthetic flow log which is reassuring to compare with the actual log when available.

10.4.7   Advantages and Limitations of the RHP

Advantages of the RHP:
- The continuous transmissivity profile obtained with a liner allows one to determine the transmissivity of any interval in the borehole which then allows the RHP measurement.
- The RHP is easily performed with the same liner already in the borehole immediately after the T profile.

- The formation head profile is useful for identification of aquitards and serves as a refinement of the transmissivity profile.
- Like the T profile, there is no need for large heavy equipment to be moved to the site, especially if scaffolding is not required.
- The time required to do an RHP is usually much less than a full day. It depends on how many stopping points are selected.

Limitations of the RHP:
- An extremely high flow zone in the bottom of the borehole will reduce the resolution of the T profile and therefore of the RHP.
- The RHP for a borehole with many high flow zones may include more stopping points and therefore more time spent waiting on the approach to equilibrium.
- The RHP and the T profile do not map the contaminant distribution. That requires the NAPL/FACT or the MLS measurements for a more complete site characterization.
- The RHP does not map the head in intervals of low transmissivity. Nor is it practical in boreholes of low total transmissivity where the T profile is also limited.

10.5   FLUTe MLS (MULTILEVEL SAMPLING) SYSTEMS

10.5.1   Water FLUTe

10.5.1.1   History of Water FLUTes

As described in Chapter 2, the early FLUTe flexible liner methods were focused on vadose zone applications, but there was not much interest in the vadose zone contamination or measurements therein. It was reasoned that the same continuous sealing liner advantage could be useful to ground water measurements. It was also easier to install an everting liner with water than with air as described in Section 3.2. With water, gravity provides the long-term pressurization of the sealing liner. The main challenge was how to pump the water sample from the sealed borehole at discrete elevations. It was easy to draw vadose zone pore gas samples with a vacuum pump at the surface, but for a water table more the 20 ft. below the surface, a positive displacement pumping system was required. The first FLUTe design used a Grundfos II pump with pneumatically activated valves to each port. It did not work well for slow flowing water samples. A much more simple easy to use pumping system was needed. That was designed essentially as currently in use.

In 1994, the first WF-like systems were deployed at an Air Force base near Riverside, CA. The pumping system was complex and unsatisfactory. An interesting automatic water level maintenance system was used as the liner was deployed. (That worked very well.) The current concept for pumping of water samples was used at Hudson Falls, NY and later at a site in Cambridge, Ontario for the University of Waterloo. It worked very well and many of those designs were ordered by Boeing for the SSFL site in

FIGURE 10.49 Early Water FLUTe installation at Cambridge, Ont. Site, ca.2001. Installation was paced by the water addition rate. The sheathed tubing bundle is on the reel. The end of the black/yellow liner is nearing the borehole.

California (2001–2002). Figure 10.49 shows the early installation at Cambridge with such ease that the installers were able to discuss kayaking as they everted the liner into place with water. This installation also used simultaneous purging of 12 sampling ports. It was also recognized that there was no borehole water to be purged prior to the sample collection since all the water was inside the liner and the samples were drawn directly from the formation. The simplicity of installing by just adding water to evert the liner was a remarkable advantage over grouting of casing into place. The WF concept with refinements has continued with installations in many places to depths as great as 1400 ft. with 11 ports with air-coupled water table measurements and nearly continuous sampling and head monitoring (Hampton Rhodes, VA, 2018).

Improvements in pump design, sampling interval spacers, and system materials have continued. Water table measurements and head histories at each port during sampling have always been an advantage. The recent major improvement in head history measurements has been the air-coupled transducer (ACT) system with the transducers easily accessible at the surface for repair or reuse. The ACT method is described in Section 9.5.

The advantages of the WF design have always been the continuous sealing liner, the simultaneous purging and sampling capability, and the removability of the systems since WFs are seldom grouted into place. WF systems have been in place for up to 17 years. The SSFL systems installed in 2001 were finally removed 15 years later. The main functional problem was that all of the downhole transducers had failed long ago. That is a problem which is now addressed with the ACT refinement. Some liners had been damaged by nearby pumping with the burst of the liner. However, there was surprisingly little evidence of deterioration with age in the SSFL liners, where not exposed to the sun. The use of stronger liner fabrics has reduced the risk of liner damage due to nearby pumping or large aquifer head differences.

Not all FLUTe MLS systems as of 2019 are installed by the eversion technique used for WFs. However, the lack of abrasion of an everting liner is still a major advantage of the WF method. FLUTe has several patents on all refinements of the WF system.

10.5.1.2   The Geometry of the Water FLUTe Design

10.5.1.2.1   The Liner Geometry

Figure 10.50 shows the main features of the WF system. The first and foremost component is the continuous sealing liner (in red). The entire borehole is sealed with the pressurized liner to avoid the influence of the borehole on the natural flow field. The spacers prevent that seal in the interval from which the water sample is drawn. An attempt to sample from a single port placed at an impermeable elevation in the formation is not useful. The spacer should be located as desired to intersect at least one flow feature in the formation. The location can be determined based on numerous FLUTe measurements such as the NAPL/FACT (Section 10.2), the T profile (Section 10.3), or the RHP (Section 10.4). The advantage of the FLUTe measurements is that they can resolve individual flow features. Those locations can also be verified with a high-quality set of geophysical tele-viewer measurements. The port behind the spacer draws through the spacer from flow features intersected by the spacer.

The port through the liner connects to the "port-to-pump tube". That tube is incased in a continuous sleeve welded to the inside surface of the liner. The port-to-pump tube extends from the port to the bottom of the inverted end of the liner, called the EP. From that point, the tube extends upward to the pump system dedicated to each port. The pump consists of a U-shaped tube formed by the "pump tube" and the "sample tube" as shown in Figure 10.50. The flow from the port passes through a first check valve where the port-to-pump tube meets the bottom of the pump tube. That flow fills the pump tube and the sample tube to the elevation of the water table at the port.

The spacer, port, and tubing with the pump system are duplicated for each sampling interval. At the bottom of the liner, the port-to-pump tubes for the several ports are gathered into a sheath that contains the entire system of tubes from the bottom of the liner to the surface including the pumps. This tubing bundle is supported by a

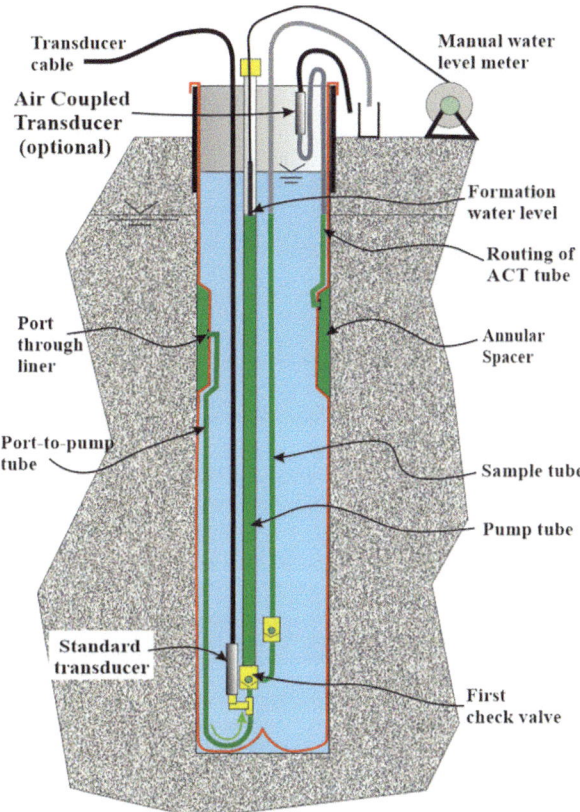

FIGURE 10.50 The geometry of the Water FLUTe system. Only a single port is shown for clarity. The red curve is the sealing liner. The standard transducer is now often replaced with the ACT transducer.

tether that is anchored at the surface to a "tether bar", and the tether is connected also to the inverted end of the liner at the EP. Several Kellum grips on the tubing bundle transfer the bundle weight to the tether. The 1/4″ tether has a tensile strength of 8000 lb, far more than the weight of the tubing bundle even though the PVDF tubing has a density of 1.7 g/cc and does not float. The tubing bundle is installed in the borehole following the everting liner. Only the port-to-pump tubes in the sleeves are everted with the liner installation.

10.5.1.2.2   The Installation Hardware Geometry

Figure 10.51 shows an especially large liner system being installed from a shipping reel over a Green Machine used as a wellhead roller at the borehole. Figure 10.52 shows the tubing bundle of port-to-pump tubes following the everting liner into the 600-ft. borehole.

For very deep boreholes and many ports, the PVDF tubing bundle may weigh over 100 lb. This is especially true for deep water tables. In that case, it is necessary

FIGURE 10.51   A 600-ft. Water FLUTe with 12 ports being installed over a Green Machine at the wellhead.

FIGURE 10.52   The bundle of port-to-pump tubes being lowered into the borehole following the end of everting liner.

FIGURE 10.53 The braking system used to support a heavy tubing bundle.

to provide a mechanical braking system to support the bundle weight plus the towing force of the everting liner. If the differential pressure across the EP is more than 10 psi, the total tension on the tubing bundle can exceed 200 lb. A mechanical braking system was designed, which uses a disk brake attached to the shipping reel (see Figure 10.53). The braking force is applied by the hydraulic pressure on the disk brake caliper. The braking tension is monitored with a load cell connected to the braking system. The photo in Figure 10.54 shows the braking system in use as the tubing bundle in Figure 10.52 is being lowered into the inside of the liner.

Once the WF system is in place and supported by the tether, the pump tubes and sample tubes are assembled in an array as shown in Figure 10.55. This clockwise array of quick connects on the pump tubes allows the convenient connection of the gas drive pumping system as describe in Section 10.5.4.2. The tubes to each port are numbered in a clockwise manner from the red capped tube, "tag tube", which is used to measure the water level in the liner.

10.5.1.2.3 The Spacer Design

The spacers are formed of a cylindrical concentric geometry as follows:

The outer layer of the spacer is a fine filter fabric passing particles less than 400 μm. The next layer is a coarse mesh of LDPE that is very permeable in the plane of the mesh to allow flow from fractures, for example, to pass through the filter fabric, through the mesh, to the port. An inner layer is a diffusion barrier of an aluminized laminate that isolates the spacer from the liner. The top and bottom of the filter fabric and diffusion barrier are welded to the liner to allow the flexible spacer to be everted

FIGURE 10.54   Braking system as used for lowering the tubing bundle of Figure 10.52.

FIGURE 10.55   Wellhead showing 13 pump tube separations and with the slender sample tubes in a subsurface vault.

into place with the liner. The mesh design is similar to the common leachate collection system beneath a landfill, allowing easy flow in the plane of the mesh.

The water fill of the liner presses the spacer against the hole wall, and the dilated liner also isolates the spacer from those other spacers above or below a port location. The water volume inside the spacer mesh is less than 300 ml in a 5-ft. spacer in a 4″ borehole. That is only about 8% of the volume of the pump for that spacer. In other words, one pump stroke would purge about 12 spacer volumes of water so that most of the purge stroke is drawing water from the formation. For comparison, a 2″ sand pack in a 4″ 5-ft. screened well contains about 12 liters of pore water between the screen and the borehole wall plus another 12 liters inside the 5-ft. screen interval. With the liner, the water inside the liner is not included in a purge during sampling. Therefore, the WF spacer requires about 3% of the purge volume needed in a 4″ screened well in order to reliably draw water from the formation.

10.5.1.3 Function of the Water FLUTe

10.5.1.3.1 Purging and Sampling

The primary function of the WF is to draw water samples from the formation. The head (i.e., the water table) at each port is measured with a slender tag line lowered into the pump tube (see Figure 10.50). Because of the valve at the bottom of the pump tube, the pump should be purged for one stroke to obtain the current water table. The head history is available with downhole recording pressure transducers as shown in Figure 10.50 or with the ACT transducer at the surface, also shown.

The WF pump system dedicated to each port is used to drive the water in the pump with a nitrogen gas. Figure 10.56 shows the main features as a gas bottle at the surface with a regulator, a connection to a higher resolution pressure gauge (the regulator pressure gauges are for extremely high pressures and difficult to read), a three-way valve, and a quick connect fitting for the connection at the top of the pump tube. The assembly from the regulator to the quick connect is called a "sample train".

The regulator is adjusted to the "purge pressure" for that system with the three-way valve closed. That pressure is typically the depth to the bottom end of the pump tube plus 10 psi. The quick connect of the sample train (the regulator, gauge, three-way valve, and quick connect) is connected to the female quick connect at the top of the pump tube for that port. The three-way valve is opened to apply the set purge pressure to the pump tube. That pressure is sufficient to close the check valve at the bottom of the pump tube and to expel all of the pump tube water through the second check valve and sample tube to a purge water container at the surface. When all the water of the pump tube and sample tube have flowed to the container, the sample tube vents the drive gas. At that time the three-way valve is closed and the pressure in the pump tube is dropped to atmospheric pressure. As the pressure in the pump tube drops, water flows from the spacer and port-to-pump tube to refill the pump tube and the sample tube to the level of the water table in the formation at the port. Opening the three-way valve expels

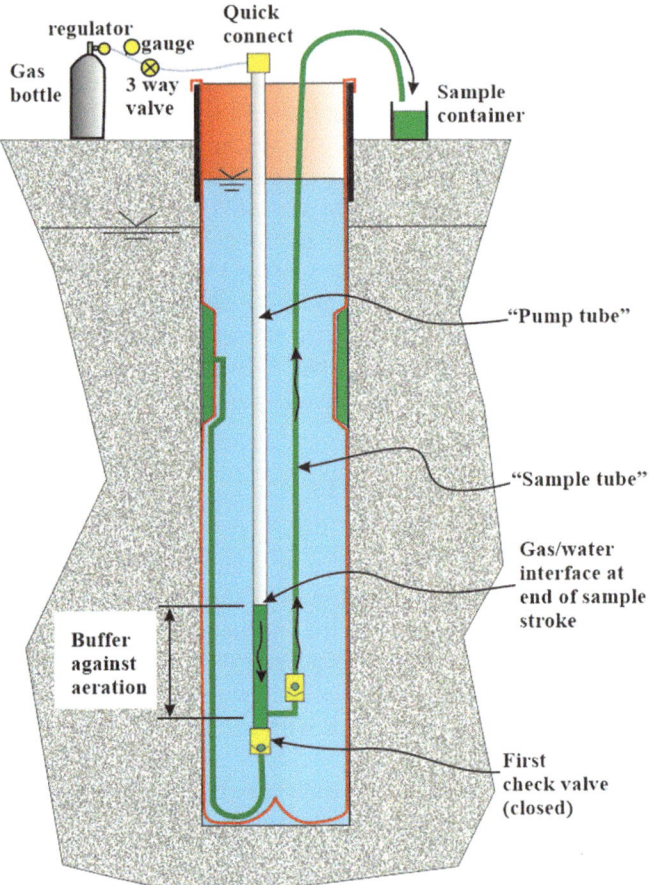

FIGURE 10.56   Pumping geometry showing buffer left at end of sample stroke.

another pump tube volume to the surface container. The pump volume is slightly more than 1 gal. for the typical pump tube extending 100 ft. below the water table in the borehole. The completion of one water expulsion is called one "stroke" of the pump.

The purge stroke is repeated until 4 gal. of water is collected in the surface container. After the pumps have refilled at the end of the purge sequence, the regulator is set to the "sample pressure" which is typically the depth of the pump tube minus 10 psi. Application of that lower sample pressure expels only ~3/4 of the pump tube volume. As the water is expelled, the flow rate decreases as the water level in the pump tube approaches the equilibrium level in the pump tube. As that flow slows, the sample is easily collected from the sample tube. With the gas/water interface above the "buffer", the water in the sample collection is more than 125 ft. typically from the gas/water interface in the pump tube.

The sample train is then connected to the next pump tube for the next port to be purged and sampled.

10.5.1.3.2   Simultaneous Purging and Sampling of All Ports

Using the manifold, which is now supplied with the WF system, the pressure source can be connected to all the pump tubes at the same time. The manifold is a compact assembly available from FLUTe and includes the pressure regulator. Figure 10.57 shows the FLUTe manifold used with a nitrogen bottle for the simultaneous purging procedure. For clarity, only two ports are shown. The simultaneous purge and sampling procedure greatly reduces the sampling time and also provides enhanced isolation of the formation volume sampled by each port. The three-way valve either connects the gas pressure to expel the water or it vents the gas pressure in the tubes to allow the tubes to refill. The purge at all the ports therefore takes the same time as the purge of a single port.

Since the pump tubes are nearly all the same length, the purge pressure is the same for all pumps. (Note: A complete purge of all the ports for systems with tubing only to the depth of the port does not provide the same purge volume for each port.) Opening the three-way valve allows all the pumps to be purged of one "stroke" of each pump. A container is needed for each port to assure that the full 4 gal. (~15 liters) is purged from each port. Figure 10.58 is a photo of an array of buckets used to collect the purge water from each port for a 12-port WF. When gas is vented from each sample tube, the valve for that port, on the manifold, is closed to allow the other ports to complete their purge. When all the valves are closed at the end of a purge

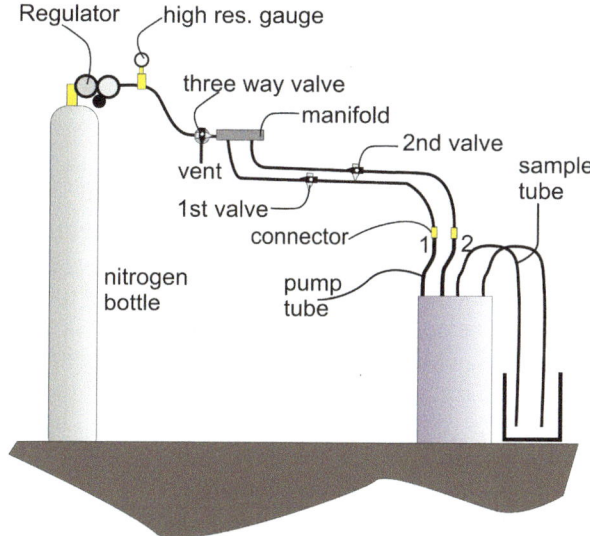

FIGURE 10.57   The manifold system attached to the gas bottle and to the connectors to the pump tubes. When the three-way valve applies the gas pressure, water flows out of the sample.

FIGURE 10.58   An array of purge water containers to monitor the purge volumes produced simultaneously before sample collection at each port. The sample tubes are secured to the buckets to prevent the flow propelling the sample tube out of the bucket.

stroke, the three-way valve is opened to drop the pressure to atmospheric and the individual valves for each port are opened to allow recharge of all pumps. Then the cycle is repeated until the recommend 4 gal. has been purged from each port. As shown in Figure 10.58, it is helpful to have each sample tube securely clamped to its numbered bucket since the flow from each sample tube is usually sufficient to flip the sample tube out of the bucket.

The sample collection should be done immediately after the purging is completed, and all pumps have refilled. The gas pressure is reduced at the regulator to the "sample pressure", which is usually the depth of the pump tube minus 10 psi.

At this time, one has the option of pressurizing each pump individually to collect a sample or to apply the sample pressure to all the pumps simultaneously. If not in a hurry, the common mode is that once the pumps have refilled, one can apply the gas pressure to one pump at a time to collect the sample. In that case, the valves to all pumps are closed and opened one at a time when each sample is to be collected and the valve is closed when the sample flow is complete.

The initial flow of this sample stroke should be discarded. That is about half a liter. The first flow contains aerated droplets left in the sample tube at the end of the last purge stroke. The sample tube flows strongly upon the first opening of the valve. However, as the water level in the pump descends, the flow slows as the differential pressure decays to the equilibrium level because the sample pressure is insufficient to expel all the pump water.

As the flow slows, it is possible to collect several samples before the flow nearly ceases as the level in the pump equilibrates to the applied pressure. After each sample is collected, the valve for that port can be closed if desired and the samples collected from the next port by opening that valve.

With practice, one can collect a sample from all the ports before the flow essentially stops. In all, the procedure is very efficient. If more samples are needed, the three-way valve can be turned to vent the gas pressure to allow the pump tubes to refill. Application of the pressure again allows more sample water flow, and no discard is needed.

It is possible to connect the sample tube to a water quality monitor as for the low-flow procedure. However, experience has shown that by the fourth stroke of the 4-gal. purge that the water quality parameters are stable for each port. It may be useful to collect one more sample stroke for a water quality measure for each port since the water quality at multiple ports is often different between ports.

There is another advantage of the simultaneous purging system as described in detail here. That is that each port is purged at a very similar draw down level at the same time. This fact discourages the sample at one port from being drawn from a volume sampled by an adjacent port. This is due to the low-pressure sink formed at each port during the recharge flow to the pump and the formation of pressure ridges between the ports. The pressure contours and flow field illustrated in Figure 10.59 shows how the simultaneous purge can develop the isolating pressure ridges. If the sampling ports were located in a continuously screened well with a sand pack such

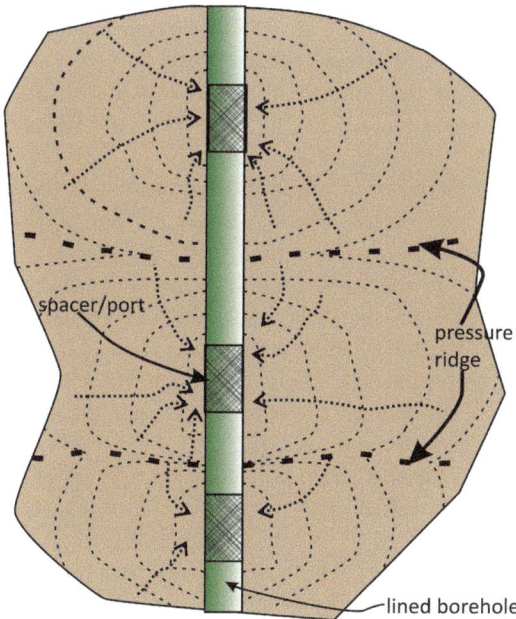

FIGURE 10.59 Illustration of the separate flow fields associated with the simultaneous purge of multiple ports in a sealed borehole.

as the CSC design, Chapter 14, the simultaneous purging would reduce the connection most likely in the sand pack. This isolation is the opposite of a straddle packer purge which is likely to draw from the open borehole above and below the straddled interval.

The T profile described above allows a calculation of the fracture aperture which allows a radius of extraction estimate associated with the 4-gal. purge of each sampling interval. In that manner, the aperture and measured purge volume allows an estimate of the range from which the sample is drawn. Of course, that requires a relatively simple 1D assumption for the fracture geometry. However, for monitoring wells, it is useful to estimate to what range from the borehole the monitoring is possible. For the large pore space in the sand pack and the open casing, a very large purge is necessary to reach a significant distance into the formation. The simultaneous purge flow field should enhance the lateral extent of sample extraction since the vertical flow is limited by the pressure ridges.

10.5.1.4   Transducer Options for Monitoring Head History

The early WF systems used a transducer located below the first check valve for recording head histories (see Figure 10.50). However, due to the finite lifetime of those transducers, FLUTe developed a transducer system for monitoring the head history with a transducer at the surface. That method called an ACT allows the head history to be measured through a slender tube. See Section 9.5 for a description of the ACT systems and utility. That tube as of 2020 is usually connected to the spacer instead of to the port-to-pump tube near the pump (see Figure 10.50). The ACT theory and validation are described in detail in Section 9.5. The ACT has the advantage of its location at the surface where it is readily accessible for repair, replacement, or reuse on other WFs. Because of the location at the surface, ACT systems are often rented at a much less cost than the dedicated downhole transducers. The downhole transducers can only be replaced with the removal of the WF system, an expensive procedure.

The ACT system is being used to monitor continuously the head at many ports in a 1400-ft. WF in Hampton Rhodes, VA for an aquifer injection project. FLUTe suggested to In-Situ that it would be ideal to do the conversion of the ACT data to a water table history in the In-Situ onboard firmware. However, that was considered too specialized a change. The pressure data of the ACT is entered into a spreadsheet provided by FLUTe to convert the data to water table depths with associated plots of the barometric changes and surface temperature history. This provides the option of correction for barometric changes as deemed appropriate (e.g., for shallow aquifers) for the several ports of a WF. The ACT system is also used on the FLUTe pdCHS and cased hole sampler (CHS) MLS systems described later for recording water table histories during pumping tests and remediation injections.

10.5.1.5   The Tracer Monitoring Capability of the Water FLUTe Design

When the pump tube has been evacuated and allowed to refill, the volume extracted is about a gallon of water. However, if the sample pressure applied is much less than the typical sample pressure, a very small amount of water will be expelled (e.g., 0.27 l with 7 ft. of pump stroke). In this manner, the simultaneous purge method has another use for monitoring first arrivals of injected tracers at each WF spacer. The tracers can be injected at the natural head or under a higher head. By using a low pumping pressure on a WF pump of the normal design, the pump can be "short stroked" to allow a much lower ejection and recharge volume. An interesting fact is that the small ejection volume is from the bottom end of the pump tube because the sample tube is connected at the bottom of the pump tube. Therefore, the small recharge through the check valve fills the bottom end of the pump tube, and the water expelled with a low driving pressure is also only from the bottom end of the pump tube. This allows one to operate the normal WF system as though the pump tubes were of much smaller volume. (Of course, such a smaller volume system is also easy to build.) With the simultaneous sampling capability and a short stroke of the system, the draw-down at all spacers is small, equal, and local to each spacer. The net flow is small, so that the perturbation of the natural flow field is minimized. Using the simultaneous short stroking of several WFs in nearby holes, one can better detect the propagation of a tracer like rhodamine dye or potassium permanganate at many elevations over a long time. Such a simultaneous short stroking of one well, or several wells, is possible with several manifolds and tubes extending from one gas driver to several wells. The driver may be as simple as a bladder pump driver connected to a gas bottle. The bladder pump driver typically controls the time the pressure is applied and the time the pressure is not applied. In that manner, it is easy to sample at many ports and several boreholes for the tracer arrivals and concentrations in time and space with both horizontal, azimuthal, and vertical resolution. Fortunately, the short stroking also reduces the gas consumption. This simultaneous sampling of many ports in several boreholes was done to monitor a surfactant injection at a superfund site in Milford, NH.

The driving pressure for a short stroke must be more than the pressure equivalent depth to the water table plus the pressure of the length to be displaced in the pump tube. With the application of several different test pressures, it is easy to adjust all ports to expel the desired amount with each stroke of the pump (i.e., each pressure application).

An additional advantage is that the above procedure can be performed while monitoring and recording the head at each port with the ACT, because the sampling procedure does not interfere with the head measurement. This is most useful with high-pressure injections of the tracer or remediation fluid.

A precaution: If the tracer injection pressure is likely to collapse the WF or the fluid injected is highly reactive, such as potassium permanganate, the liner should be first filled with a weak grout and is therefore no longer removable, but it can be drilled out for removal.

10.5.1.6 Materials in the Water FLUTe Construction

The flexible liner is normally a urethane-coated nylon fabric of 400 denier and ~200 lb/in. tensile strength. The urethane coating is on both sides of the fabric. Other coatings such as PVC are less expensive, but not as tough nor so well bonded to the fabric. It was learned very early that a poor bond to the fabric leads to a zipper-like series of perforations in the coating, which leak. The tubular plastic films tested were especially susceptible to that failure mode as well as lacking sufficient strength for the pressures applied to evert the liners or seal the boreholes. Other fabrics have been tested, but they do not bond as well to the urethane coating. RF welding is used to form the tubular liners and most attachments.

Each lot of liner material is tested to assure that it does not contain conflicting VOCs with those compounds of interest. One VOC does appear in the liner tests, toluene, because it is used in the urethane coating procedure. Based on water samples from WF systems, the toluene dissipates in ~3 months. It appears at very low levels in water samples at 10–50 μg/l, well below the drinking water MCL of 1000 μg/l. If there is no flow past the liner, it takes longer for the toluene to dissipate, but that is less concern for water samples.

Toluene does appear in FACT carbon samples where the toluene level is used to detect active flowing fractures as in the daFACT design (Section 10.2.2.5). Since FACT measurements are performed immediately after the borehole is drilled, the leaching of toluene from the liner by fracture flows transports toluene to the FACT. Since the FACT is isolated from the liner by a diffusion barrier, toluene peaks in the FACT provide a possible means of identifying active fracture flows if toluene is not a contaminant of concern.

Whereas it is known that toluene can enhance the degradation of TCE, the fact that it dissipates not long after emplacement makes it less of a concern. Urethane-coated fabrics manufactured without the use of toluene were found to be mechanically inferior and FLUTe has decided to accept the temporary concern about the toluene reaction with chlorinated solvents for the advantage of a high-quality urethane coating.

For situations requiring a stronger liner, an 840-denier ballistic nylon, coated on both sides, has a much higher tear strength and double the tensile strength of standard 400-denier fabrics. The 840-denier material is surprisingly flexible and more elastic then the 400-denier fabrics. It is more costly and only used when considered necessary.

The tubing in the WF system has been PVDF (Kynar) tubing as superior to the more common LDPE and HDPE plastic tubing, which is more commonly used for ground water sampling. The superiority of PVDF has been demonstrated by Parker et al (1997–1998) for absorption and leaching of VOCs. PVDF is actually superior to Teflon for TCE. It is also more expensive than the polyethylene tubing. There is a concern because PVDF is a fluorinated compound that it could confuse PFAs measurements. However, week-long leach tests show minimal PFAS leaching from WF materials, and sampling of DI water through the FLUTe MLSs of current design with PVDF tubing using the prescribed purging and sampling procedure show non-detects (ND) for PFAs in the samples. Other tubing materials can be used in WFs, but sampling for VOCs and even such caustic materials as potassium permanganate make the PVDF tubing the preference.

Brass fittings are used in a few locations in WFs, but Kynar fittings or stainless steel are an option, if preferred. Some fittings are of urethane. PTFE sealing tape for fittings is no longer used in WFs due to PFAS concerns.

All other materials in WFs including tether, spacer materials, and even tapes have been tested in week-long leaching tests for PFAS and VOCs with NDs for those materials at the very low detection limits normally used (0.25 ng).

10.5.1.7   Installation and Removal Procedure for Water FLUTes

The WF system is normally installed in a borehole already sealed with a blank liner. The blank liner is inverted from the borehole and the WF everted into the borehole by FLUTe-trained personnel. Figures 10.51–10.55 show a typical installation procedure for an especially large WF. Figure 10.49 shows a more typical WF of 300 ft. depth and shallower water table. However, shipping reels are smaller than in 2001.

For shallow water tables, it is useful to add a short interval of weighted mud to the interior of the liner. A more attractive practice is to install a water removal tube that is used to pump the water from beneath the liner during installation. Once in place, the liner is deflated by pumping water from inside the liner and removing the water removal tube in the borehole and to reinflate the liner to the recommended level. The same practice is used if the borehole has a very low transmissivity in the lowest interval of the borehole. (Note: The water removal tube is not to be confused with the pump tube name in the WF system.)

If there is any possibility that the highest head in the borehole is not identified in one of the ports, one can measure the highest head in the borehole as described in Section 3.4.2 by incrementally lowering the water table in the liner until it does not stay down. That identifies the highest head in the hole to assure that the water level in the liner is sufficient to seal the borehole everywhere. This overcomes the criticism offered that an unknown high head in the formation could prevent a seal of the borehole.

The WF system has often been removed when monitoring was to be discontinued. The removal is by inversion from the borehole or by removing all the liner water and lifting the liner from the borehole. The removal procedure also assures that if there is any malfunction covered by the FLUTe warranty against defects of manufacturing or materials, FLUTe can remove the system and repair or replace the WF. This is typically done at FLUTe expense if it is warranty work.

Installation and removal procedure of a WF are usually performed by FLUTe personnel. An advantage of the CHS and pdCHS described later (Section 10.5.3) is that they can be installed by other than FLUTe personnel and very quickly.

10.5.1.8   Advantages and Limitations of Water FLUTe System
Advantages of water FLUTe system:
1. Flexible liner seals the entire borehole
2. Quickly installed without field construction of the system or the associated space requirements, and no large crane or drill rig is used
3. Very simple and efficient sampling procedure
4. Simultaneous purging capability commonly used for labor and sample quality advantages
5. Removable
6. High-quality tubing for minimal contaminant perturbations

7. Especially useful for tracer tests
8. Allows easy head measurements at each port with several head recording options without conflict with the sampling method or number of ports available
9. Can be used in cased or uncased stable boreholes
10. Are especially effective at isolation of flow zones in karst formations

Limitations of Water FLUTe system:

1. Requires a stable borehole or a cased hole with screened intervals
2. Difficult to address artesian head conditions without a grout fill of the liner
3. Other situations where a grout fill is advantageous prevents removal of the liner other than by drilling
4. Chemical attack of the nylon liner by potassium permanganate requires a grout fill of the liner and special design features such as PVDF fittings and a stainless steel filter fabric (May require a special CHS drop-in system)
5. The interior water level must be maintained for the liner seal if not filled with sand or grout
6. Extreme gradients or head differences can be a hazard to the liner burst limit
7. Sharp rocks in the borehole are dangerous, but surprisingly seldom a hazard
8. If a perched aquifer exists, the water level in the liner must exceed the perched aquifer head. A stronger liner may be required to support that high excess head.

10.5.2   THE SWF (SHALLOW WATER FLUTE)

The SWF is similar to the CHS described in Section 10.5.3. The main difference from the CHS, which is lowered into the hole, is that the SWF is everted into the borehole. Since the tubing extends from the surface to the port, there is no pumping system. Unlike the WF and the pdCHS (Section 10.5.4), the SWF requires peristaltic pumping. Due to the simplicity of the design the price is substantially less than the WF but somewhat more than the CHS system. The additional cost over the CHS is that the tubing lies in sleeves welded to the interior of the liner like the WF. The purpose of the SWF is to provide a more economical MLS than the WF with advantage of the eversion installation.

10.5.2.1   The Design and Function

Figure 10.60 shows the SWF geometry. The tubing from the surface runs in interior sleeves, welded to the liner, to the port. Since the liner is everted into the borehole, the tubing is relatively slender (e.g., 1/4″ OD). With the connection of a peristaltic pump to the sample tube at the surface, the samples are drawn directly from the formation. All of the water in the borehole is inside of the liner and need not be purged. This provides a limited amount of IDW (investigation derived waste). The popular LFS (low flow sampling) method can be used with monitoring of the water quality as the tubing and nearby formation are purged prior to the sample collection. With experience, the recommended 3-gal. purge from each port prior to sample collection may be found

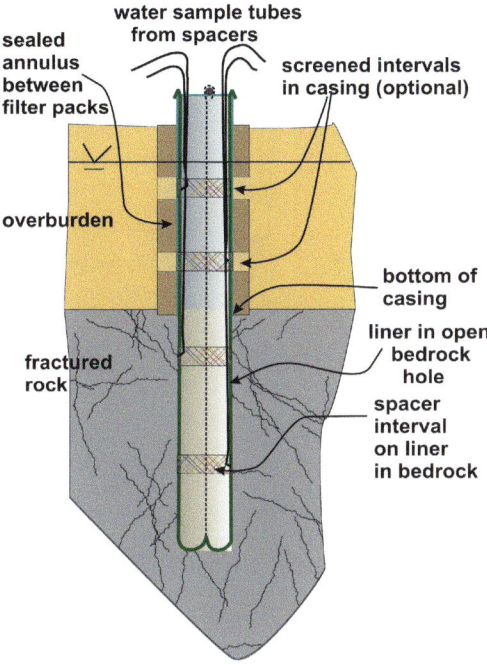

FIGURE 10.60 The shallow Water FLUTe system with sampling intervals in the saturated and vadose zones and also inside a cased interval with matching screens.

sufficient without the need to monitor the water quality as for the LFS instrumentation. A disadvantage is that the system cannot be simultaneously purged as with the WF and the pdCHS. However, one can connect several peristaltic pumps to the sample tube and reduce the time to purge accordingly. The SWF was used for the samples collected for comparison with the FACT results at the ten ports described in Section 10.2.2.4.5.

10.5.2.2 Other Advantages and Limitations of the SWF

The ACT can be connected to each sampling tube and stored inside the liner for continuous monitoring of the head at each port. The installation procedure is like that of several FLUTe systems with the liner deployed off a shipping reel with the water addition to evert the liner. In most cases, a separate pump tube (also called a water removal tube) is first lowered into the borehole before deployment to remove the water from beneath the liner. If the bottom of the borehole is of very low transmissivity, a special long vent tube is added to the design to allow water addition beneath the liner for removal by inversion from the bottom end of a tight borehole. See Section 3.3.3 for the long vent tube description and use. Removal of the liner is by inversion or if not to be reused, the liner can be emptied of water and lifted from the borehole.

The number of ports available depends mainly on the borehole diameter. If the primary function is to monitor the head history, many very slender tubes can be used to the ports with ACTs at the surface. The water table at each port requires the

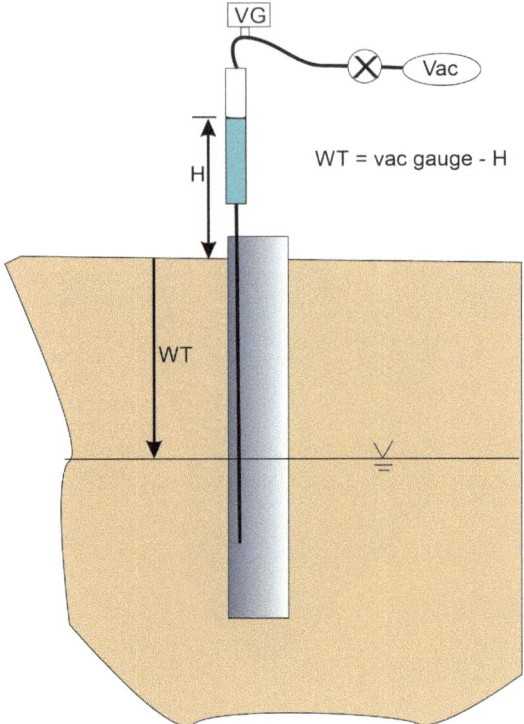

FIGURE 10.61 Vacuum water level meter. WT = VG − H, the water table depth BGS. VG is the vacuum reading on the gauge in water column height units. H is the height of the meniscus above the GS.

VWLM (vacuum water level meter) described in Section 9.5.2, because the tubes are too slender for the electric water level meter. Figure 10.61 shows the basic VWLM available from FLUTe for use with the SWF. That is an advantage of the CHS over the SWF. The CHS does allow the use of the slender water level meter for water table measurements in larger diameter tubing.

The SWF is more attractive for installation in marginally stable boreholes, since a slough of the borehole is unlikely to entrap an everting liner.

Obviously, the system is only practical for water tables at each port of less than 25 ft. below the surface due to the limit of peristaltic pumping. FLUTe has designed a system for simultaneous purging of the SWF or the CHS, but the use of several peristaltic pumps is more practical due to the general utility of additional peristaltic pumps. Because the water samples are drawn directly from the formation, LFS-type monitoring of sample quality is not needed unless preferred with the collection of the sample from each port after purging. Experience with the use of the LFS will demonstrate whether a larger purge volume than the recommended 3 gal. should be collected from each port.

The SWF can be installed in cased holes in unstable formations as easily as in an uncased borehole. Most installations are in uncased boreholes where abrasion or hole stability would be a concern for the CHS system. However, use of the SWF has been displaced, in most situations, for the cost advantage of the more recent design of the CHS system.

It is noteworthy that the sample tubes can be used to collect gas samples and to monitor barometric changes in the vadose zone even if at some times in the year the water table is above the port. This is possible because there are no valves in the system.

10.5.3 CHS (Cased Hole Sampler) Systems

10.5.3.1 Background and History

10.5.3.1.1 Design Objectives and Criteria

In 2018, it became clear that the need for FLUTe to install WF systems was more difficult for sites distant from the FLUTe field offices such as Sweden and Australia. It was also obvious that the WF was expensive for short holes from 50 to 150 ft. It was also clear that it would be advantageous to use FLUTe MLSs in unstable media requiring casing. WF systems had already been installed in cased holes to 1400 ft. with 11 screened intervals. The deep installations were to monitor water quality and head histories continuously during potable water injections into critical aquifers. The shallow CHS system described above is a more simple, less expensive system well suited to open or temporarily stable boreholes.

Given that shallow installations were often cased with single or multiple screens through unstable overburden, FLUTe designed a simple liner MLS system for cased holes. The system was named a CHS (cased hole sampler) system.

A third application desired was a liner MLS that could be used in casing as small as 2″ in diameter. This was especially true for 2″ continuously screened cased holes with a continuous sand pack. With the FLUTe capability of simultaneous purging and sampling, one could expect reasonable sample isolation despite the relatively permeable sand pack of CSC. Then in the typical FLUTe evolution, it was reasoned that the continuous sand pack did not need to be continuous to provide useful measurements. This concept led to the CSC design described later in Chapter 14.

In summary, the simple design objectives for the CHS systems were:

1. Nearly full function of a WF system including:
    a. Liner seal of the borehole
    b. Sample extraction from the formation, not the borehole water
    c. Simultaneous purging and sampling of all ports
    d. Manual and recorded head history measurements at each port
2. Less expensive
3. Easier to install

The design objectives have led to the two versions of the CHS system with the addition of the pdCHS (positive displacement CHS). The pdCHS meets two requirements. The first is to replace the peristaltic pumping use, which is convenient for shallow water tables but which was not acceptable for volatile contaminants like TCE. The second is to overcome the depth and flow limitations of peristaltic pumping. Only positive displacement pumping allows flow from deep water tables and in the long slender tubing needed for many ports in small casings and deeper ports.

Finally, another difficulty with past procedures was the requirement that FLUTe install all WFs systems. While a few properly trained individuals have installed WFs, the eversion procedure is not easily done so well as FLUTe staff can do it. Therefore, the preferred flexible liner would be installed as easily, and more quickly, than even a simple system like the Solinst CMT (continuously molded tubing) design. A CMT system is best not constructed in the field under adverse weather conditions (e.g., far below freezing) and requires that ample space be available for construction on the surface. More quickly and less expensive is possible for the CHS systems by avoiding the backfill procedure in open stable holes needed for the CMT system.

With these design objectives, FLUTe designed two CHS systems according to the criteria:

1. Much less expensive per foot than short WF systems
2. Less expensive to install, no FLUTe travel expenses to the site or FLUTe labor of installation
3. Less costly to manufacture
4. Less costly to ship (smaller and lighter)
5. Can be installed in cased holes with multiple screens or in shallow uncased holes, and holes which can be thoroughly developed in the usual manner, prior to the liner installation
6. Can be easily installed by anyone in a short time with only written instructions
7. Uses the same simultaneous purging and sampling method as for WFs
8. Has use in unstable media
9. Can be used in both forms (CHS or pdCHS) in 2″ or larger casing
10. Requires no field construction
11. Useful in both discrete screened casing or in FLUTe CSC (continuously screened casing)
12. Fully removable

10.5.3.1.2 Some Other Benefits
Some other special benefits are:
1. Can be installed in a cased hole in 10–15 min
2. Can install two to three systems in one day with testing with full dilation of the liner
3. Works well with manual water level measurements
4. Connects directly to FLUTe ACT systems for head histories at each port
5. Does not require scaffolding for shallow water tables or artesian conditions
6. Does not need heavy mud to install
7. Is easily grout filled to resist collapse of nearby high-pressure injections or extreme water table drops from nearby pumping
8. Can be installed in shorter open stable boreholes with no casing, or cased with screens in the unstable upper portion, and open stable boreholes in the fractured rock below the cased interval
9. Can accommodate much larger diameter tubing which cannot be everted in a WF or SWF

Despite the recent CHS invention (patented of course), there have been >50 installations already by early 2021 at six sites in cased and uncased boreholes. One of the first installations was in a 2″ continuously screened well with four ports. It was a pdCHS system that was less compact than a CHS system. The only problem was that the well had not been developed and the sample water contained excessive sediment at some ports. A special advantage of the use in screened casing or open stable holes is that the well can be rigorously developed before the installation.

An obvious question is why is the price much less than for a WF? The answers are that it takes less labor to build the system and fewer materials. The liner material is the same urethane-coated nylon fabric. The typical FLUTe tubing is PVDF which is far superior to LDPE or HDPE for TCE absorption or leaching. However, if desired, an HDPE tubing version is available at somewhat lower cost, but not as well suited for TCE or PCE as tested by Parker et al (1997) at CREEL.

The depth of installations has been increasing. The deepest pdCHS installation to date was in 2019 in a 3″ casing with six screened intervals and ports to 255 ft. The screens were well developed with a surge block and air lift pumping. The use of ACT measurements of head histories during a pump test was done in 2020. That site has 12 pdCHS systems in place in 4″ screened wells to 150–200 ft.

10.5.3.2   Geometry of the CHS

The CHS geometry is shown in Figure 10.62. The liner is continuous to seal the borehole. The spacer defines the unsealed interval from which the sample is drawn. The tubing from the port to the surface is called the sample tube. The central tube is used to add water to the interior of the liner after it has been dropped into place. The bottom seal of the liner supports the weight suspended from the bottom seal. The weight is adjusted to be sufficient to sink the liner in the borehole water and to overcome the buoyancy of the system when not filled with water. The spacer is constructed of an open LDPE mesh which lies over the port and allows water flow to the port when the liner is dilated against the borehole wall or screen. A filter fabric surrounds the liner and covers the spacer to prevent sediment from entering the tubing to the surface. The sample tubing from the port to the surface is sufficiently large (3/8″ or larger OD) to allow the use of an electric water level meter to measure the water table level in the formation at each port. Only a single-port system is shown for clarity. The multi-port system has an additional spacer, port, and sample tube for each port. The tether is connected to the bottom seal and supports the weight below the liner and the Kellum at the top of the well. The Kellum supports the tubing bundle. The tether is connected to the bottom of the Kellum. The upper tether is connected to the top of the Kellum and supports the tubing bundle and the weight via the bottom seal. In an uncased hole, there is no backfill expense.

10.5.3.3   Installation Procedure for CHS

The liner arrives in a shipping box with the weight and cap separate. The shipping container (a box instead of a reel) is placed near the borehole and the cover removed. If the top of the surface casing is not smooth and fair, a temporary protective edging is placed on the top of the casing. The water table depth and total borehole/casing depth is checked. The bottom of the borehole is tagged for the actual depth.

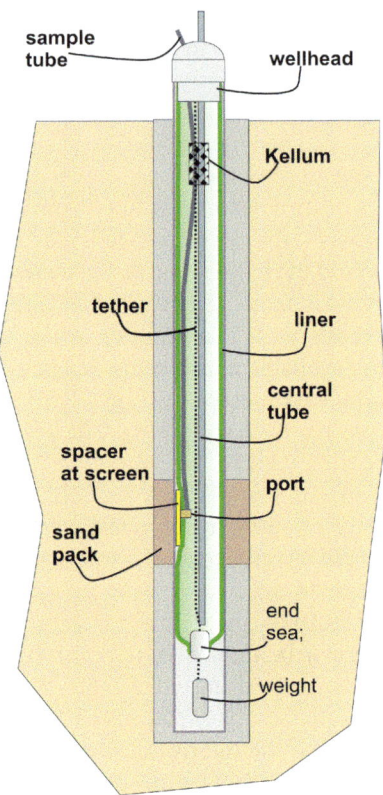

FIGURE 10.62 The CHS geometry in a cased hole (single port for clarity). Allows water level measurement in sample tube and peristaltic pumping. Weight adjustable on tether to allow for sediment in casing. Also used in open uncased holes.

(Any sediment accumulation in the bottom of the casing or the borehole should be removed with air lift pumping as described in Section 3.2.4.7) Simply pumping the well for development will not remove the sediment. The bottom end seal is above the bottom of the hole with a tether connected to the bottom of the seal to support the weight. If the borehole depth is less than specified, the bottom tether should be shortened by the difference before attaching the weight to the tether.

The weight is attached to the tether and lowered into the casing (see Figure 10.63). The liner is lifted from the box and lowered, following the weight, into the hole until the wellhead is seated on the top of the casing. If a protective edging was used on the casing, it must be removed before the wellhead is seated on the casing. The tether in the slot of the wellhead is now supporting the full weight of the system except for the liner. The top tether knot is usually formed at the proper elevation in the fabrication procedure.

The central tube is connected to a water source such as a pump submerged in the water tank adjacent to the borehole. It is prudent to check that each sample tube has filled with water before dilating the liner. The liner is filled with water through

FIGURE 10.63 CHS installation from the shipping reel into a 100 ft. 4″ casing with four ports on the liner. The white intervals are the spacers for each port. This installation took less than 10 min not counting setup at well or water addition to seal the liner. The pdCHS system installation is the same.

the central tube until the liner water level is 10 ft. above the water table. The water can be added through the wellhead before the cap is attached. The liner will dilate to seal the casing or borehole. As the liner dilates, some water will be trapped between the casing or borehole wall and the liner. As that water bleeds through the screen or into the formation over time, the water level in the liner will drop. Refill the liner to the prescribed depth. If it is still descending, refill again. This can be a slow process. With each refill, the rate of the drop of the water level in the liner will decrease. When the water level stabilizes, thread the central tube and each sample tube through the numbered holes in the cap. Slide the cap down until fully seated on the wellhead piece. Tag the water level in each sample tube. Record the water levels.

CHS systems can be filled with mud or grout through the central tube as described below in Section 10.5.4.4. The mud fill forces the borehole water outside the liner to the surface as mud is added. The water between the liner and casing can be pumped out through a slit in the liner below the wellhead. This speeds the dilation process. CHS systems can also be installed in uncased holes of limited depth as described in Section 10.5.4.5 below using a protective sheath or a stronger/thicker liner to protect against abrasion. The mud can later be pumped out to remove the liner, but the weak grout fill requires the liner be drilled out of the borehole.

10.5.3.4   Purging and Sampling

To test the purge of the CHS ports, connect a peristaltic pump to each sample tube and purge at least one sample tube volume from each tube. If a 3/8″ OD sample tube, it is 1/4 gal. per 100 ft. of tube submerged. Check the water level in each sample tube with an electric water level meter. Thereafter, check the water level in the central tube. The water level in the liner should be 10 ft. above the highest head in any port. If the water table is too shallow, the water level should be at least 5 ft. above the highest port level.

   At this time, the installation is complete. Three gallons are purged from each port and then the sample is collected. It is best to purge all the ports simultaneously with a peristaltic pump on each port. If that is impractical, each port can be purged separately and a sample is collected from each.

   For a 3-gal. purge of each port, one may not need to perform the LFS procedure. However, if desired, connect a water quality monitor to the sample tube downstream from the peri-pump and record the water quality for that port. If you continue to pump, there should be little change in the water quality after the recommended purge volume from each port. Because there is no connection of the sample water with the water inside the liner, you are not purging the well as with an open cased hole. One can easily monitor the entire purge procedure to assure that the recommended 3-gal. purge leads to a steady state of water quality parameters. If this is the first sampling episode, one may be seeing drilling effects. For later episodes, that should not be significant, especially if the well was thoroughly developed before the liner installation.

10.5.3.5   The Removal Procedure

The removal is relatively simple for the CHS system. The liner is pumped empty through the central tube, the cap removed, and the liner system is lifted from the casing. For larger than 2″ casing, a more aggressive pump may be preferred to remove the water from the liner.

   When lifting the end of the liner above the water table, the water in the interstitial space of the tubing will accumulate at the bottom of the liner and dilate the liner making it very difficult to lift the liner. The remaining water is simply pumped from the liner with the central tube. Then the liner is lifted out of the casing.

10.5.3.6   Special CHS Design for Potassium Permanganate

Installations of CHS systems where potassium permanganate, or a similar strong oxidizer, is to be injected, can damage the liner by the contact with the oxidizer. Also, the injection pressure may collapse the water-filled liner. A third concern is that the injected fluid can flocculate or generate sediment that can clog the standard

CHS sampling system. For that situation, a special design has been developed. The changes are as follows:

1. The liner is filled with a weak grout that will still seal the borehole even if the thin liner (20 mils) is damaged. The grout fill also prevents collapse of the liner with injection pressures.
2. The tubing size is increased to 0.5″ OD or larger to prevent clogging of the tubing.
3. There is no valving in the system except for the pdCHS version when sampling is to be performed.
4. The spacer filter is constructed of a fine stainless steel screen to prevent damage by the oxidizer.

These changes are only possible with the CHS drop-in design. These changes cannot be made in everted liner systems.

The positive displacement system of the permanganate version is very similar in function to the pdCHS system described next. The permanganate version of the pdCHS system uses a small central tube in the larger sample tube of the CHS to allow a pressurization of the outer larger sample tube to force water up the small central tube for collection. A ball attached to the bottom end of the small central tube on a short tether allows the seal of the port fitting during the pressurization of the sample tube and allows recharge when the pressure is dropped to atmospheric. When pressure is applied to the outer tube, the ball seals the port fitting and the water fill of the sample tube flows through the scallop in the small central tube to the surface for collection. When sampling is completed, the central tube can be removed for the water level measurement or left in place until the next sampling event. Raising the small central tube prevents the ball sealing of the bottom fitting and then allows an ACT monitor of the injection history through the small central tube. The tubing is of the same PVDF material as most FLUTe systems. The tubing should be capped to prevent injection pressures forcing fluids out of the sample tube.

These changes allow the high-pressure injection of a strong oxidizer without impairing the function of the CHS liner in either a cased hole or uncased borehole.

10.5.4   The pdCHS (Positive Displacement CHS)

10.5.4.1   The Design of the pdCHS System

The pdCHS geometry is shown in Figure 10.64. The liner and exterior features are the same as the CHS system shown in Figure 10.62. The main difference is that the pumping system is built into the liner for positive displacement pumping. A single-port system is shown for clarity. The main features are the sealing liner, the spacer with the outer filter fabric and the coarse LDPE mesh, and the port in the liner behind the mesh. The pumping system is a U-shaped tube. The right-side tube in the drawing is the pump tube, the left side is the sample tube. The sample tube has a tee connection to the port tube. The port to sample tube extends from the tee upward to the normally open check valve to the port.

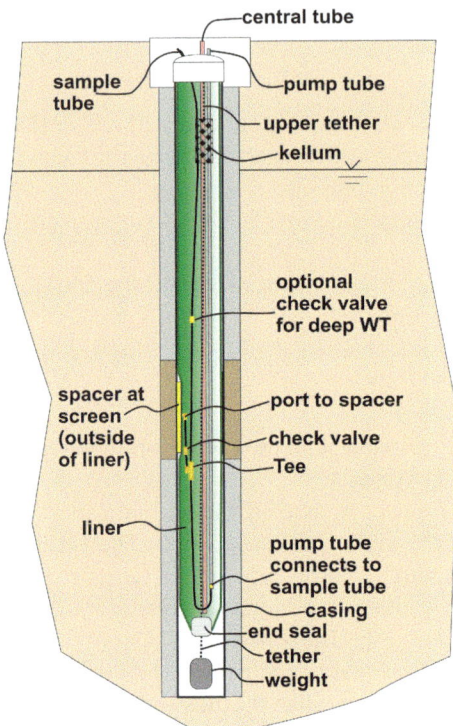

FIGURE 10.64   The pdCHS system geometry with the sample and pump tubes forming a U extending to the bottom of the hole. Normally open valve allows positive displacement pumping and the use of ACTs on all ports for water table history. Multiple ports can be pumped simultaneously because each port system is nominally the same length.

When the liner is installed, the spacer conducts water to the port, through the check valve and filling both sides of the U tube to the water level in the formation at the port. The pump tube side is usually 3/8 OD tubing which allows the water level to be measured with a slender electric water level meter and enhances the purge volume per stroke. Applying a specified gas pressure to the pump tube closes the normally open valve and drives all of the water to the surface out of the sample tube. Dropping the pressure allows the system to refill. Applying the "purge pressure" again to the pump tube expels another U tube volume of water out of the system. These are called purge strokes.

When the prescribed 3 gal. of water have been purged from the U tube, the gas pressure is reduced to the "sample pressure". Now the gas pressure applied to the pump tube does not displace all of the pump tubing water. Therefore, the aeration of the sample water is prevented. As the water level in the pump tube drops to the depth of the equivalent gas pressure, the flow slows to a stop. The simple procedure is to allow recharge of the pump system and after applying the sample pressure again, collecting a larger water sample. The first one fourth of a liter of sample water should be discarded since it can contain aerated droplets from the previous purge. If a larger sample is required, the sample cycle is repeated at the sample pressure with no discard needed.

10.5.4.2 Simultaneous Purging and Sampling of the pdCHS

The advantages of simultaneous purging and sampling are described in Section 10.5.1.3.2. For the pdCHS system, the method is the same as for the WF system. The gas source is connected through a manifold, available from FLUTe, which connects the gas source to each pump tube (see Section 10.5.1.3.2 for the description of the manifold). The purge pressure is set at the gas source and a three-way valve applies the same pressure simultaneously to all the pump tubes. The effect is that the entire volume of water from each U tube is expelled simultaneously through the individual sample tubes. Figure 10.65 shows the flow from four sample tubes as gas pressure is applied to all pump tubes. That purge volume should be collected in a separate container for each port. Dropping the pressure to ambient allows recharge of the system and applying the gas pressure to all ports again with the three-way valve causes another purge volume to be ejected. When 3 gal. have been expelled from each port, the purge is complete for all the ports. Valves on the manifold allow individual tubes to be purged if some tubes refill more slowly as controlled by the formation conductivity. The advantage of the manifold is that the time to purge and sample all the

FIGURE 10.65 Showing the simultaneous flow from a four-port pdCHS system in a 2″ × 100′ casing using a four-port manifold. Operation of one valve causes flow and recharge of all ports for purging. Individual valves allow sample collection from each port after purging 3 gal. Note the central tube used to fill the liner with water.

ports is not much longer than the time to purge and sample one port. If some tubes refill more slowly, the wait for recharge can be increased to allow a more uniform fill of all tubes. That allows the purge volumes to accumulate more equally in time. More details of the simultaneous purge procedure are provided in Section 10.5.1.3.2.

10.5.4.3   Installation and Removal of the pdCHS

These procedures are the same for the pdCHS as for the CHS system.

10.5.4.4   Installation of CHS and pdCHS with Mud or Grout-Filled Liner

An advantage of the CHS systems is that they are easily installed with very shallow water tables or even artesian conditions. However, the liner must be at an internal pressure above the formation water to obtain a good seal. There are at least two simple options. The easiest is to fill the liner with a heavy mud instead of water. The mud fill is discussed in detail in Section 3.5.7, but such an advantage for CHS systems is also described here. A particular advantage of the mud fill is that is pushes the liner against the casing/hole wall as the mud level rises and does not entrap water between the liner and casing which can take a long time to move to the permeable intervals allowing full liner dilation. With that advantage one does not need to wait for the head to stop dropping in the liner as the liner dilates. However, the mud overpressure depends on depth as explained here. For very shallow water tables, or artesian conditions the mud fill does not seal to the surface with a significant overpressure. The mud excess head above the same water elevation is $\Delta H = (\rho - 1)H$ where ρ is the density of the mud in g/cc and H is the height of the mud column in water. Typical mud densities are as high as 1.8 g/cc. The mud density is gained by mixing a bentonite powder with a barium sulfate powder marketed as Baroid. The Baroid is commonly used to weight a mud mixture in mud drilling procedures. The table in Figure 10.66 shows the mixtures as a function of density, and Figure 10.67 shows the mud overpressure with depth in the column as compared to the water fill pressure with depth.

Mud mixture for Liner fill

Density gm/cc	Bentonite #/gal	Baroid #/gal	pressure with depth of mud (PSI) if filled to the water table.												
			20	40	60	80	100	120	140	160	180	200	250	300	350 ft of mud
1	0	0.00	0	0	0	0	0	0	0	0	0	0	0	0	0
1.05	0.3	0.32	0	1	1	2	2	3	3	3	4	4	5	6	8
1.1	0.6	0.66	1	2	3	3	4	5	6	7	8	9	11	13	15
1.15	0.6	1.24	1	3	4	5	6	8	9	10	12	13	16	19	23
1.2	0.6	1.85	2	3	5	7	9	10	12	14	16	17	22	26	30
1.3	0.6	3.11	3	5	8	10	13	16	18	21	23	26	32	39	45
1.5	0.6	5.89	4	9	13	17	22	26	30	35	39	43	54	65	76
1.8	0.6	10.83	7	14	21	28	35	42	48	55	62		87	104	121

Add half a teaspoon of soda ash per gallon. (three teaspoon Blowouts have occurred at 25 psi for 400d DC
    (16 table spoons per cup, or 48 teaspoons per cup)   However, that is ~1/2 normal burst pressure
    (i.e., one cup of soda ash is good for ~96 gallons) relatively comfortable level of high pressure
7-10 ft. preferred , 5 ft minimum (~2 psi)
Too low for sealing need excess head above water table

FIGURE 10.66   For each density desired and the bentonite weight (lb/gal.) shown, the Baroid column provides the Baroid weight in lb/gal. to be added for that density. The rest of the table shows the overpressure in psi. developed with mud depth. The color highlights indicate which depths for a given density are sufficient or excessive for the liner.

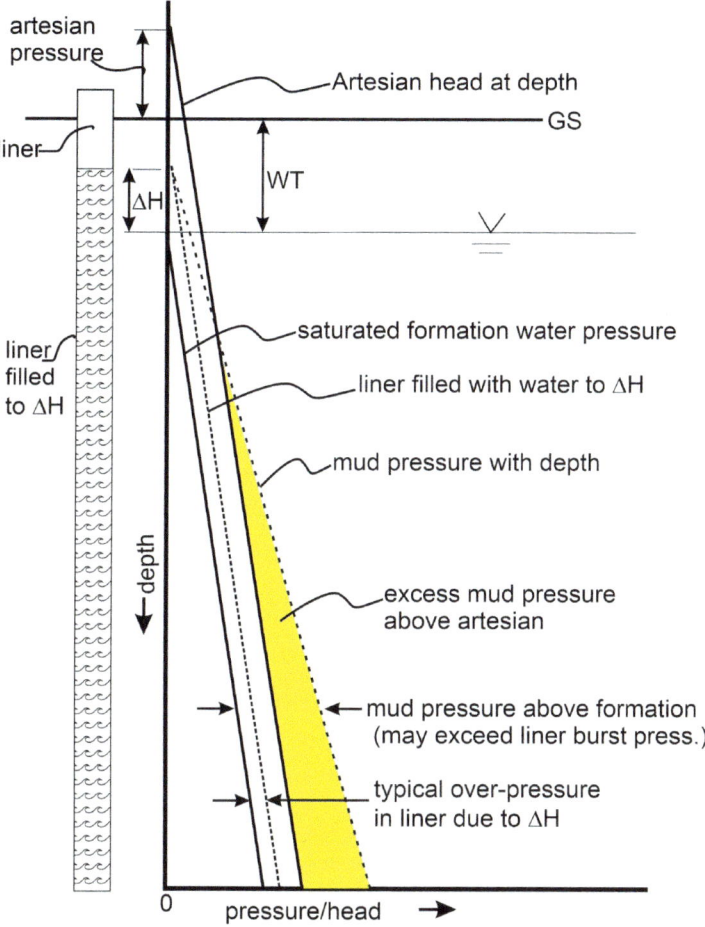

FIGURE 10.67 Comparison of normal water pressure gradient to that with an elevated head, ΔH, and a heavy mud gradient. The gradient of an artesian head is also shown assuming it exits throughout the formation.

Note that for some mud columns the overpressure can be dangerous to the liner. For example, at 100–300 ft. a mud density of 1.8 is certain to burst the liner except in a cased hole. Reducing the density of the mud may lead to an insufficient overpressure at shallow depths. Stronger liner materials such as 840-denier ballistic nylon can withstand much higher head differentials in breakouts of the uncased hole wall. It is good that the high mud pressure is not a concern when the liner is fully supported by a casing.

A second option for a fill other than water, if the mud pressure is not sufficient, for example as with high-pressure injection of remediation fluids which are likely to collapse the liner, the fill of the liner can be a bentonite grout mixture. The grout density of about 1.35 g/cc is not as high as the mud, but the advantage is several-fold. The liner cannot be collapsed and chemical attack of the liner by oxidizers like potassium permanganate are not a concern because the sample tubing system is not affected. The disadvantage is that a grout-filled liner system must be drilled out for abandonment or at least the tubing filled with a sealant. Drilling a grout-filled liner is easier than drilling a water-filled liner system. For deep open boreholes to be filled with grout, the grouting of the liner should be limited to 120 ft. lifts between cures unless the stronger liner material allows higher lifts of density 1.35. Tests have shown that the bentonite grout mixture recommended does not overheat the liner system in a 6″ hole below the water table. Larger diameter boreholes above the water table should be further tested since the heat of hydration can damage PVC casing. A higher content of bentonite is recommended since the grout strength is not a concern as with a grouted empty casing. It is also useful that an overheated casing cannot collapse a grout-filled liner. It is noteworthy that the grout mixture can also be weighted with Baroid if the grout density is insufficient to seal the artesian intervals.

10.5.4.5   Installation of CHS Systems in Uncased Holes

As described above, CHS systems have been installed in uncased holes. The main concern about lowering a liner system into an uncased hole is the abrasion of the liner leading to many small leaks which violate the main requirement of a liner sealed borehole. Liners if pumped empty and dragged from the uncased borehole can create such leaks. In a PVC-cased borehole, the abrasion is much less than in a drilled open hole, so the CHS is relatively immune to serious abrasion. The fact that CHS systems are urethane coated on the inside and the outside surface of the fabric is also helpful. However, it is best to not risk the abrasion damage in open-hole use. However, that depends on the hole depth and the nature of the hole wall. CHS systems have been installed to 100 ft. in a 6″ hole in shales of NJ without leakage. A five-port pdCHS system was lowered into a 4″ cored borehole to 115 ft. in Guelph, Ontario without damage. However, there is some risk of abrasion in an open hole. In contrast, a 3″ pdCHS liner with six ports was installed to 255 ft. in PVC casing without damage.

A solution to the hazard of abrasion in open holes is to provide a protective sheath, which is a patented solution for several FLUTe systems which are lowered into the borehole. The protective sheath must meet several criteria:

1. The liner must not slide out of the sheath as the sheath is being lowered.
2. The sheath must be very abrasion resistant.
3. The drag of the sheath on the liner must not be so great that the liner system rises with the sheath as it is being removed from off the CHS.
4. The drag of the sheath on the borehole must not be so great as to prevent the liner being lowered to the bottom of the hole.

Those criteria have been met as follows:

1. The sheath is temporarily attached to the bottom end of the liner so the weight cannot pull the CHS out of the sheath. The attachment is releasable.
2. The sheath is of very strong nylon fabric.

3. An anchor is provided at the bottom of the liner by inflation of a water balloon inside the liner which anchors the liner in the borehole before the sheath is lifted. Or the bottom of the liner below the bottom of the sheath can be filled with a heavy mud which also provides an anchor of the liner in the borehole.
4. The nylon sheath is relatively slippery and the weight used for the CHS must overcome both the buoyancy of the liner and the drag on the hole wall.
5. A fifth practical criteria met is that the addition of the sheath must still allow a reasonably priced system. It does.

It is interesting to note that criteria no. 4 was not met for an early installation and the sheath was removed and the CHS installed without the sheath and yet survived without serious abrasion. However, that is not expected. The sheath material is now much more slippery and has been used in subsequent installations.

It is important to remember that there is no concern with an overpressure of the liner in a cased hole because the liner is supported by the casing even when it is mud-filled. In an uncased hole, the liner is not supported in enlarged intervals in the borehole, so that a water fill to the surface or a mud fill may exceed the liner tensile strength. However, the cost reduction with no need to backfill outside a casing makes installation in an uncased hole very attractive. A CHS system in wire-wrapped screens can be a concern because of the sharp edge common at the bottom and top of the wire-wrapped screened interval. This increases the concern with overpressure of the liner on that edge. Only one liner has had a problem with wire-wrapped screens near Amarillo, TX, in the early years of FLUTe development. An alternative to the sheathing has been to use an extra strong and thicker liner of 840-denier instead of the standard 210- or 400-denier liners. However, the sheath is the better protection for a more hazardous borehole whose caliper shows an irregular and abrasive condition.

10.5.5 USE OF ACT SYSTEMS WITH THE CHS SYSTEMS

Since the pdCHS system uses a normally open check valve which only closes with a strong pressure application to the pump tube, the water levels in the pump and sample tubes of the two CHS systems always follow the formation head. The ACT system is described in detail in Section 9.5. The connection of the ACT transducers to the sample tubes of either The CHS or pdCHS system allows one to monitor and record the water table variations at each port in time (note the pdCHS optional second check valve for very deep water tables requires the ACT be attached to the pump tube). The ACT transducers can be located in the vault at the surface or in an enlarged extension of the casing. Another option is to lift the cap from the wellhead piece and to store the Act transducers in a vertical string inside the surface casing inside the liner. FLUTe usually provides rented ACT assemblies in a vertical sleeve for ease of storage inside the surface casing. The ACT data can be downloaded as needed and converted to water table histories. Each time the CHS system is to be sampled for water quality, the ACT is disconnected and the sample is collected from the sample tube. The water tables are remeasured and recorded, and the ACT connections are redone. The water table must stabilize before the water table depth is measured and the ACT reconnected. The wellhead cap must allow the ACT cable to be accessed or the cap removed to download the data if the ACTs are stored in the surface casing. A water table measurement for each port is always required before the ACTs are connected to the tubes.

Act measurements are useful for seasonal changes, pumping tests, and for events like the removal of blank liners in nearby wells for assessment of cross well connections.

10.5.6 Depth Limitations for CHS and pdCHS Systems

10.5.6.1 Depth Limits for CHS Systems

In cased holes, the depth limitation for the CHS system is the length of the tubing to be effectively pumped with peristaltic pumping. For very long tubes small diameter tubing (1/4″ ID) a high vacuum must be applied to draw the water sample to the surface. The deeper the water table, the more limiting that flow rate and the higher the vacuum commonly applied. With larger diameter tubing that is less limiting, but still a practical limit. The difficulty of peristaltic pumping increases with deeper water tables to a practical limit of ~25 ft. below which the water vapor pressure plus TCE vapor pressure competes with the peristaltic pumping. For deeper water tables, the flow rate for a given vacuum decreases to insufficient to drive flow in long tubes. Therefore, only for very shallow water tables, large diameter tubing and low vacuum values would the concern about volatilization be reduced to insignificant. Since those are not easily controlled, especially the vacuum applied, many regulators do not allow peristaltic pumping of VOCs. For that reason, FLUTe does not recommend a CHS for VOC sampling or for a depth greater than 300 ft. for 1/4″ ID tubing for other applications. The pdCHS does not have that limitation.

Since CHS systems have no valves in the tubing, they can be used for injection and extraction and the tube diameters can be much larger than for everted liner systems. In that case, only the liner interior pressure differential with the formation water and injection pressure limits the length of tubing or depth in the borehole. In extreme cases, the liner can be filled with grout and injection pressures can be very high without fear of collapsing the liner. Injection or extraction of fluids is better addressed with the DEIL (discrete extraction injection liner) system described in Section 10.5.11. The DEIL is an extension of the CHS design, but with a different set of advantages and limitations.

At some depth and number of tubes and the amount of tubing curl, which is related to the diameter and material of the tubing, even the abrasion in a cased hole is a concern. In wire-wrapped screens with typically razor-sharp steel edges at the top and bottom of the screen, only a sheathed version should be considered. However, the use of the heavier 840-denier liner material may provide sufficient protection. It is useful to note that curl of the tubing as it is lifted from the shipping container is more significant in very cold weather and more significant for larger diameter tubing. For that reason larger tube diameters of the order of 1.5–2″ diameter and made of HDPE are very awkward to handle in cold weather and are not used with FLUTe systems even for the CHS design.

10.5.6.2 Depth Limits for pdCHS Systems

For the pdCHS, the pumping system for sampling is limited to the pressure capacity of the tubing. To purge the system, the pressure applied must be greater than the depth of the liner because the U-shaped tubing extends to the bottom of the borehole (unlike the WF system). After that limit, there is the pumping rate through extremely long tubing. There is also some concern about sediment accumulation in the bottom of the U due to the long water column above the bottom of the U. The water column and sediment load above the NO check valve is very small. However, a hybrid system has been designed that can be used at greater depths. That design does not exceed the pressure capacity of the tubing, but that design has not been tested except that it is very similar to the WF design and probably works equally well, but with a much smaller purge volume per stroke.

These same limits for the pdCHS are not true for a WF system. The effective U of the WF pumping system only extends to the first check valve which is usually far above the bottom of the borehole. So, there is no purge or sampling pressure limit for the WF. Also, the WF is everted into the borehole and does not have the hazard of abrasion that exists for a drop in system. Furthermore, the pump geometry of the WF can re-suspend the sediment accumulation above the first check valve because the flow is upward through any accumulated sediment during recharge of the pump with the finest component on top of the settled sediment. For the pdCHS, the flow on the pump tube side pressures the top of the sediment settling from the larger pump tube. That is not the preferred flow direction to clear the sediment. Because there have been no reports of clogged pdCHS tubing except with one 100 ft. continuously screened well with extreme sediment content in the bottom end of the screen, it only suggests that all wells should be well developed before installing a sampling system. Even with the mud clogging of the one continuously screened well, the ports were functional with peristaltic pumping of the pdCHS system.

Any sediment accumulation is proportional to the height of the water column above the point of potential accumulation and the diameter of the tubing above the accumulation point, since it is the sediment per unit volume that controls the accumulation. Therefore, the WF system, with a fixed limit of 100 ft. of water above the first check valve, is likely to have less sediment accumulation than any pdCHS system of more than 100 ft. submergence. Many years of experience with the WF system supports the belief that sediment accumulation is not a common problem. The shallower the borehole, the less the potential height of sediment accumulation for the pdCHS.

CHS systems are at their best for shorter boreholes than are common for the WF system.

A depth limitation for unsheathed pdCHS systems is the amount of abrasion that can occur for unsheathed systems and for sheathed systems the amount of drag that is encountered in lowering the system into place, or in removal of the sheath. Experience will quickly define those limits which depend on formation and borehole wall characteristics, number of ports, tubing diameter, and hole diameter. At this time a 300-ft. cased hole installation is relatively simple for both systems in a 3″ casing. For cased holes without a sheath, 150 ft. is the conservative limit for less than five ports in a 6″ hole for a pdCHS and ten ports for a CHS, both with 1/4″ tubing and 3/8″ pump tubes. Check with FLUTe's latest experience. A six-port pdCHS in a 3″ casing to 255 ft. was easy. A CHS system has half as many tubes as a pdCHS system. If uncertain, a WF is the conservative choice with known advantages under a wide variety of conditions.

10.5.7   Relative Cost of the CHS Based Systems

CHS systems are much less expensive than the WF systems for two reasons. The amount of materials is less than in WF systems and the labor of construction is less than for WF systems of equal depth and no. of ports. However, that leads to depth limitations and somewhat slower sampling procedures. CHS systems do not have the diffusion barrier to isolate the liner from the sample water because the presence of toluene in the sample water is usually not important, and it decays in time if from the liner material. The diffusion barrier can be added to the CHS system if desired at additional cost.

The price of the CHS system is about one-third that of the WF but is limited to shallow water tables and peristaltic pumping. The pdCHS system is about half the price of the WF, but limited in depth.

10.5.8   Advantages and Limitations of Both CHS Systems

Advantages of both CHS systems:
1. Cost of system
2. Low cost to install
3. Full function
   a. Two modes of sample pumping
   b. Simultaneous purging (mainly for the pdCHS)
   c. Manual water table measurement or water table history with transducers for each port
   d. Removable unless grout-filled
   e. No field construction
   f. Continuous liner seal of borehole
   g. Used in slender casing of 2″ or larger
4. Can be installed by the customer
5. No special equipment to install (e.g., no crane truck or drill rig)
6. Easily removed
7. Cased or uncased boreholes
8. Use of ACT systems for formation head histories
9. May be used in continuously screened wells, especially CSC systems
10. Can use much larger diameter tubing than WF and SWF

Limitations of both CHS systems:
1. Hole depth and number of ports versus hole diameter
2. Smaller purge volume per stroke compared to WF
3. Driller must install multilevel screens if a cased hole installation
4. Wire-wrapped screens require heavier liner material or sheath use

5. Limited depths in uncased holes without sheath
6. Less immune to sediment in the sampling system than the WF system
7. Requires grout fill of liner for high-pressure remediation injections
8. CHS is limited to peristaltic pumping.
9. Abrasion is a concern in open holes

10.5.9 Use of FLUTe MLS Systems in General

As for most MLS systems, the FLUTe systems are useful for collection of water samples at discrete elevations in a borehole. Since the liner seals the entire hole except at the spacers which define an unsealed interval, it is best to not try to sample over extremely long intervals. The standard spacer length is 5 ft., but 10 ft. is common at extra cost. All FLUTe liners can be inflated with water, sand and water, mud or grout.

FLUTe systems are either everted into the borehole (the WF and Shallow WF) or lowered into the borehole (the CHS and pdCHS). Some differences in the two systems are as follows:

10.5.9.1 Water FLUTe (In Use Since 1996)

1. Used for boreholes more than 150 ft. deep typically
2. Everted into place and therefore preferred for
   a. Angled holes
   b. Highly enlarged boreholes such as karst
3. Large volume pumping system as used for deep boreholes
4. More expensive than CHS systems
5. A positive displacement pumping system
6. Simultaneous purging and sampling
7. Installed by FLUTe-trained personnel
8. Installed in both cased or uncased boreholes
9. Not normally installed in holes less the 3.5″
10. ACT head history measurements
11. Very economical sampling costs

10.5.9.2 Shallow Water FLUTe (SWF) (In Use Since 2014)

1. Peristaltic pumping system
2. Everted into the borehole
3. 4–10″ hole diameter
4. May be best used for many ACT measurements with slender tubing connections to many ports

10.5.9.3 CHS Systems (In Use Since 2018)

1. Both CHS systems are drop in place systems and therefore not suited to angle holes or cavernous holes
2. Less expensive due to material, labor and installation costs
3. Limited in depth usually to 200 ft. (300 ft. in cased boreholes with special construction)
4. Peristaltic pumping (CHS) and limited to shallower water tables

5. Positive displacement pumping (pdCHS) for any water table depths
6. Simultaneous purging and sampling (pdCHS only)
7. Easily installed with minimum training of the customer or driller
8. Installed in cased holes to 300 ft. or uncased holes to <200 ft.
9. Easily installed in 2″ casing (no. of ports depends on which system and hole diameter)
10. ACT head history measurements

All FLUTe MLS systems are well suited for cross-hole measurements as described in Section 10.5.9.4, using the ACT head measurements on each port. All systems (SWF excepted) allow manual water level measurements at each port. All systems allow head histories at each port. WF and pdCHS systems allow manual water table measurements at the same time as recording head histories.

10.5.9.4   Mapping Cross-Hole Connection with FLUTe MLS Systems

10.5.9.4.1   Background

Many customers use the "FLUTe sequence" of NAPL/FACT, T profile, and RHP profile to measure the borehole properties of contaminant distribution, conductivity, and head. However, those measurements are near the borehole and are not easily related to the larger scale flows between boreholes. The borehole measurements are used to design the MLS system sampling intervals for water samples and head measurements. Cross-hole connection information is more useful for designing remediation injections and estimates of contaminant migration.

A method to assess cross-hole connection has been reported by Persaud et al (2018). However, in that study, the method used the driving pressure of FLUTe blank liner installations in boreholes and head measurements of head changes in nearby open holes. As expected, the deliberate change in head in one borehole was seen as a change in head in nearby boreholes. This is the basis of pumping tests. The same general effect was observed while performing a RHP measurement in New Bedford, MA. The typical RHP measurement of borehole equilibration with the flexible liner inverted to different elevations produces a head history like that in Figure 10.68 as measured by a transducer in the bottom of the borehole

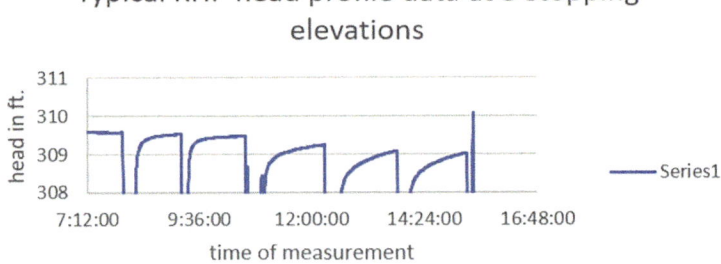

FIGURE 10.68   As the inverted liner is stopped at higher elevation in the borehole, the head beneath the liner is allowed to equilibrate. The data is used with the T profile to assess vertical formation head distribution.

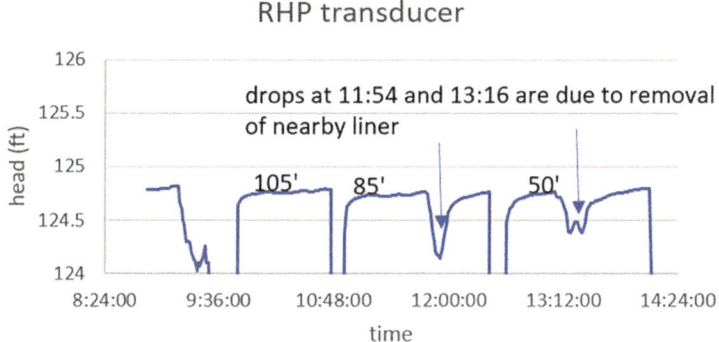

FIGURE 10.69   A RHP head measurement impacted by the removal of a blank liner from a relatively distant borehole. It was surprising that the distant borehole was so well connected as to produce such a large drop in the RHP data.

(see Section 10.4 for the method). Such measurements are usually done with no nearby open holes or activities like liner installation or removal nearby to perturb the data collection with cross-hole connections. The same precaution applies for the T profile measurement. However, the RHP head history in Figure 10.69 shows the effect of a liner removal being performed in a nearby hole. But the nearby hole at a distance 305 ft. was not considered very nearby. The customer at this site said that effect must be due to a scarp known to exist at the site. None of this is new information and cross-hole connections are the subject of many pumping tests with monitoring of nearby open boreholes. Such open-hole measurements do not identify individual flowing fractures and are subject to the connection of the open hole with many fractures in the same hole as well as borehole storage effects. However, it is reasoned that it would be easy to alter the installation of the typical FLUTe MLS systems described in this chapter to gain even better information on the flow paths between boreholes.

   Head measurements in open holes do not identify the cross-connection effect except as an average of individual fractures intersected by the borehole. Also, the borehole storage masks the head change in an open hole. Other nearby open holes with similar measurements of head change allow short cuts to other fracture systems not normally connected by the open holes. In summary, it would be better to have all boreholes sealed with one borehole as the source, or sink, when assessing borehole connections. Furthermore, it would be best to have individual fractures monitored in boreholes at the same time as an isolated distant source or sink is driving the flow field, in order to assess cross-connection to individual fractures in nearby boreholes. That was attempted in a test at the NAWC site in NJ by the USGS using several packers in a borehole and tomographic techniques, and other methods, to study cross-connection in a very small volume of rock surrounded by a 30 ft. circle of boreholes with a central borehole (Tiedeman et al, 2015). However, such a small volume assessment of interconnections is not useful for a larger site. The application of tomographic assessment techniques to the data set from the proposed measurements could be very useful.

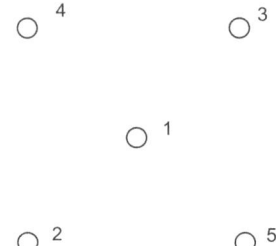

FIGURE 10.70   Example of simple borehole geometry to illustrate the cross-hole connection measurement concept.

10.5.9.4.2   A Simple Cross-Hole FLUTe MLS Measurement

The typical FLUTe sequence described in Section 13.1 leads to the design of MLS systems to be manufactured for installations of the several MLS systems at a site. The FLUTe sequence often ends with a T profile of each borehole with the transducer used, left at the bottom of the hole. When an MLS system is to be installed, the blank liner is removed, the transducer recovered and the MLS installed. Two simple changes can make the procedure a very effective cross-hole mapping of flow paths with very high spatial resolution.

As a simple example, assume a regular (albeit unrealistic) borehole pattern (Figure 10.70) for a site characterization after the FLUTe sequence is completed with sealing liners in all the holes and transducers at the bottom of each hole. The MLS systems are shipped to the site and ready to be installed by removing the blank liners and installing the MLS systems. A first difference from the normal procedure is that sufficient ACT systems (Section 9.5) can be rented to instrument most of the ports of the MLS systems. A second difference is that a FLUTe linear capstan is on hand to perform the blank liner removals. The linear capstan (Figure 10.71) is often used to quickly remove blank liners. The linear capstan has the ability to monitor and record the tension applied to the liner, control the speed of removal, and record the position of the liner in the borehole

The cross-hole measurement is performed as follows:

The first blank liner is removed from hole no. 1 (it could be any of the several holes). The MLS is installed and each port is fitted with an ACT to monitor the head history at each port. The second hole blank liner is removed with a special procedure. The downhole transducer is set to record the head in the borehole on 5 second intervals, for example. The liner capstan is connected to the tether to remove the liner with a constant tension on the tether and on the inverting liner. That constant tension causes the liner to invert and also drops the head beneath the liner by an amount $\Delta P = 2T/A$, where T is the tension applied to the liner and A is the cross-section of the borehole. As the liner inverts, it sequentially uncovers flow paths which, with a constant tension on the liner, also maintains the same ΔP beneath the liner. The drawdown of ΔP is applied to each new fracture abruptly when it is uncovered. The linear capstan is instrumented for measuring and recording the liner position in the borehole and the tension on the liner/tether. The speed is variable and can

FIGURE 10.71 The linear capstan removing blank liner from borehole. Tension on the liner is monitored and the speed is controlled using a variable speed DC motor. The liner depth can also be monitored/recorded at the first roller.

be controlled to provide a constant tension as the liner removal rate increases with the added flow into the borehole of each fracture uncovered. Eventually, the removal rate needed for a constant tension can exceed the capstan maximum rate and the ΔP beneath the inverting liner will decrease as the tension decreases.

The recording transducer in the borehole will monitor the head beneath the liner in time and it can be compared to the recording of the linear capstan tension and liner depth in time. This procedure for removal of the blank liner provides a draw down interval of increasing length and known drawdown as a function of the fractures being unsealed by the inverting liner. Discrete fracture locations would be known from the T profile and tele-viewer images commonly available. The ACTs at each port in the first MLS will provide the head change in time as the blank liner is being withdrawn in the second borehole. The ACT head data, the borehole draw down in the second borehole, the liner depth and individual fractures uncovered with liner depth can all be plotted on one time scale graph.

Then the second MLS is installed and the ports instrumented with ACTs and the third borehole liner is withdrawn with the linear capstan and transducer recordings. Now both the first MLS and the second MLS are recording the effect at each port of the blank liner removal in the third hole.

The result is a head history at each MLS related to the timing of the blank liner withdrawals and the direction of the liner removal location from each MLS already in place. The data set has vertical resolution, directional resolution, and head history resolution. And only one borehole is not sealed when the data is collected from all the MLS systems. This should be far more than is normally available from a pumping test. There are no borehole storage effects or open borehole flow effects. One advantage of the method is that the drawdown can be controlled by the tension applied. In a tight borehole, the drawdown can be as high as the liner can withstand or a 50–70 ft. of head reduction, or more, depending on the hole diameter and liner strength. The challenge is to use the wealth of information in the data sets.

An enhancement and simplification of the procedure would be to do a steady-state pumping test in each borehole after each blank liner is removed and before each MLS is installed. The steady-state data may be easier to analyze, but without the variable fracture information. The steady-state data would still show the response at each spacer. In theory, one can also deduce the vertical connections in the formation because of the sequence of fracture exposures with the liner removal. The steady-state pumping may provide better detection at distant boreholes.

The effect of the 300-ft. distant borehole liner removal on the head measured with the downhole transducer in Figure 10.69 shows that the cross-connection head changes would be easy to measure with the ACT systems. Since the transducers are at the surface, they can be returned after an economical rental period. The downhole transducers are used with the T profile. The downhole transducers are another available measurement set as the single blank liners is removed until all the blanks have been removed.

10.5.9.4.3 In Summary

The sequence above is similar to the normal installation of several FLUTe MLS systems at a site. The only change is the addition of the ACT systems and the recording of the linear capstan data as each blank liner is removed according to a prescribed procedure to make the data interpretation easy. The adaptation of the normal FLUTe MLS installation does not involve an extraordinary cost for the extra data gained. With the best forethought, the location of the boreholes might be optimized to gain the best data on lateral and vertical flow paths. One can use the T profile and fracture aperture estimates already available at a site to compliment the cross-hole path identification. Any information on geologic structure should also be helpful, including fracture orientations and other information from geophysical measurements in the individual boreholes.

The ultimate advantage would be in the design of the site remediation with expectations on where the flows are likely to propagate. The MLS systems can be used to monitor the remediation and transparent liners could be very useful in evaluating injected tracers or remediation fluids.

10.5.10   COMPARISON OF FLUTe MLS SYSTEMS WITH OTHER MLS SYSTEMS

The summary below is a list of the characteristics of the FLUTe MLS systems. A primary difference from other MLSs is the continuous seal of the borehole with the flexible liner. This continuous seal is a significant advantage over the use of straddle packers for isolation of the several intervals to be sampled. The large interior volume of the liner allows the use of relatively larger pump tubes and transducers. Other MLSs are often constrained to the interior of a 2″ ID casing. The ability to remove the WF MLS for warranty repairs, or at the end of use, is another significant advantage. Because of the liner seal of the borehole, there is no need for the expense, or potential sample contamination, of using an annular grout seal outside of a casing in an open stable borehole. Because the FLUTe liner system is pressure tested in the factory fully assembled before shipping, there is less concern than with any field assembly in inclement weather or contamination of the system in the assembly in the outdoor environment.

Vadose gas sampling ports can be added to a WF above the water table in the formation. Those are described in Chapter 11. Following is a general comparison of FLUTes with other systems.

FLUTe MLS systems:
- The flexible liner MLS seals the entire borehole except for the defined sampling intervals. No sealing grouts are used.
- The sampling intervals are defined by an exterior spacer with a port in the liner which draws the sample directly from the formation.
- All systems have been installed in boreholes with or without artesian conditions.
- All ports can be purged simultaneously (CHS and SWF used special equipment).
- Purge volumes are typically 3–4 gal., but much larger volumes are easily obtained.
- All systems are fabricated and leak checked in the factory fully assembled.
- All systems are usually installed in less than one day.
- The WF systems are installed by FLUTe-trained personnel.
- All systems are removable except for liners rarely filled with grout to withstand extreme pressure differentials in the formation or to withstand high-pressure remedial injections.
- Head measurements are performed manually or with recording transducers.
- Vadose pore gas sampling intervals can be added above the water table.

Other MLS systems in general:
It is usually considered impolite to criticize alternative methods unless it can be done objectively. It is the author's intent to be objective/factual. Shapiro (2002) describes

the difficulties in obtaining a representative sample in an open uncased borehole and also in a cased hole with multiple screens and even in a long single screen in a cased borehole. The problem he describes is the lack of isolation of the individual flow zones with different levels of contamination and different conductivities. The FLUTe MLS systems described above address that problem. Non-FLUTe systems in general suffer from some of these difficulties:

- Require backfill of the borehole (e.g., sand, grout, or bentonite)
- Or, use packer seals which seal only a small part of the otherwise open hole
- Cannot practically be purged simultaneously, because the sampling system tubes only extend to the port and are not of the same length
- Cannot be purged at all (e.g., Westbays)
- Are not removable and packers can get stuck in open holes
- Are not deployed in a single borehole (e.g., cluster wells)
- Have limited numbers of ports (e.g., nested wells)
- Cannot sample and monitor head in the same port at the same time (may be possible in some)
- Are more expensive or less expensive depending on the system and number of ports
- Cannot produce a large purge or sample (e.g., 1–5 gal.) in a reasonable time
- Require field construction in whatever weather exists
- Require a large area at the well for components or assembly
- Are expensive to ship (large diameter coil or tubing bundle or many pipe sections)
- Require large equipment (e.g., drill rig or crane truck) for installation (CMT excepted)

Obviously, these points do not fit all systems, but should be considered in a selection. Those systems used for measurements at different elevations with no seals at all in open holes or open casings are not considered reliable or to be actual MLS systems (e.g., diffusion bags, LFS, suspended carbon blocks, …) even though positioned at different elevations.

10.5.11 The DEIL

10.5.11.1 The Purpose and Design of the DEIL (Discrete Extraction and Injection Liner)

10.5.11.1.1 Background

In response to a question of whether FLUTe can seal a borehole and extract or inject in discrete intervals of the borehole, the DEIL design was developed. Because extraction or injection procedures usually use larger flow rates than water sampling procedures and more viscous fluids than water, they require larger diameter tubing than our usual designs. Such large diameter tubing cannot be everted in the usual FLUTe installation manner. Check valves cannot be used in the injection mode. Therefore, we developed a new design that allows one to lower the liner down the borehole with the larger tubing interior to the liner and with discrete intervals defined by exterior spacers where the extraction or injection would occur with the rest of the borehole sealed by the flexible liner. The liner is then dilated with a water fill or a different fill such as sand or grout for injection sites. Removal of the sand fill material allows the liner to be removed.

10.5.11.2 The Geometry of the DEIL Liner

The geometry of the FLUTe DEIL is shown in Figure 10.72 for only two extraction intervals. The large tube from the port extends to the bottom of the borehole where it connects with the bottom end of an even larger diameter tube to the surface. There are a variety of extraction options. If the water table is shallow, some surface pumps

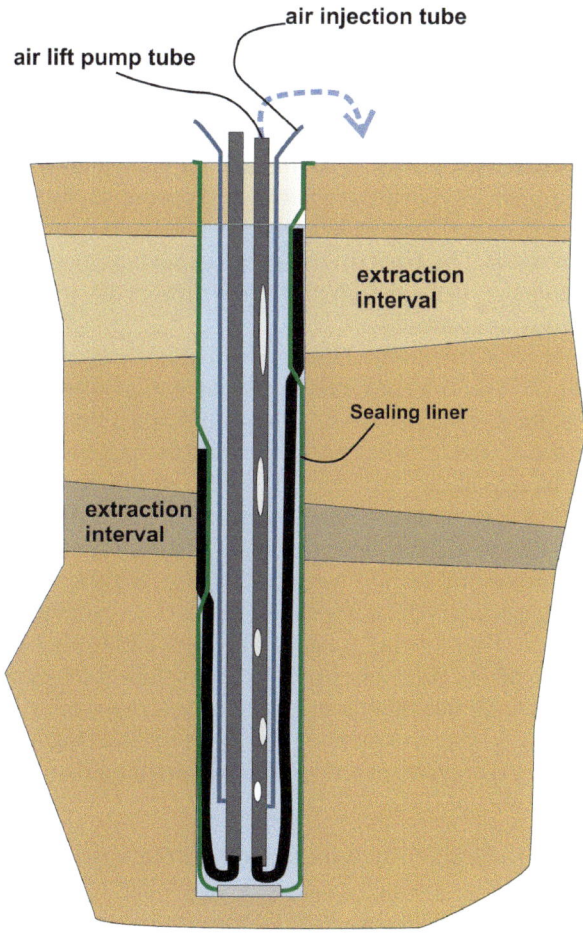

FIGURE 10.72 The DEIL geometry showing the large tube from the extraction interval and the larger air lift pump tube. Air injected near the bottom of the pump tube reduces the density of the water column and causes it to be driven to the surface. Water table depth at the port should be less than half the liner depth.

can draw the formation fluids to the surface. This is not an option for deeper water tables. Another attractive pumping method for extractions is an air lift pumping as shown in Figure 10.72. For an air lift pump method, an air injection line connects as shown to the bottom of the larger air lift pump tube. The air injection, at a surprising small rate, causes the large tube to act as an air lift pump carrying the extracted water to the surface where it can be treated as desired. In one possible application, the outlet can be directed into an activated carbon filter to remove contaminants.

The air lift pump is essentially gravity driven, and it provides a gallon per minute flow from 65 ft. at an air flow rate of ~1 standard cubic foot of air per hour. That is a rate that can be supplied by a relatively small compressor as used for common work like spray painting, although at far less than the air pump capacity of those inexpensive compressors.

The same tubing can be used to inject remediation fluids into the spacer intervals. This method does not require more than an injection pump or gravity driven system at the surface to inject through the illustrated tubes into discrete intervals. However, the injection pressures are often higher than the excess head inside the liner which provides the liner seal. In that case, the injected fluid would collapse the liner, violating the seal. Two attractive options are to fill the liner with sand or with a weak grout. The sand fill of a liner has been removed with the same air lift pumping method applied to the inside of the liner which then allows the liner removal by pumping the liner empty of water. The grout fill can only be removed by drilling out the liner. There are no components in the design that would prevent the drilling of the liner out of the hole if filled with grout.

Liners were not normally lowered into open boreholes because of the concern of abrasion of the liner. Therefore, a protective sheath is included in the DEIL design as is described for the CHS systems which are also lowered into open boreholes. The sheath is removed after the liner has been emplaced in open uncased holes. In PVC-cased holes with screens at each extraction level, the sheath is not needed. In open or cased holes, the intervals of interest should first be well developed.

Since there are no valves in the design of Figure 10.72, it is easy to collect samples at the adjacent intervals not being pumped. It is also possible to monitor the head history in the adjacent intervals as fluids are injected or extracted to determine to what extent the intervals are being treated. The head monitoring can be done with ACT systems which do not need tubes so large as to allow the insertion of transducers. During an injection procedure at one or more intervals, the adjacent tubes can be sealed with the ACT system to monitor injection pressures higher than the surface elevation. The seal also prevents overflow of the injected fluids.

10.5.11.3   The DEIL Design Advantages and Limitations
Advantages
- The liner can usually be installed or removed by the customer.
- The entire borehole is sealed except for targeted intervals.
- The pumping rate is very efficient.
- The cost of the pumping system is not much more than the tubing to the surface and less than most electric downhole pumps.

- One is not treating water that is free of contaminants from uncontaminated intervals.
- A water sample is easily drawn from the bottom end of the pump tube to assess contaminant levels unaffected by the aeration of the pumping mechanism. Head histories can be monitored at intervals not being pumped.
- A special design can include slender tubing for monitoring more elevations between the high flow zones.

Limitations
- The number of intervals available depends on hole diameter and tubing diameter.
- The air lift pumping mechanism is less effective if the water table is relatively deep compared to the borehole depth. A different pumping system is needed for deep water tables. (i.e., if WT > borehole depth/2).
- The large tube diameters may require a protective sheath as used with the CHS systems to avoid abrasion of the liner with the more stiff tubing.
- As with any large diameter tubing system, the installation in very low temperatures is more difficult due to curl of the tubing.

10.5.12 OTHER SPECIAL CHS SYSTEMS

10.5.12.1 Many Head Measurements in a CHS
The drop in design allows the use of tubing as small as 3/16″ OD to monitor formation head changes. Such small tubing allows far more monitoring intervals to be included in a single liner. That is because the ACT design can be used at the surface to monitor each slender tube. As many as 20–30 monitoring ports are possible. The use of so many tubes would make the eversion of a liner difficult, but the drop in design of the CHS allows so many tubes mainly limited by the borehole or casing diameter.

10.5.12.2 Hybrid pdCHS for Deep Boreholes
In consideration of how one might pump a very deep pdCHS system with a deep water table, a simple design was developed. It is shown in Figure 10.73. It is a geometry similar to a WF but does not require the tubing to be everted with the liner since the system is dropped into place and filled with water like the CHS systems. Because the pumping system only expels the water in the upper portion of the pair of tubes above the "U", there is no need for the high pressure of the complete expulsion of the pump tube volume as is the case for the more typical operation of the pdCHS. The standard pdCHS system uses a purge pressure greater than the depth of the borehole. Unlike simply short stroking the pdCHS, to obtain a sample or purge at lower driving pressure, the water of the deep pdCHS does not oscillate in the pdCHS lower portion of the tubing between pressure applications. Rather there is a continuous but interrupted flow from the port to the surface. The water in the tubing in the lower portion of the system is displaced completely by the inflow from the formation between pressurizations. The valved portion of the tube is tied to the supporting tether to assure the valve is vertical as required to function as a normally open valve.

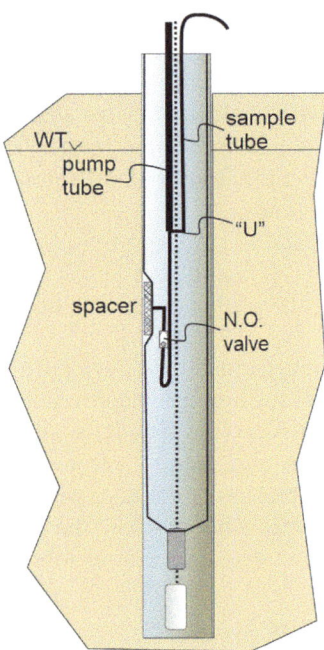

FIGURE 10.73   Pump geometry for deep ports of pdCHS system. Applying gas pressure to pump tube closes valve and expels water in pump and sample tube only. Recharge fills pump and sample tubes again.

The pump tube and sample tube pairs for each port can be of the same length which then allows the simultaneous purging of equal volumes from all ports with equal draw down at each port. The standard FLUTe manifold can be used. The length and diameter of the pump tube defines the volume of water produced with each pressurization. This design has not yet been fielded, but can be lowered into a cased hole of 300 ft. with a water table and ports well above the bottom the junction of the pump and sample tubes, called the "U". The depth of the U beneath the water table defines the volume of a purge stroke. The tube diameters can be optimized to improve flow rates and the volume produced with each pressurization of the system. This is especially attractive for monitoring tracer arrivals as described in Section 10.5.1.5.

10.6   STRETCH OF LINERS AS IMPORTANT TO FLUTE METHODS

Stretch is elongation of the liner from its original length lying on the table in the fabrication facility. Numerous FLUTe methods are measurements of borehole characteristics that are relevant to the depth in the borehole. The liner depth in the borehole is used as the reference for the depth of those measurements. For that reason, the subject of stretch of the liner during emplacement is addressed here for each of the several FLUTe methods that are described earlier in this chapter. It is useful to compare the depths of the FLUTe measurements with the depths measured using other methods.

In Chapter 3 is described the end load on the liner due to the pressure difference between the fluid inside the liner and the fluid outside. That leads to a tension in the liner Ti. Other factors can cause tension in the liner such as when the liner is inverted from a borehole with low transmissivity causing the fluid pressure outside the liner to be greatly reduced. When the liner is subjected to any tension, it tends to elongate. The liner is relatively elastic, as most are, when the tension is reduced the liner will contract. The cover materials used on blank liners described in Section 10.2 can also be elongated by tension during installation and during the process of being pulled from inside an inverted liner after removal from the borehole. Since the cover in the inverted liner is compressed by the liner interior pressure Pl, the cover moves intimately with the liner.

It is important to recognize that liners stretch and that the elastic behavior is different for different liner materials. The elongation of a liner is nearly directly proportional to Ti. Ti is the inverted liner tension which depends on the differential pressure ΔP as explained in Chapter 3. However, the tension above the EP at the surface in a vertical installation is also dependent on the drag and the hanging weight of the liner system including the tubing in sleeves in the liner. The drag may be so high as to reduce the tension at the surface to near zero or the liner so heavy as to greatly increase the tension at the surface to more than the towing force Ti at the EP. However, most of the stretch of the liner during installation is due only to the tension at the EP because once the liner is everted against the hole wall the friction on the borehole wall prevents further elongation or contraction of the liner unless the ΔP drops and allows the liner to slide on the borehole wall. There are reasons to avoid that drop in head in the liner during an installation as described later.

The amount of elongation of the liner depends on the length of the liner the diameter of the liner, the driving pressure, ΔP, and the stretch of the liner occurs both in dilation or elongation. The liner has different properties depending on the yarn and whether the strain of the liner is in the circumference or longitudinal. That is whether the tension is in the direction of the "warp" or the "weft", the weave perpendicular to the warp. The weft is the primary resistance to the radial dilation of the liner. When supported in the borehole, the dilation is near zero except in breakouts. As described in Chapter 3, the stress in the longitudinal direction, the warp direction, is less than the hoop stress. The stretch of the liner in length is the primary feature of interest. Dilation can allow the liner to seal the borehole better, but liners are deliberately sized to be slightly larger in diameter than the nominal borehole diameter to better seal small enlargements. There is also less hoop stress in enlargements if the liner is larger than the borehole diameter. When unsupported, the dilation of the liner can be quite large as the ΔP approaches the burst pressure (e.g., a ballistic nylon may dilate to nearly twice its diameter before burst). However, most nylon liner fabrics used do not dilate so much at normal driving pressures.

In a typical installation, the liner is stretched an amount ΔL by the end load at the EP according to the relationship: $\Delta L = E\ L\ Ti/C$, where E is a measured coefficient for a particular liner material, L is the length of the liner under load, T is the tension, and C is the circumference. However, in Chapter 3, it is shown that Ti depends on the cross-section of the liner or Ti = ΔP A/2 = ΔP π r²/2, where r is the borehole radius. C = πD, where D = 2r. Therefore, ΔL = E L ΔP r/4 gives less stretch for smaller diameters. If the situation is a small borehole with a larger diameter liner, the Ti depends on the borehole diameter and C is the circumference of the liner which may be more than the hole diameter.

From the expression for ΔL, it is easy to determine E = ΔL × C/(Ti L) for a variety of materials by measurement of ΔL, T, and C in a pull test of a known length of liner at various tensions, T. Those measurements have been done for the range of materials used in FLUTe liners. It is noteworthy that the liners are always welded into tubular form with the warp of the fabric in the direction of the liner axis since the liner rolled bolt stock is long in that direction.

Stretch of the liners is not important for sealing blank liners. However, the stretch of the cover on a blank liner is important for the NAPL FLUTe and FACT covers in order to assign the stains and carbon analysis results as accurately as possible according to the depth when in the borehole. The cover can be marked for the original depths when fabricated on the blank liner and the liner is marked for location in the borehole assuming no stretch of the liner. However, the liner does undergo some stretch during emplacement due to the tension on the liner at the EP when driven by a particular ΔP mainly due to the excess head in the liner whether water or mud. For shallow boreholes of less the 50 ft. depth, the stretch is not significant as less than half a foot depending on the driving head in the liner. But for a 200-ft. liner and 20 ft. of driving head, the stretch is 2.3 ft. for the single coat 400-d. liner. That is 1.1% of the liner length. It has been noted that the location of fractures based upon the FACT profile are somewhat deeper than seen on the optical tele-viewer log. At 10 ft. of driving head, the stretch is half as much. A total of 10–20 ft. of excess heads in blank liners are normal. With deep water tables and wet film adhesion drag, the driving head may need to be higher, causing more stretch. An example of an extreme case of stretch is for a 900-ft. liner with a 280-ft. water table depth and 20 ft. of excess head. The stretch accumulated at the 900 ft. depth is ~13 ft. for a ballistic nylon liner which is more elastic than other liner materials. For that much stretch, a 10 ft. spacer at 900 ft. on the liner without any stretch adjustment would not even be everted if the liner design and emplacement were not adjusted to accommodate the liner stretch. The stretch calculation model described above has been compared to the actual stretch observed in some liner emplacements with good agreement.

It is useful to determine the actual stretch that has occurred in liners to estimate the correction needed for the depths measured next to the NAPL FLUTe or FACT cover (Section 10.2) with a tape measure extended adjacent to the cover. That is possible by measuring the actual depth of the EP in a borehole with a tag line and recording the offset of the nearest liner EP mark from the ground surface. From that measurement, the difference between the current liner depth in place and the amount of liner everted into the borehole can be determined. That depth must be measured with the current excess head similar to the emplacement excess head or with the liner supported by the end of the borehole (rarely the case). The ratio of the liner current depth to the length of unstretched liner everted is the fractional increase that has occurred. It is necessary to divide the EP mark distance from the GS by two when comparing the unstretched liner length to the tagged depth of the EP.

   The cover material can be stretched by both the liner stretch during the emplacement and the liner removal and also by the tension during the cover removal from the inverted liner. If the cover removal from the liner stretched the cover, the stains and the FACT sample depths after removal would be slightly greater than the actual in situ depths. The tape measure distance from the GS mark to the deepest depth mark on the cover is another measure of how much the cover has been stretched. The mark on the cover should be less than the tape measurement value to the mark from the GS mark. The ratio is the stretch that has occurred in the cover. However, the stains on the cover and the depth of the FACT carbon samples measured with the tape should be the actual depths in the hole unless the cover has been stretched by the removal tensions. However, the removal from the inverted liner tensions is usually not as high as the end load on the EP during the installation and removal from the borehole. Measurements have shown that the cover material is not as elastic as the liner and only recovers some of the stretch length change when not under load.

   Since some stretch is unavoidable with liner installations, FLUTe has developed several methods for compensation or correction of depths important to the measurement. For T profiles, the actual depth of the liner is measured frequently and recorded at the exact time during the measurement. In the data reduction, those depths are compared to those calculated from the encoder measurement. They are often quite close, but in very cold weather, the effective diameter of the liner on the encoder roller is different than on a hot day. For that reason, the encoder roller diameter is adjusted in the calculation to match the measured liner depths during the T profile. Comparisons with the tele-viewer images of the geophysics measurements have shown the resulting T profile depths to be very comparable (i.e., within a few inches typically). For other depth-dependent measurements such as NAPL/FACTs, the depths are not determined from a liner measurement but from a comparison with a tape measure at the surface. Some offset has been noted, but not as large as the estimated liner possible stretch. One reason is that blank liners are not installed with excess heads in the liner much more than the minimum needed to evert the liner which results in minimum stretch. Also, unless used to emplaced a NAPL/FACT there is no concern about the exact location of a blank liner in the borehole to seal the borehole since the liner is continuous.

   For the emplacement of multilevel water sampling systems, there is a concern about the exact location of the liner in the borehole. For shallow boreholes, the concern is not that great, because the spacers are long enough to straddle the interval of interest such as a screen or a fracture zone. However, for deep boreholes with deep water tables, several adjustments are made to assure the sampling intervals are properly located. The first precaution is to calculate the stretch expected based on the measured properties of the liner material and the planned installation excess head to be used. In everting to the water table, the head in the liner is measured as it is added to the liner and maintained. Therefore, the driving pressure is essentially constant and based on comparisons with known installation depths, the calculations are very accurate. Fortunately, only the excess head affects the liner stretch and not the drag term, the tension at the surface or the weight of the liner. That stretch in travel to the water table is offset by a shift of the liner at the surface at the start of the installation. Therefore, the uppermost sample interval depths are not significantly affected by the stretch when the liner is fabricated according to the specifications of the order form. For those sampling intervals at depths where accumulated stretch is to be expected, the spacers are shifted upward by the calculated stretch for a known excess head. That head used in the calculation is the head used in the installation. This is most important for installations greater than several hundred feet. Another accommodation is to use longer sample spacers extended above the planned depths to assure that a screened interval, for example, will be completely or sufficiently covered by the spacer as emplaced. Fortunately, the sand pack high conductivity behind a screen makes even a small overlap of the spacer with a screen a good connection with the entire sand packed interval. These adjustments to the MLS construction are done for deep boreholes. The CHS systems are lowered into position and the stretch of the liner is insignificant and the weight on the bottom end of the liner is supported by the interior tether which has very little stretch.

It is very fortunate that comparison of MLS liners removed from boreholes has shown that screen impressions on the liner have confirmed the ability to reasonably predict the liner shift due to stretch.

Liners removed by inversion from very low transmissive holes are often subjected to high tension on the liner with significant stretch of the liner. Once removed the liners contract to the initial length, but a cover on the liner may be stretched. For that reason, the use of a long vent tube described in Section 3.3.3 reduces the tension of removal from a tight borehole of low transmissivity.

11 FLUTe Vadose Multi-Level Measurements

Herein are described a variety of vadose zone measurements of pore fluids with both gaseous and liquid sampling methods.

11.1 PORE GAS SAMPLING

11.1.1 THE GEOMETRY

The in-place geometry of the vadose sampling system is shown in Figure 11.1. The installation is completely above the water table. The liner can be everted into the borehole with a variety of driving fluids from air, to water or mud. The usual installation is done with water to the bottom of the borehole and the liner is then dilated with air to urge the liner against the borehole to isolate each sampling interval. An airtight seal is required at the top of the liner if the liner is to be dilated with air. In most cases, a small air pump is used to maintain the liner interior pressure in combination with a pressure relief valve set to the desired pressure. The air pump is often powered by a solar panel for remote locations. For short boreholes, it is easy to use water as the fill and pressurizing fluid. A water fill is reasonable for boreholes up to 40 ft. in depth in the vadose zone. Vadose liner systems have also been installed in driven casing (e.g., in sonic casing) in unstable media. In those situations, the interior of the liner is filled with dry sand and the liner system is not removable by inversion until the sand is removed with a method devised by FLUTe with a vacuum style system with the liner inflated. If the sand is wet, it is difficult to remove.

The number of gas sampling intervals is only limited by the space for tubing in the interior sleeves of the liner. A total of 10–15 ports are common. Vadose sampling ports are sometimes included above the water table on a Water FLUTe system to combine both water and pore gas sampling in open stable boreholes. Pore gas sampling systems have been installed to over 1700 ft. in Nevada.

11.1.2 THE GAS SAMPLING PROCEDURE

A representative pore gas sample can be drawn to the surface after the interstitial volume of the spacer and the tube volume have been purged. A typical design for a vacuum pump and gas sample collection volume are shown in Figure 11.2. The system includes a flow meter to measure the flow rate in order to determine when a sufficient gas volume has been purged. A simple venturi vacuum pump provides a steady gas flow that greatly aids the flow meter reading versus the positive displacement vacuum pumps which can produce pulsating flows. Once the required volume of gas has been purged, the gas sample is drawn into the gas collection volume. That

DOI: 10.1201/9781003268376-11

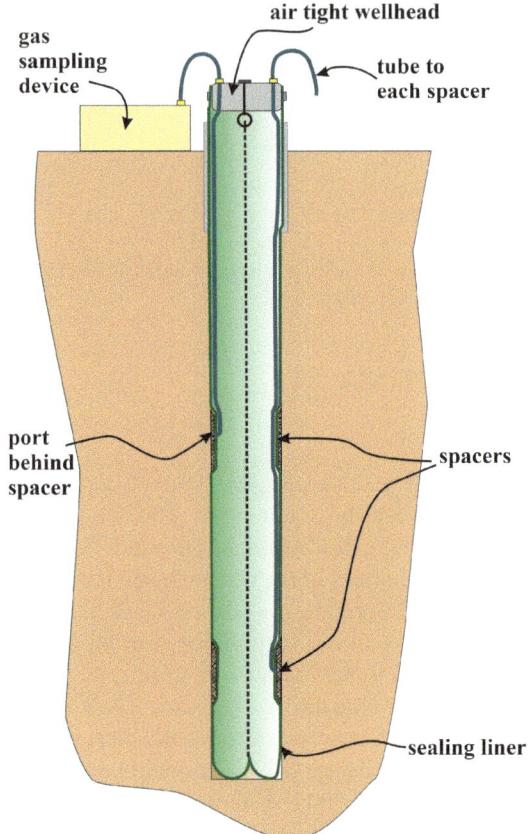

FIGURE 11.1   The vadose gas sampling liner system.

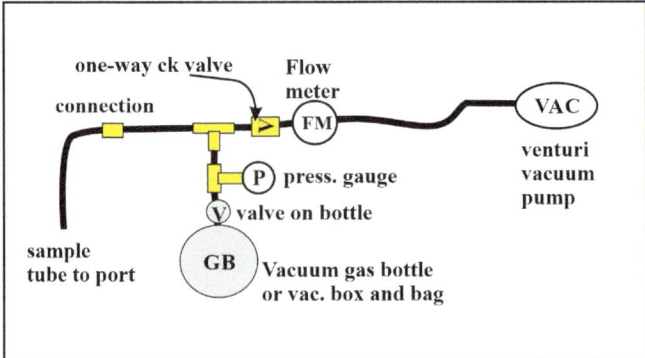

FIGURE 11.2   Pore gas sampling system. Vacuum pump purges tubing and sample is collected by opening the valve to the gas bottle. Flow meter allows determination of sufficient purge volume. The check valve prevents backflow when vacuum pump is stopped before gas collection.

collection volume can be either a vacuum bottle or a flexible bag housed in a vacuum box. An alternative approach can use a peristaltic pump for extraction of the gas from the formation and for filling a *Tedlar* sample bag. The flow meter is still desirable, but a balloon volume can also be used as a flow measurement determination. The tubing volume is usually small.

A novel gas sampling procedure was used to detect tritiated water (water containing tritium). The procedure was to extract water vapor from the pore space at discrete intervals and flow the gas with vapor through a silica gel container to collect the water vapor in the gas flow. The silica gel was then analyzed for tritium.

11.2   PORE LIQUID SAMPLING IN THE VADOSE ZONE

11.2.1   THE USE

The pore water sampling liner system is installed in the same manner as the pore gas sampling system. Pore water can be obtained from the vadose zone if the capillary tension is not too high (i.e., the pore space has relatively high water saturation). The samples are collected into absorbent covers on a liner everted into place. Keller et al (1993) describe the wicking process and the characteristics of a variety of absorbent materials for a range of vadose conditions. Sometimes a wire pair with two contacts is positioned behind the absorber to monitor the increase in saturation during the absorption process. When the absorber has gained as much water as possible (i.e., the capillary tension of the absorber equals that of the formation), the electrical resistance between the two contacts stops decreasing. The resistance measurement should not be done with a DC circuit, but rather with an AC voltage source to prevent charge accumulation in the measurement system.

11.2.2   THE GEOMETRY OF PORE LIQUID SAMPLING

The geometry of the absorber covered liner, Figure 11.3, shows the absorber geometry used for the collection of pore water samples. The liners are usually everted into open stable boreholes. In some situations, the vadose liner with absorbers is everted into a borehole already containing a pore gas sampling liner system. That method is called a "Duet". The Duet is described in Section 13.6. The smaller everting absorber collection system is driven with a higher pressure than the original liner such that the second liner displaces the first liner. When in place, the driving pressure (usually air pressure) of the second liner is dropped, allowing the absorber carrier liner to collapse and restore the borehole seal of the first liner and to then wick a sample from the borehole wall. The second liner is then removed by re-inflating the liner and inverting it from the borehole, allowing the original liner to dilate back against the borehole wall.

11.2.3   THE SAMPLING PROCEDURE FOR PORE WATER

When the absorber has been left in place for a sufficient time period, the carrier liner is inverted from the borehole. The carrier liner is then everted horizontally into a

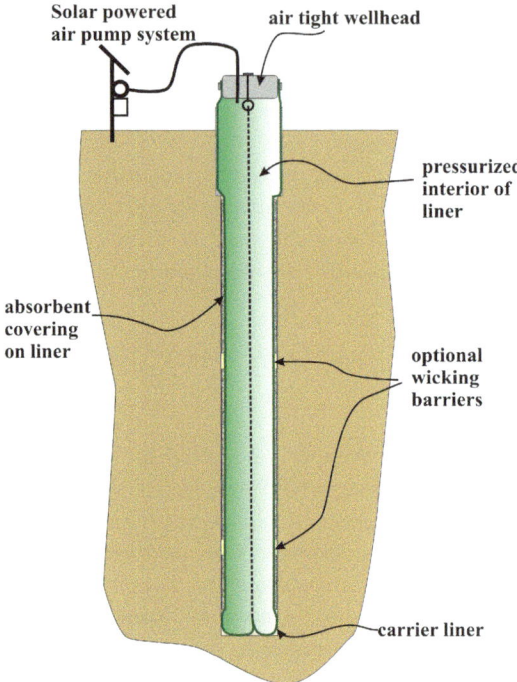

FIGURE 11.3   Pore water absorber system on carrier liner. Covering can be instrumented sections or continuous. Absorbers are recovered after liner is inverted from the borehole. Typical solar powered air pump shown.

tubular poly-film in the same manner as it was everted into the borehole. A continuous absorbent covering can be disconnected from the bottom of the carrier liner and the carrier liner is then inverted from the interior of the absorbent covering. The collapsed absorber can then be sealed in segments by elastic bands on the poly-cover. The poly-cover is to reduce evaporation before the absorber is analyzed for the distribution of contaminants. The absorbent covering often contains wicking barriers to prevent longitudinal migration of the absorbed liquid in the absorbent covering to improve spatial resolution. An extensive investigation of the wicking capabilities and capillary tension curves versus saturation was done by Keller et al (1993) to assess the potential measurement of soil capillary tension with absorbers on flexible liners.

Other kinds of attachments to liners include electrical contacts on the outside of the liner connected with wires to the surface for tomographic resistance measurements. Various instruments have been lowered down the borehole and isolated by everting a liner over them. If the liner seal is important, the suspension of several instruments can be done on thin steel cables.

11.2.3.1   Other FLUTe Measurements in the Vadose Zone

Both NAPL FLUTes and FACT systems can be installed in the vadose zone for mapping the contaminants. Air pressure is also used to seal the liner against the hole wall. The protection of logging tools from contamination or borehole collapse is also reasonable methods using blank liners in the vadose zone. A simple air lock design at the wellhead can allow the logging tool installation while maintaining the air pressure in the liner.

11.2.3.2 In Summary

- Both pore gas and pore liquid systems are available using flexible liners for pore fluid collection in the vadose zone.
- The pore gas sampling liners are usually removable by inversion unless sand filled.
- It is noteworthy that the vadose sampling systems were the original Flexible Liner systems and have been in use for several decades.

12 The TACL (Traveling Acoustic Coupling Liner)

This chapter treats how flexible liners are used to enhance acoustic coupling of geophysical measurements.

12.1 THE TACL METHOD

In 2018, the USGS asked FLUTe to build a 5″ liner to be filled with water to allow a sonic transducer to be well acoustically coupled to the unsaturated formation. However, the water table was greater than 400 ft. below the surface in an essentially basalt formation with large borehole enlargements in brecciated layers of the basalt flows.

But, a 400 ft. water fill of a liner above the water table has a differential pressure in the liner of over 170 psi, far above the burst pressure of the liner. In the breakouts of the borehole, such a differential pressure is certain to burst the liner. It was also known that the basalt formation has sharp edges in the vesicular breccia intervals. The initial conclusion was that it was not possible.

But in consideration of the objectives, FLUTe suggested an alternative approach. If the liner was installed with a shorter water fill of perhaps 50 ft., the liner could be everted with the sonde submerged in the water-filled interval with good acoustic coupling. Then everting the liner another 50 ft. sealed the next interval of the borehole. In that manner the liner could be everted in increments to the water table or below. The drawing in Figure 12.1 illustrates the traveling acoustic coupling liner (TACL). As the liner is everted to a lower elevation or inverted to a higher elevation, the acoustic sonde is moved so as to remain in the water-filled interval where it has a good coupling to the hole wall. Further consideration of the borehole environment led to the suggestion of the use of the stronger 840 denier ballistic nylon liner.

The installation at the site in Idaho of the recommended liner provided the coupling needed. However, the long water column with the high pressure at the bottom of the water fill causes the liner to stretch and to curve as it propagated through the long enlargements of the borehole. The liner curvature caused the liner EP (eversion point) to be misaligned with the open borehole beneath the long enlargement, halting the liner propagation. That required that the water column be reduced to reduce the curvature of the liner in order to enter the borehole beneath the enlargement. Figure 12.2 shows a drawing of the liner trajectory through a cavern (e.g., in a karst formation or a large breakout of the borehole) with the curvature under a high driving pressure, if the liner is not supported. The reason that the liner curves under a high differential pressure is that the seam of the tubular liner enhances the tensile strength on the seam side of the liner. The end load of the differential pressure at the

DOI: 10.1201/9781003268376-12

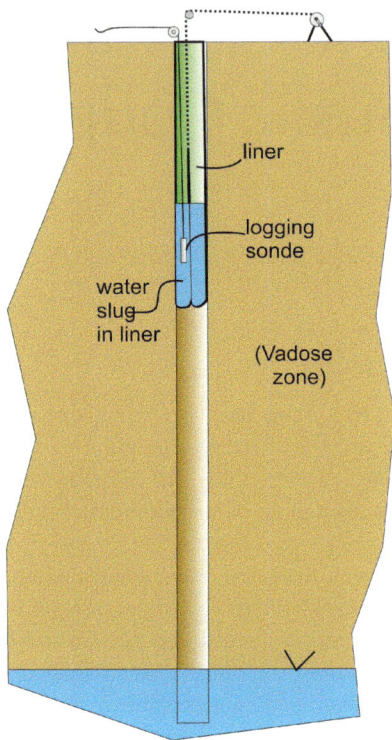

FIGURE 12.1 The TACL moving water-filled portion of an everting liner in the vadose zone. This allows a sonic sonde or an exterior optical cable to be acoustically coupled to the formation.

EP causes the liner to stretch. The enhanced strength of the double thickness at the seam causes the liner to curve in the direction of the seam.

The solution was simple. FLUTe welded a second simulated seam strip to the interior of the liner opposite of the seam. A helpful change in the procedure was to propagate the liner by eversion to the water table with a shorter water column and then to fill the liner with the longer water column. The liner was then raised by inversion in successive steps to the surface using the linear capstan (Section 9.2).

During the eversion to the water table and then while the liner is being inverted, the liner is dilated with an air blower to support the hole wall and to reduce the drag due to wet film adhesion as described in Section 3.2.4.2. It is common for an open hole to "breath" in or out depending on the barometric pressure history. During a high pressure period, the flow into the open borehole pressurizes the permeable layers such as the brecciated basalt layers in the vadose zone. During a low barometric pressure, the borehole expels the air from the permeable porosity in the formation. The flow of air into the borehole during a low pressure period tends to collapse the liner in the borehole. The blower pressure in the liner keeps the liner dilated above the water-filled interval and makes access for the sonde easier through the interior

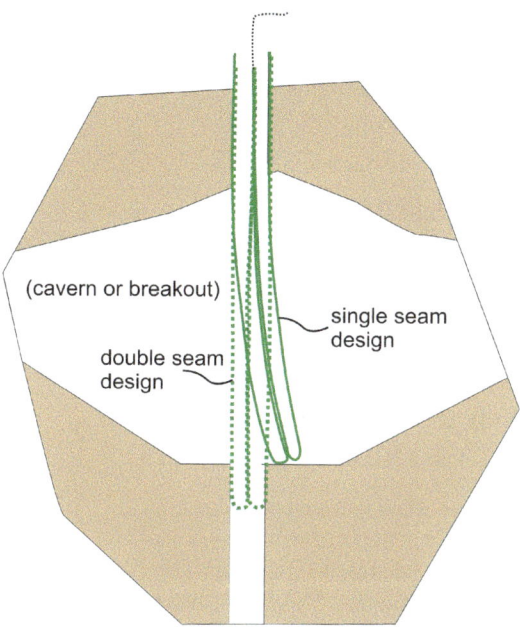

FIGURE 12.2   Effect of asymmetric seam on liner propagation. Seam on one side causes curvature under high differential pressure in liner.

of the liner. It is impressive how much additional drag there is on the liner removal if the wet liner is collapsed on the inverted liner in the borehole. The drag effect is addressed in the liner mechanics description of Section 3.2.4.2. The liner interior is necessarily wet with water during the eversion into the borehole using the pressure of a slug of water inside the liner. This method is patented.

This technique will work equally well with a water-filled liner pressing a fiber optic cable to the borehole wall as described by Munn et al (2017) in the coupling of a cable to the formation for seismic measurements. The water column can be as long as the liner strength and the linear capstan load limit allow in order to increase the efficiency of the process with the moving water column. Note that the burst pressure of the liner is inversely proportional to the liner diameter and the tension on the liner is proportional to the liner cross section. This makes more slender holes attractive, but a borehole less than 3-in. diameter with an exterior cable is perhaps a limit on how small a borehole is practical for inverting a liner. Everting a liner in a small borehole is made easier by just increasing the pressure.

Advantages of the TACL method

1. Provides acoustic coupling in the unsaturated zone where filling the liner with water is not possible
2. Protects the sonde from potential slough of the borehole
3. Enhances liner propagation through large breakouts or caverns
4. Uses the linear capstan for high tension removal of long liners

5. The same system can be used in angled or horizontal holes in the vadose or saturated zone
6. In horizontal or high angle holes, the sonde can be towed by the everting liner as described in Section 3.5.1

Limitations

1. Requires an open stable borehole
2. Does not seal and support the borehole throughout its entire length like a complete water fill
3. Slough of the hole wall on the liner above the water fill could entrap the liner

12.2 USE OF THE BLANK LINER TO PROVIDE COUPLING OF FIBER OPTIC CABLES

The paper by Munn et al (2017) describes the use of the FLUTe blank liner to press a special fiber optic cable against the borehole wall for seismic measurements of the geologic structure in a saturated formation. The paper describes the method and the results of such measurements. The blank liner in those applications is usually installed by other than FLUTe personnel, but preferably by those who have been trained in the procedure.

The blank liners have also been used to isolate fiber optic cables in boreholes for measurement of temperature distributions (Munn et al 2017).

These few applications were not developed by FLUTe but use the advantages of the FLUTe sealing blank liner for both sealing of the borehole and the compression of the cables against the hole wall. FLUTe has developed designs to allow the installation and removal of a variety of flexible liners in boreholes of very low transmissivity that are of use in the above installations. One example is the long vent tube (Section 3.3.3).

13 Application of Combinations of Liners and Other Methods

Combinations of several FLUTe methods are used for augmentation of horizontal drilling, a unique progressive packer method, and a sequence of FLUTe methods to obtain a set of hydrologic measurements to characterize the full hydrologic state of a subsurface situation.

13.1 THE FLUTe SEQUENCE

The FLUTe sequence is the usual installation of several FLUTe measurements as follows:

1. A sealing NAPL/FACT covered blank liner is installed to prevent cross-connection immediately after the hole is drilled and to obtain a map of the contaminants in an open hole.
2. The use of the same blank liner to map the flow zones using the T profile.
3. The removal of the blank liner after the T profile while performing the RHP method to map the head profile.
4. The blank liner is used to seal the borehole until the FLUTe multilevel sampling system is constructed.
5. The MLS systems are installed to draw water samples and measure the head at each port based on the above measurements.

This sequence has become the most common application of FLUTe methods. Occasionally the sequence is reduced to fewer measurements, but the combination seems to be a very economical site assessment measurement. For both the RHP and the MLS emplacement, the procedure can be adjusted to provide useful cross hole connection data as described in Section 10.5.9.4.

The high-resolution measurement of the contaminant distribution (both NAPL and dissolved) is of obvious value. Coupling the high-resolution measurement of the flow paths and the head distribution allows better guidance for selection of the water sampling measurement intervals for the regulators and a better design of remediation systems.

The same sequence is available in unstable formations using the continuously screened casing (CSC) design in Chapter 14.

DOI: 10.1201/9781003268376-13

13.2   LAHD (LINER AUGMENTATION OF HORIZONTAL DRILLING)

The everting liner has been used beneath a landfill in Indiana to emplace eight sampling intervals in a horizontal hole. The method is described in detail in Section 15.1. The everting liner in following a reamer on the drill rod supports the hole wall against collapse and forces the mud and cuttings of the reamer out of the drill hole entrance at the drill rig, rather than leaving the cuttings in the borehole. Details are provided in Section 15.1.1.

13.3   PROGRESSIVE PACKERS

13.3.1   Purpose of Design

The concept is a traveling "straddle packer" but with the passage sealed with a liner in both directions from the straddled interval. This method illustrated in Figure 13.1 was conceived for use in horizontal boreholes. It has not been used. It would also be useful in testing for leaks in horizontal pipes. The requirement is for access at both ends of an open pipe or borehole. It could also be used in detection of leakage beneath a landfill if the pipe through which the liner is traveling is perforated or permeable as for bisque-fired clay pipes. Such a system is described in a paper presented at a landfill monitoring conference (Keller, 1996). The design is called a "Progressive Packer". The advantage is a sealed pipe or borehole over its entire length with only the extraction/injection interval open. This is like a straddle packer without the leakage past the packers and without the heavy equipment needed. Also, the liners employed can traverse turns in the pipe impossible for inflatable straddle packer systems. An advantage is that one does not need to select the sampling interval as with the FLUTe MLS systems of the liner augmentation of horizontal drilling (LAHD) concept. Another advantage is the high rate of injection or extraction possible. The straddled interval travels continuously through the passage and can be halted anywhere. Finally, straddle packers are not easily installed in horizontal uncased boreholes.

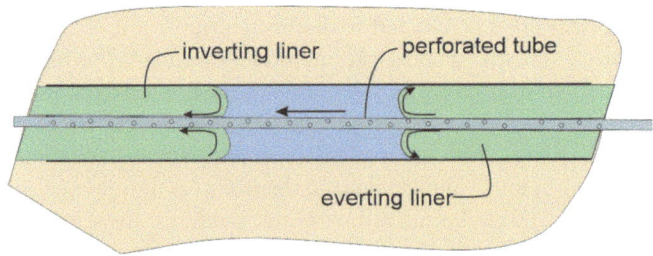

FIGURE 13.1   Progressive packer concept. The perforated tube is everted from advancing liner and inverted in the retreating liner maintaining a constant gap between them. The open gap is in fluid connection with the perforated tube for either injection or extraction in the interval.

13.3.2 THE METHOD

The interval for extraction is between two flexible everting/inverting liners. The liners are connected to cause them to invert/evert at the same rate with a constant separation (see Figure 13.1). The fluid between the two liners is drawn into a perforated tube that is everted from one advancing everting liner and inverted into the second retreating liner. As the perforated tube travels along the passage between the liners it is isolated and sealed by both liners except in the interval between the liners. The friction of both liners on the perforated tube causes them to invert and evert with a constant interval between them. As the liners propagate around turns in the pipe, the tube aids the bend of the liners in the turns.

13.3.3 EMPLACEMENT TECHNIQUE

The first liner is everted through the entire hole/passage and exits at the other end. The end of the first liner is then attached to the perforated tube interior to a second liner on a reel. (Note: both the liners are pressurized.) As the first liner is inverted with the tension on the tether or liner, the second liner everts off the reel extruding the perforated tubing that is pulled into the interior of the first liner. The tube everted from the second liner can be connected to a hollow axle on the second reel for a continuous connection of the advancing tube to a pump extracting or injecting. The inverting liner is followed by the everting liner with a moving gap between the two liners (Figure 13.2). In this process, both liners are pressurized above the pressure in the perforated tube. If the perforated tube is under vacuum, the friction of the tube on the inverted liners is enhanced.

An alternative is to have the tether on the first liner to be a tube that is rolled onto the reel of the first liner. That tube is not perforated but maintains a connection to the perforated tube after they are connected. The flexible liners are under pressure and therefore seal the perforated tube everywhere but in the interval between the two liners. The perforated tube must be deployed from the second liner for this design.

13.3.4 THE MEANS OF KEEPING THE LINERS PRESSURIZED

It is useful to have the liners deployed from sealed canisters such as the air pressure canister described in Figure 13.2 (and in Section 3.5.1) in order to maintain the

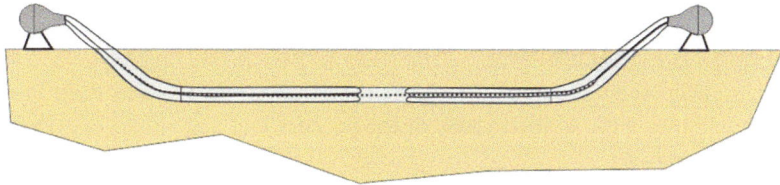

FIGURE 13.2 Air canisters and traveling packer interval. Can be used in any direction upward or above the surface in piping. Would be well suited for use with the "magic gland" described in Section 3.5.4.

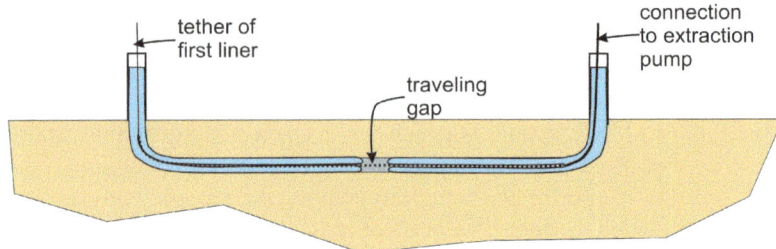

FIGURE 13.3 Water-driven liners with traveling gap allowing injection or extraction in the movable interval. The differential pressure, DH, depends on subsurface passage to allow for a water fill.

interior pressure of both liners. The everting liner friction on the tube assures that the liners travel together. The liner pressure must be greater than the pressure in the tube. If a vacuum is applied to the tube as during extraction from the interval, the necessary friction is easily maintained.

A relatively easy method for keeping the liners inflated which does not require a pressure canister is to use an excess water head in the liners with a standpipe at each end of the borehole as shown in Figure 13.3. The second liner can be deployed from a reel into the top of the standpipe.

13.3.5   Other Concepts of Potential Use of The Progressive Packer

If the everting liner from the second reel is covered with a diagonally woven cylindrical cover, the perforated tube can be feeding a resin to fill the interval between the two liners. The resin saturates the woven cover of the everting liner as it enters the traveling interval. The saturated covering is pushed against the hole wall as the central perforated tube is everted. When done, the everting liner holds the saturated cover (e.g., resin-soaked fiberglass as shown in Figure 6.2) against the hole wall. (The passage may be a leaking sewer pipe.) Flowing hot water through this liner can initiate the cure of the resin. When the resin is cured, the carrier liner can be inverted leaving the fiberglass liner. This assures that the cure-in-place liner is in intimate contact with the hole wall sealing any permeable intervals such as a hole in the original pipe. This is an alternative to the common sewer relining procedure where the cover on the liner is already saturated and emplaced by the everting liner while towing a hose to provide the hot water flow through the liner. In that case, the resin tends to be squeezed out of the cover where it enters the pressurized system which may prevent a sufficient saturation of the cover in a permeable passage.

The traveling interval can also be used for discrete remediation fluid injections in horizontal holes without the bypass of the packers that one would experience in a permeable formation.

The traveling interval can be used to stabilize an unstable formation with grout injection if the first liner is emplaced in an unstable horizontal hole using the LAHD technique described in Section 15.1. An attractive alternative is to emplace an undersaturated diagonally woven cover which when cured supports the hole wall but is permeable for access to the formation. This avoids the gap between the slotted screen casing installed in horizontal boreholes and the hole wall without any seal of the open annulus outside the screen. In that situation, the full range of FLUTe measurements can be performed in such a permeable cure-in-place casing in intimate contact with the formation. The method has been patented.

13.4 TOWING SONDES AND SUPPORTING BOREHOLES FOR LOGGING

To tow a sonde through an everting liner, the sonde is simply attached to the sealed end of the liner and the tether is the cable to the sonde. The method (Section 3.5) was used for towing a neutron moisture sonde beneath a simulated landfill. A design for monitoring of leakage from a layered bed beneath a landfill was proposed to the California regulatory agency for landfills. It was so well received as to be offered by the state as a reduction of the tipping fee state tax used to address landfill eventual leakage. However, due to the cost of the layered design and the use of today's money rather than future revenues, no landfill company accepted the tax reduction by construction of the monitoring design. The design is described in an invited paper presented by Keller (1996) at a GSA landfill monitoring conference in Austin, TX. The landfill design used permeable bisque fired clay pipe to maintain access to the monitoring layer. The pipe also provided access to the subsurface for gas monitoring for leakage and for potential remediation of the leak if detected. The test samples of curved bisque fire pipes were obtained from a clay pipe manufacturing company in Ohio. A perforated PVC pipe may be preferred for the permeable passage, but clay pipe was inexpensive.

A demo of the sonde towing method was done in a vertical upward borehole from an INCO mine tunnel in Sudbury, Canada. It also was not pursued by INCO. The traditional method was to use push rods for installing logging tools from the mine tunnels. The liner protection of the tool from entrapment in a borehole slough was apparently not compelling. A complication was the inflow of water into the vertical holes in the deep mine which could frustrate the liner eversion without a perforated vent tube in the hole. The liner was driven with air pressure from a canister as illustrated in Section 3.5.1.

13.5 TRANSPARENT LINER

Transparent liners are of use with a wide array of other methods. Cameras inside such liners can map NAPL FLUTe stains as they develop and are especially useful for assessing the progress of a newer more absorptive NAPL cover material for the process of removing NAPL from a formation. The NAPL absorption design includes a dye striped cover to develop a stain as the NAPL penetrates the absorptive cover, and visible through the transparent cover, to aid in the decision of when to remove the cover from the borehole for disposal of the NAPL absorbed. The use of a UV light, with a UV camera, can detect the arrival of tracers behind the liner or the presence of fluorescing liquids such as fluorescein dyes or petroleum products behind the liner. As described in Chapter 5 the presence of potassium permanganate can be observed, but that remediation fluid can also quickly attack the transparent liner. Dark contaminants like coal tar can be seen through the liner. A limitation is that the visible resolution is usually limited to materials in contact with the liner rather than at a distance behind the liner. The liner is more nearly transparent when saturated and only coated on one side. A test was performed by Dakota Technologies demonstrating the ability to see with UV light the emissions from petroleum products against the liner using a camera inside the liner.

13.6   DUET METHOD

The DUET is the method of installation of a second liner in a borehole already lined with a liner. The drawing in Figure 13.4 shows the displacement of the first liner by the second liner. This is only possible if the second everting liner is driven with a higher fluid pressure than the first liner fluid fill (either air or water). The method has

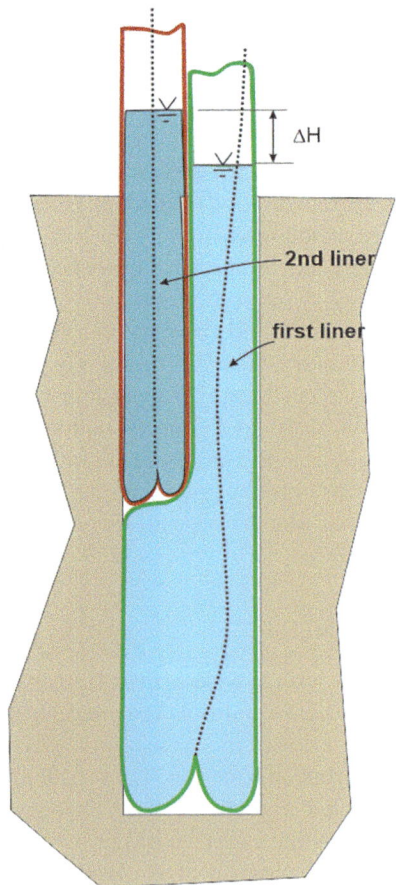

FIGURE 13.4   Duet method with second liner displacing the first liner. The second liner can be deflated to allow the first liner to reseal the borehole.

been used to install a second liner with absorbent coverings into a borehole already occupied by a first liner with gas sampling tubing. Upon the full extension of the second liner, the second liner internal pressure is dropped to atmospheric allowing the second liner to be collapsed by the first liner, restoring the seal of the borehole. The method has the advantage of not interrupting the first liner support of the borehole or a serious leak path along the borehole due to the second liner presence. This method was used in Yucca Mt. for monitoring of pore gas and liquids near a heater test simulation of a nuclear waste canister.

Once the second liner absorbers have absorbed the pore fluids of the formation, the second liner can be reinflated at a higher pressure again and inverted from the borehole to recover the absorbers. As the second liner is inverted, the first liner will expand to reseal the borehole and support the borehole. This technique was used at LLNL in the vadose zone in California and also at the Yucca Mt. site in Nevada. At the Yucca Mt. site both liners were of a high-temperature silicone rubber. The absorption in the absorbers was monitored by the conductivity change between two electrical contacts on the exterior of the liner at each absorber passing through the liner. A pair of electrical leads connected to the contacts extended through the interior of the liner in sleeves to the wellhead for monitoring as described in Section 11.2.

A further possible application of the method is to raise the pressure in the two liners to a level sufficient to exceed the tensile strength of the formation causing a fracture to form in the borehole wall parallel to the contact of the two liners. This would be most useful in weak formations at shallow depth and may be used to initiate a vertical fracture in the formation instead of the probable horizontal fractures common in stratified sedimentary formations. Figure 13.5 is an illustration of such a DUET geometry. The friction of the liners on the hole wall tends to reduce the hoop stress at the liner contact. Each liner of the pair is of smaller circumference than the hole. This approach avoids the loss of the pressurizing fluid into the formation.

It is well known that overinflation of straddle packers can also initiate a fracture parallel to the hole causing a leak past the packers. Ironically the leak path can develop due to an attempt to better seal the packers to a rough hole wall. In contrast, thin flexible liners do not need a high pressure to seal well.

The second liner can also be used to tow various kinds of sondes, or a borehole camera, into unstable boreholes that sonde or camera can then be retrieved by inversion of the second liner from the borehole. In theory, the first liner can support the borehole against collapse and yet allow access with the second liner.

In 1997, the DUET approach was successfully tested for eversion of a second liner into a 6″ pipe with a 90-degree elbow, through the elbow, just to see if it could be done. It worked easily. It displaced the first liner through the elbow and propagated further past the elbow.

13.7 VERTICAL CONDUCTIVITY MEASUREMENTS USING FLUTe MLSs

As described for cross hole connection assessment, one can use a FLUTe MLS equipped with pressure transducers for assessment of the horizontal connection in a formation. Whereas the vertical conductivity in stratified formations is usually low

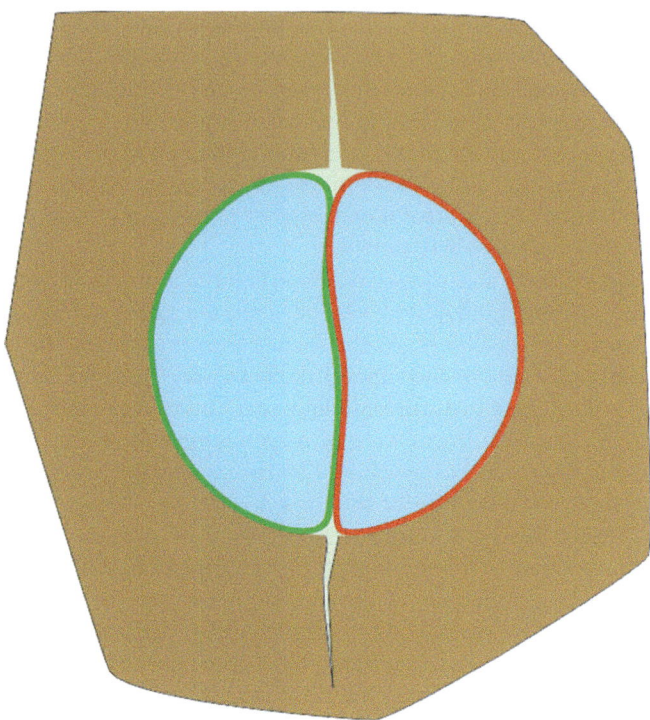

FIGURE 13.5   A pair of liners when highly pressurized can exceed the hoop stress in the hole wall to cause a rupture in the plane of the liner junction.

compared to the horizontal conductivity, it is useful often to assess the vertical connection. That can be done as follows:

With transducers connected to ports at different elevations, one port can be purged in the usual manner which upon recharge drops the pressure at the spacer to a relatively low level compared to the head at spacers above and below the purged port. By monitoring the pressure at a port, or ports, vertically distant from the purged port, one can deduce whether there is a vertical connection in the formation between the ports and the purged port if a drop in pressure is observed. This is possible because the liner seals the borehole between the sampling intervals. This was done at the SSFL site north of Los Angeles, CA and it was found that some parts of the formation between the ports were more well connected than at other elevations in the sandstone formation.

A simple 1D spherical calculation model of the test geometry is useful for determination at what vertical conductivity the measured pressure history could be detected. With a low conductivity or high conductivity, at some separation, the pressure drop cannot be detected depending on the resolution limits of the monitoring transducers, and the limited volume usually purged with the declining gradient driving the system during refill. The small volume purged reduces the pressure pulse propagation due to purging. A high horizontal conductivity can deplete the vertical pressure change over a short vertical distance. However, a vertical fracture connection would be seen more easily. If head histories are part of the monitoring, the test is relatively easy at little additional cost. Only a significant vertical connection can be expected to be detected with the method. The advantage with the FLUTe MLS liners is that there is no borehole storage to mute the head change measurement. It is noteworthy that the simultaneous purge procedure would prevent such a vertical connection between ports. That is one of the advantages of the simultaneous purge for sample isolation.

13.8 LINER PRESSURIZATION FOR SHALLOW WATER TABLES OR ARTESIAN CONDITIONS

13.8.1 The Problem Addressed

Liners have been everted into artesian wells using a variety of FLUTe methods from heavy mud to scaffolding to obtain a driving pressure greater than the formation or borehole head. Shallow water tables also require a means of increasing the head in the liner above that of the highest head in the borehole. However, once the liner is in place and the scaffold removed, there is a need to maintain a higher head in the liner.

Filling the liner with a heavy mud is one method that has been used, but it is limited due to the uncertainty of which portion of the formation contains the artesian head conditions. If it is a shallow artesian aquifer, the mud density needed to seal that shallow interval can produce an overpressure at depth that may be a hazard to the liner. For example, a mud density of 1.5 g/cc will produce an overpressure above the normal hydraulic gradient in a 300 ft. hole of 150 ft. at the bottom of the hole. This is too near the typical liner burst pressure. However, the overpressure at 40 ft. is only 20 ft. and may not be sufficient to seal the artesian interval. Some FLUTe lined boreholes have had 30 ft. of artesian head with successful installation by FLUTe personnel.

A grout fill of the liner described in Section 10.5.4.4 has some of the limits of a mud fill for shallow artesian aquifers.

A preferred solution to an artesian condition is to pressurize the liner uniformly over its entire length as needed to exceed the artesian head above the surface. Several methods have been used to achieve that, but they are less than satisfactory under a variety of conditions. One method of pressurization requires an airtight seal of the liner, which is not easy to achieve for long periods (note the deflating of a balloon in time due to diffusion through the balloon). If the liner has a very slow leak rate, the drop of the water head in the liner can also compromise the sealing pressure. The simple device of a water-filled standpipe above the surface is usually impractical.

Most liner installations occur without any leakage, but since some very slow leakage may occur in a liner, it is desirable to have a method that allows a uniform differential pressure increment along with the entire liner and that can also be easily adjusted if some leakage does occur over periods of months to years. Two methods have been designed. One is the WILD method (Section 13.8.2) and the other is called a submerged standpipe method (Section 13.8.3). Only the latter has been used. Each has benefits and limitations as described.

13.8.2   FLUTe's Weighted Inverted Liner Design (WILD)

13.8.2.1   The WILD Method

It is well known that a flexible liner can be everted with an excess head inside the liner. The excess head, ΔP, provides a tension in the inverted portion of the liner. As the ΔP increases, the tension increases according to T = 1/2 A ΔP. Where A is the cross-section of the liner. Using a tension to generate a ΔP across the end of the liner is a common experience in a borehole of low permeability. Increasing the liner tension decreases the pressure/head beneath the liner especially if the flow rate into the borehole is very low due to a low conductivity.

The WILD geometry in Figure 13.6 uses a heavy weight clamped in the inverted top end of a flexible liner to generate the tension desired on the inverted liner at the top of the borehole. The pressure increase, ΔP, is generated in the sealed interior of the liner. As the weight descends due to gravity, the pressure in the liner increases until the differential pressure across the inverted top end of the liner is sufficient to support the weight, stopping its descent. Conveniently, if the liner has a very slow leak rate (very rarely seen with new fabrication methods) the pressure in the liner

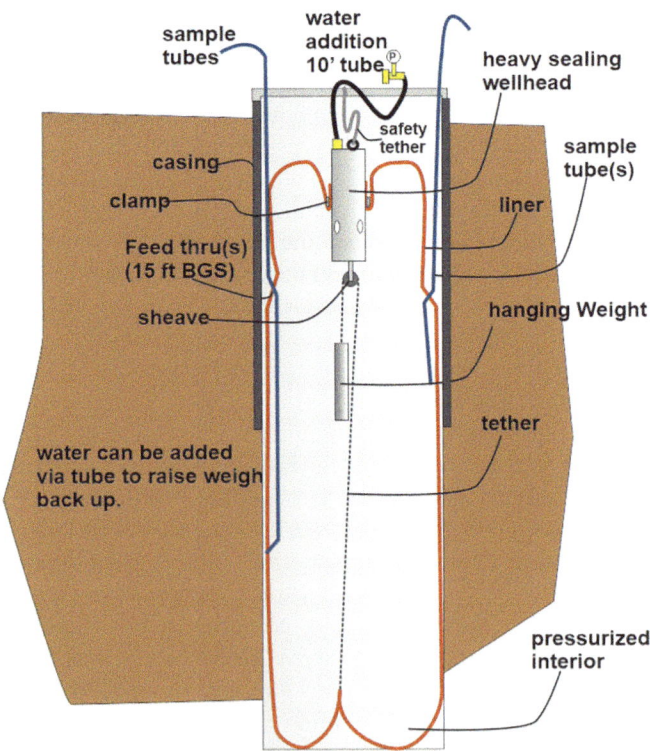

FIGURE 13.6 The weighted inverted liner design, WILD, for maintaining a uniform pressure in a flexible liner. Note, the hanging weight is amplified by 2× and keeps the tether suspended inside the liner. The sample tubing routing is only needed with a SWF system, not for a blank liner.

decreases, and the weight descends a small amount until the pressure in the liner again supports the weight.

In Figure 13.6, the additional feature of some liners, such as the Shallow Water FLUTe, is the presence of tubing used for water sampling. That tubing would complicate the seal to the weight in the end of the liner. In Figure 13.6, the tubing is shown routed through feed-throughs that are welded to the inside of the liner. The tubing is still accessible for water sampling with shallow water tables or artesian conditions. The external routing of the tubing near the surface has only been done occasionally.

Other features of the design add to the utility. The water addition tube allows one to add water to the liner causing the weight to rise, if it has descended due to leakage. The pressure gauge allows the head in the liner to be observed as water is added to raise the weight to a convenient height. The tether normally attached to the bottom end of the liner is usually anchored at the wellhead. In this design, the tether is routed through a sheave on the bottom of the weight and an additional weight is attached to the end of the tether. This smaller weight, doubled due to the sheave, is added to the downward tension on the sealing weight clamped in the liner. This also allows the tether to extend if the liner inverts downward or everts upward with water addition through the water addition tube.

The net effect of the inverted liner and weight is to increase the pressure in the interior of the liner along its full length above that pressure provided by the water fill alone. Adjustment of the two weights allows one to obtain the desired pressure increase.

13.8.2.2 Advantages and Limitations of the WILD Design

Advantages of the WILD system:
1. Provides a uniform pressure increase along with the entire liner.
2. Avoids the addition of heavy mud to the liner.
3. The system can reside below the ground surface in a vault.
4. The pressure increase is adjustable by selection of the weights used.
5. The liner tether is accessible for the liner later removal.
6. The system does not interfere with the sample tubes of the Shallow Water FLUTe or CHS system.
7. The system can be lowered below the frost line in winter.
8. There is no need of an airtight seal of any components.
9. Water addition to the liner is easy in order to adjust the elevation of the weight and the excess head in the liner.

limitations of the WILD design:
1. The weight of the wellhead required increases with the diameter of the surface casing.
2. The weight needed for high head additions to the liner may require a weight sufficiently long as to be awkward to assemble and may require a taller scaffold to achieve the excess head required during the assembly and installation of the weighted wellhead.
3. For large artesian heads, the weight needed may be difficult to handle without a crane at the wellhead.
4. The main utility is limited to modest artesian heads or very shallow water tables.

13.8.3 THE SUBMERGED STANDPIPE DESIGN

The submerged standpipe is another alternative to the standpipe or casing that rises well above the ground surface to maintain a constant overpressure with depth in the liner. As the name describes, the effective standpipe above the surface is a slender pipe below the surface that provides the same overpressure as a water-filled pipe above the surface.

The design is shown in Figure 13.7. The top of the liner is sealed with a water-tight wellhead. It is most convenient if the liner has no tubing that must pass out of the wellhead, but the only application of the system as of 2021 did have Water FLUTe

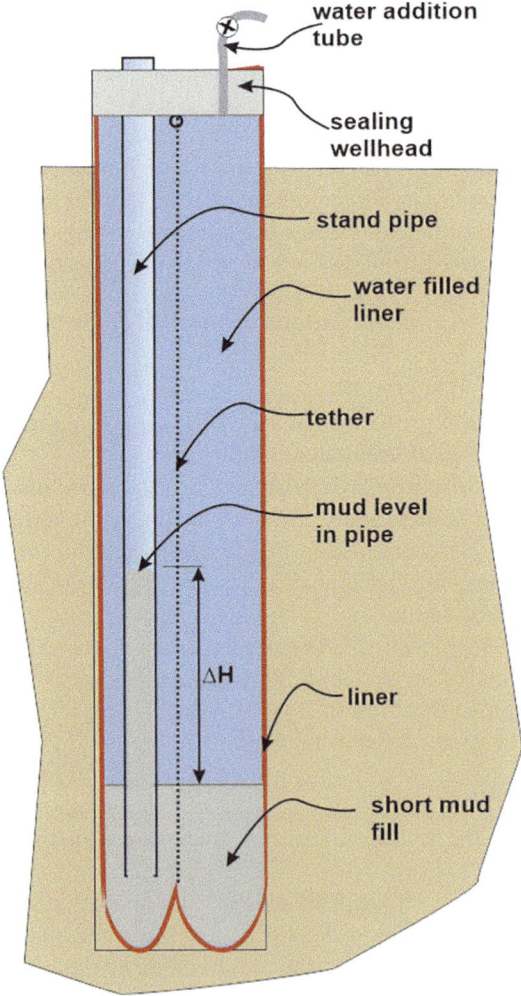

FIGURE 13.7 Geometry of the submerged standpipe inside a liner. Water pressure in liner causes the mud to rise in the standpipe, which maintains the head in the liner for the full length above the short mud fill.

tubing extending through water-tight seals in the wellhead. An additional tube called a water addition tube, which penetrates the wellhead, is equipped with a manual valve to seal the tube. Also, penetrating the wellhead is the "submerged standpipe" extending to near the bottom of the liner. An important feature is the short increment of heavy mud added to the bottom of the liner. That mud may be needed to install the liner in the borehole against an upward artesian flow or to evert the liner with a shallow water table.

13.8.3.1 The Function of the Submerged Standpipe Design

Given the initial condition of the liner installation to the bottom of the borehole with a short heavy mud fill and the seal at the top of the liner, water is added through the water addition tube at a pressure sufficient to seal the borehole adequately. That pressure/head is higher than the highest head in the formation by about 10 ft. That elevated head inside the liner is applied to the top of the mud column and forces the mud to rise in the submerged standpipe until the level of mud in the pipe produces a pressure at the bottom of the pipe matching the head applied to the liner for a sufficient seal. That height is such that $\Delta P = (\rho - 1) \Delta H$ where ΔH is the mud rise in the pipe and ρ is the density in g/cc of the mud in the bottom of the liner. Because the submerged standpipe is of a small diameter (e.g., 2″), there is not a major downward displacement of the top of the mud column in the liner for a significant mud rise in the standpipe. The mud rise in the standpipe does add a significant pressure rise in the liner. The mud rise inside the pipe causes an over flow of the small standpipe above the wellhead as the system equilibrates to the liner pressure developed by the water addition at the wellhead. At that time, the valve on the water addition tube is closed. The water pressure in the liner is now increased by the amount ΔP along its entire length.

If there is a decay of the head in the liner due to leakage, the water level in the standpipe will descend. In order to avoid freezing of the water in the small pipe, the pressure can be raised above that needed, and lowered to a new level causing some small drop in the water level in the pipe.

If the head in the liner decreases due to leakage, the mud will descend to maintain a somewhat lower head in the liner. This is better than the loss of head inside the liner without the mud in the pipe. Without the mud, and because the water is relatively incompressible, the excess head loss with a sealed wellhead is abrupt and complete with leakage. Adding water to the water addition tube will raise the mud level in the standpipe restoring the full head desired inside the liner.

13.8.3.2 Details of the Function

The rise of the mud is resisted by the viscosity of the mud that is initially small. However, when the rise stops, the mud will gel to produce a new temporary gel strength that will resist temporarily the drop of the mud to maintain an overpressure in the liner. The gel of the mud mixture is necessary to maintain the suspension of the added weighting material. Tests have been done of that process to measure the gel strength and the effective shear strength of the gelled mud. The shear strength is not excessive and is less important if the submerged small pipe is greater than 1″ diameter. It is relatively easy to make the water addition but adds to the maintenance needs for the liner system as compared to a simple water fill in the liner. It is also useful that the weighting material in the small PVC pipe does not settle in the pipe as has been observed in a mud-filled liner. The ideal mud fill would be like that of mercury (ρ = ~11g/cc), but that is not allowed. A heavy mud fill of density 1.8 would need a ΔH of 12′ for a head increase of 10′.

An interesting modification of the design is to cap the submerged standpipe and extend a tube from the sealed sample tube in an MLS liner from the artesian interval to the sealed standpipe. The sample tube end must be sealed to prevent artesian flow. The flow through the sample tube connection to the standpipe would then maintain the water head in the standpipe above the mud in the pipe. Then, if the mud level drops in the standpipe, there is no loss of water head above the mud in the standpipe. That refinement would further reduce the drop of head in the liner due to any slow leakage. The water added later would reverse the sample tube flow. The sampling procedure for a Water FLUTe system would not be affected by the connection to the sample tube. In many ways, the submerged standpipe is less complicated for a simple blank liner. Without the automatic recharge of water above the mud, which is possible with an MLS liner, the water in the standpipe must be replaced if the mud level drops and prior to any water addition to the liner to raise the head.

Advantage of the submerged standpipe:

1. Avoids the standpipe or casing rise above the surface for shallow water tables.
2. Maintains an essentially uniform overpressure throughout the length of the liner.
3. Avoids the hazard of a mud fill for deep boreholes.
4. Avoids the effect of settlement of the weighting material in the mud in lined boreholes.

limitation of the method

1. Complexity of the system.
2. Need to maintain the head with water addition to the liner in the case of possible seepage from the liner.
3. Lack of understanding by those doing the water addition thereby causing overpressure of the system.

14 CSC (Continuous Screened Casing) Design

This chapter describes a unique casing design that allows access to unstable formations for the use of the full suite of FLUTe measurements.

14.1 PURPOSE AND DESIGN

The problem addressed with this design is that measurements of contaminants in unstable formations currently done with either nested wells located at discrete elevations, or cased holes with multiple screens at discrete intervals, do not allow high spatial resolution of actual contaminant distribution or access for mapping the conductivity distribution. Those cased hole intervals are selected with less than complete information on the subsurface characteristics and contaminant distributions. To date, direct push methods are available for high-resolution mapping of some of those parameters, but only in soft soils that do not refuse direct push methods for mapping those parameters.

Currently, FLUTe high-resolution methods are mainly used in open stable boreholes. In those boreholes, the several FLUTe methods map the NAPL (NAPL FLUTe), the dissolved phase (FACT), the conductivity distribution (T profile), and head distribution (RHP) with high spatial resolution at low cost. However, it would be a great advantage to use the same methods in the unstable overburden or weak rock holes likely to slough. It would be even more convenient to use the same borehole for those measurements in the overburden and in the bedrock.

The continuous screened casing used in many old site investigations would allow use of FLUTe methods. However, the common coarse gravel packed annulus frustrates the resolution possible with FLUTe methods and more importantly the high flow in the gravel pack and the casing is a serious concern for cross-connection of contaminated aquifers with uncontaminated aquifers.

The CSC (continuous screen casing) design illustrated in Figure 14.1 allows the FLUTe methods to be used in unstable formations. The most important aspect of the CSC design is that it provides many sealed intervals in the sand packed annulus and limits the cross-connection in the annulus to potentially insignificant levels when the casing is sealed with a flexible liner. The key word is insignificant. It is suggested that "significant" is more than occurs in the natural formation or in using other methods of site characterization.

Finally, the question is whether that leakage in the annuls is milliliters/day, liters/day, or hundreds of liters/day, and is it significant? The fact that the sealed intervals are essentially impermeable allows the possibility that when combined with the permeable sand annulus, it can provide a very low effective permeability in the annulus.

DOI: 10.1201/9781003268376-14

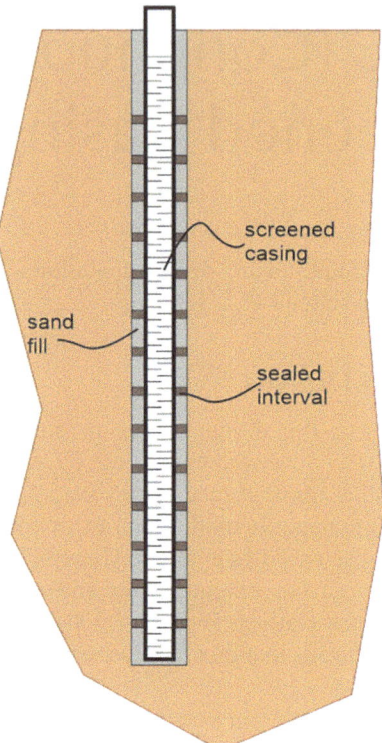

FIGURE 14.1 A layered annular fill between the screened casing and hole wall. The gray intervals are typical fine sand and the brown layers are seals preventing vertical flow in the sand-filled annulus. For 10-ft. intervals, this is a nominal 150-ft. hole.

The insignificance of flow in the CSC design is proven only with experience, but the amount of cross connecting flow can be estimated with a calculation of the flow in the annulus for a wide variety of conditions. The ability to use so many different conditions in a calculation is better than even the experience in a few sites. Those calculations have been done.

This report describes those calculations and the results which show how much flow is expected in the annulus of the CSC design. In most situations where there is insignificant flow in the annulus, the FLUTe methods can measure both the same parameters as in open stable boreholes but can also measure the significance of the actual flow in the annulus. If that measured flow is judged to be a long-term concern, the screened casing and annulus can be grouted and the borehole defined as only exploratory rather than of long-term use. It is useful that inexpensive multilevel sampling systems are available from FLUTe to measure the long-term contaminant distribution in the unsealed segments to aid that decision of whether to grout fill the casing after the measurements are completed. Such a grout fill would further seal the annulus.

The method of building the CSC backfill design is relatively easy, and not very much costly than building a casing with a single screen or multiple screened intervals in a sonic drill casing. That technique and cost are described later herein after the flow assessment. The sonic installation of a 3″ multi-screened well with six screened intervals to 255 ft. was done in Espanola, NM. (The multilevel sampling systems used in that well happened to be a FLUTe pdCHS system described in Section 10.5.4. The well development, MLS installation, and purging and sampling of all six ports simultaneously was done in a single day.) The challenge in the Espanola installation was to select the useful screen elevations based on available knowledge of contaminants, conductivity, and geology. While geologists do a multilevel screened design very well, it could be done better with nearly continuous measurements of conductivity, contaminant, and head distributions in hand. The purpose of the CSC is to provide that kind of information. Direct push methods at the Espanola site to the 255-ft. depth were not an option. It is also useful to have the core from a sonic drilling procedure to judge the utility of the sealed intervals

14.2 CALCULATION OF FLOW IN THE INTERRUPTED ANNULUS

This description is of the calculational model used to assess the flow in the CSC sand pack as a function of the important variables, some of which are not known such as the formation vertical conductivity everywhere. There are two flow regimes of interest. One is the early time-dependent flow after the construction of the CSC system. If that flow is significant, the casing and annulus may best be sealed with grout soon after the system is constructed and the measurements have been performed to map the parameters of interest. The second calculation is the steady-state flow over longtime periods to determine if the casing sealed with a liner over longtime periods is a concern for cross-connection of contamination transport.

The time-dependent calculation model addresses explicitly the question of the early vertical flow rate in the CSC design. Note the interior of the continuous screen is sealed with a flexible liner. The calculations address the questions of how much cross-connection can be expected in the unsealed portion of the sand-filled annulus. The model uses the important parameters:

1. The gradient driving the vertical flow
2. The hole and casing diameter
3. The conductivity of the formation
4. The conductivity of the sand
5. The dimensions of the annular seals

The geometry of the model is shown in Figure 14.2. The flow path modeled is the dashed line. The driving condition is a boundary excess head at the top of the model. This boundary is assumed to exist at some elevation in a shallow aquifer below the water table. The important features are a screened casing (radius of ri) sealed with a flexible liner and therefore an impermeable inner boundary at ri. The inner wall of the borehole left with the sonic casing withdrawal is at the radius ro. The grout seal of the annulus at a depth Zt_j to Zb_j of thickness $Zb_j − Zt_j$ for the jth sealed interval. The grout seal outer radius is a variable allowing some invasion of the grout into a permeable formation to a radius rg.

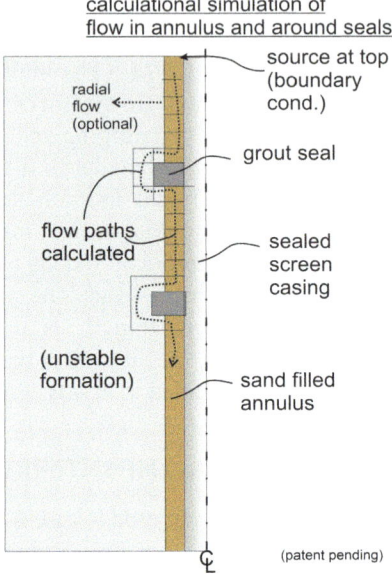

FIGURE 14.2   The flow path of interest is the dashed line through the annulus and around the seal in the formation. The seals may extend into the formation depending on the formation conductivity.

The flow path is through the sand-filled annulus of conductivity C_s in a formation of conductivity C_f. The flow around the seal is in a single-cell thickness Δh and radial width of the same Δh beyond the grout radius. The length Δh is also the vertical cell dimension of the calculation.

Any flow in the annulus, at a head above that in the formation, can also experience radial horizontal flow into the formation. That component of flow was modeled with the classic Thiem equation used for straddle packer calculations. The flows are calculated over many days. The steady-state flows are calculated later in this assessment.

The model is a simple finite differencing of the flow equations and is only expected to provide a reasonable bound on expectations given the uncertainty of the conductivity distribution of the formation without a FLUTe measurement.

It is very interesting to see how variation of the parameters in the calculation affects the results. The results plotted are the history of flow in the sand-filled annulus past the bottom of the first and second plugs. The first calculation was without the sealing plugs, called seals. Next is shown the result with the seals, but no radial flow from the annulus. The third calculation is the flow including radial horizontal flow from the annulus. Unlike a typical Modflow calculation, contours of the flow field are not available because this is a 1.5D calculation. Pressure plots are shown for the flow in the annulus for the several situations. After the three calculations just described, the conductivity in the formation was varied over a reasonable range for many formations. It is generally assumed that a preferred 20/40 sand size would be used as commonly available and yet would not allow sand flow through the screen slots. There is no assumption of how the fines in the formation would tend to clog the outer radius of the sand pack during development, but the well should be thoroughly developed for FLUTe measurements.

It is useful to remember that a precise calculation of a real situation is not done, and the point of these calculations is to bound the unbounded opinions offered so far as to the flow in the sand pack and its effect on the measurements possible in the CSC design.

14.2.1 THE RESULTS OF THE CALCULATION

Many calculations can be easily performed by simply changing the parameters in the calculation. Only four calculations are shown here. The first is to assess the flow without any seals in the annulus. The second adds the seals to the annulus for only the first five sealed intervals. The third calculation allows radial flow from the annulus into the formation. The fourth calculation shown changes the formation conductivity by increasing the conductivity by a factor of 10. In all cases, the conductivity of the sand pack is the optimum, which is the smallest grain size that does not pass through the slots in the casing. These four calculations show the effect of the major parameters on the flow past the first, second, and third seal in the annulus. Below the third seal, there is no significant flow by 6 days. Cross-connection at later times can be assessed with a steady-state flow model or more sophisticated implicit model such as Modflow. This simple explicit differencing model shows well the effect of the parameter variations. In all the calculations, the interior of the casing is assumed to be sealed with a continuous flexible liner. Flow at later times is addressed with steady-state calculations described later.

14.2.1.1 Calculation No. 1: Calculation with No Seals in the Annulus

Radius of the casing is 2″
Radius of the borehole 4″
Radius of the seal 6″
Spacing of seals 10′
Thickness of seal 1′
Cell size in the annulus 1′
Driving head at the top is 10 ft.
Conductivity of the sand 0.176 cm/s
Conductivity of the formation 0.001 cm/s
Conductivity past seal 0.176 cm/s, sand (i.e., no seal in place)

The pressure change has passed the third seal after only 3 days (Figure 14.3). The flow past the first, second, and third seals is very large as seen in the next graph (Figure 14.4). It is clear that the continuous common sand pack of more coarse sand is an extreme flow path. If the casing is not sealed with a liner, it is an even more extreme flow. Such flow is why the earlier practice of continuous screened wells is to be avoided.

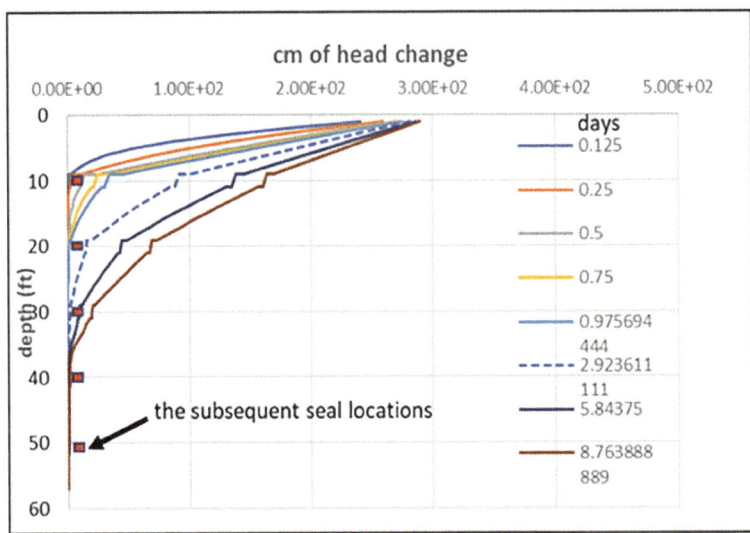

FIGURE 14.3 Calculation no. 1. The head distribution is in the annulus. The head at the top boundary is set to 10 ft. (304.8 cm). The kinks in the curve are where the seals are to be modeled in the calculation and should be ignored. After nearly 9 days, the flow has passed the first three seals. The locations of suggested seals are indicated with the red rectangles.

Flow past the first seal location (with no seal) is 8400 liters by 8 days. The flow past the third seal location with no seal is nearly 1120 liters by 8 days. The sand pack is obviously a major flow path without any seals. As the system approaches a steady state, the several curves will converge to the inflow rate at the top. The flow must displace the water in the original sand pore space to be significant.

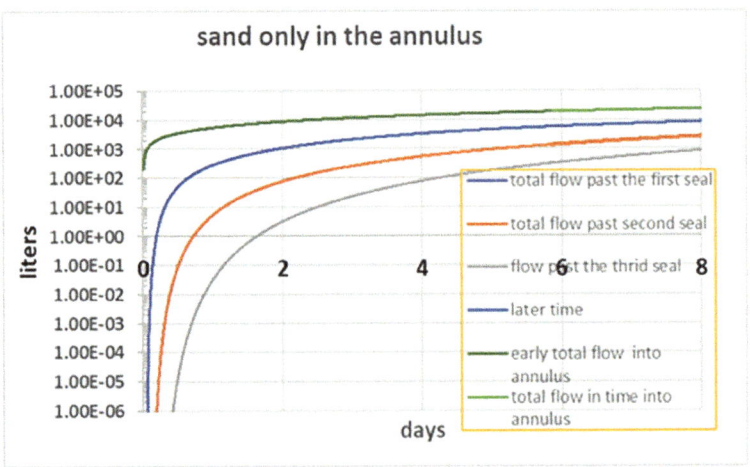

FIGURE 14.4 Calculation no. 1. Total flow past the top seals and flow into the top of the annulus (the olive curve).

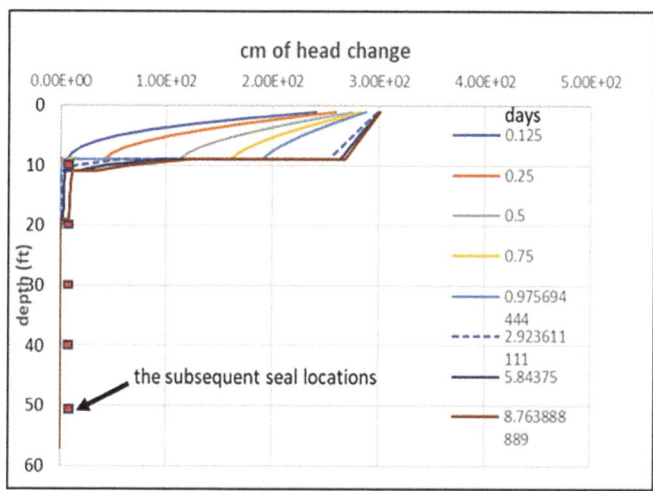

FIGURE 14.5 Calculation no. 2. The head distribution in the annulus with the seals at the indicated locations. The flow nearly stagnates against the seal with bypass in the formation at the formation conductivity of 0.001cm/s.

14.2.1.2 Calculation No. 2: Calculation with Grout Seals in the Annulus

The differences from the first calculation was to change the conductivity of the bypass of the seals to the formation conductivity of 0.001 cm/s, a very silty sand, instead of the high annular sand conductivity used in calculation no. 1.

There is only modest bypass of the first seal in 8 days as shown in the next graph (Figure 14.5) (blue line).

Bypass below first and second plug after nearly 9 days has been reduced by ~70 fold. Note the log scale.

The rate of bypass of the first seal is slowing toward a steady-state rate, but still less than a total of 122 liters after 9 days. The bypass of the second seal has dropped by 30,000 fold to only 5 ml. Bypass of the third seal is trivial yet, at this time. The steady-state calculation described later shows the third seal to be less resistant to flow than suggested at this early time.

14.2.1.3 Calculation No. 3: Calculation with Seals in the Annulus and Allowing Radial Horizontal Flow from the Annulus

Grout seals force bypass in the formation and radial horizontal flow decreases the driving pressure against the seals (Figure 14.6). The head above the first seal is greatly reduced by lateral flow from the annulus into the formation. The flow past the first seal is only 62 liters in 9 days, and only 14 ml has passed the second seal. The bypass of the first seal has also been reduced by a factor of ~2 with the reduced head above the seal. The formation conductivity of 0.001 cm/s limits the radial horizontal flow. If radial flow is a concern in the upper portion of the borehole, several adjustments to the design such as longer surface casing or a longer grout seal near the surface are in order.

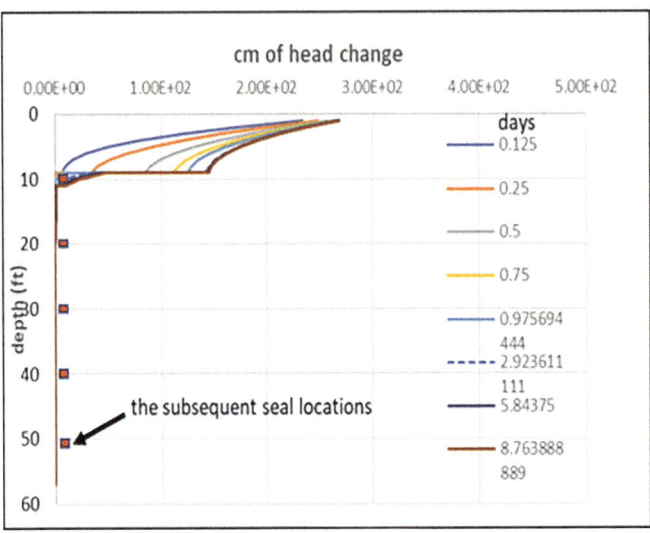

FIGURE 14.6   Calculation no. 3. The head distribution when lateral flow from the annulus is allowed to reduce the driving head in the annulus.

14.2.1.4   Calculation No. 4: Calculation with Seals in the Annulus and Allowing Radial Horizontal Flow from the Annulus and upon Increasing Formation Conductivity by Factor of 10

Conductivity of the formation 0.01 cm/s
Conductivity past seal 0.01 cm/s (same as the formation value)

The higher formation conductivity allows much more lateral flow from the annulus dropping the pressure in the annulus (Figure 14.7). It also allows easier bypass of the seals. The net result is not significant reduction in the flow past the first seal or second seal.

The ten-fold increase in formation conductivity produces a trivial change in flow past the first seal because the bypass is also calculated at the formation conductivity.

Dropping the excess head above the top boundary by half should reduce the leakage past the seals, and it does change the flow past the second seal by half. The flow past the third seal is still in the sub milliliter range.

Reducing the formation conductivity at the seals to 5e–04 cm/s cuts the leakage past the first seal by half as it should. Obviously locating the top seal in a clay-loaded interval can greatly impede the flow in the sand pack. It may be easier to identify clay-loaded intervals in the sonic core than to identify the high flow zones in the gravel and sand of the sonic core. The high flow zones may be better located with FLUTe measurements in the CSC design.

14.2.2   What May Be the Definition of Significant Vertical Flow?

One possible definition is that the flow in the annulus with the seals is less than the normal vertical flow in the formation volume removed in drilling the borehole.

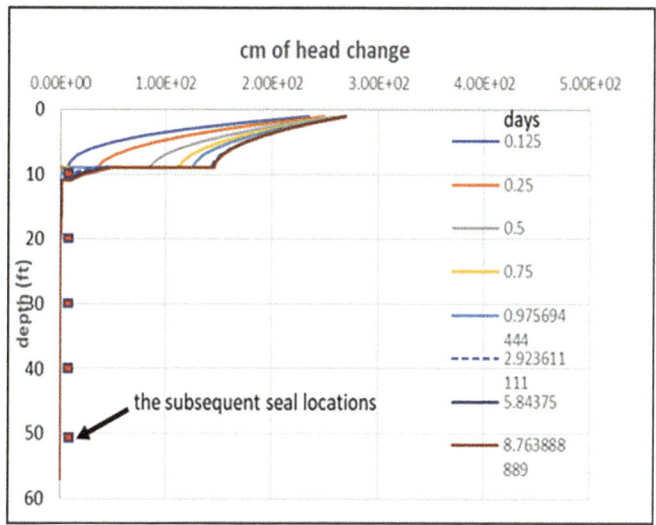

FIGURE 14.7   Calculation no. 4. The head/pressure distribution in the annulus with a ten-fold increase in the formation conductivity. The lateral flow is much higher with the increased formation conductivity. The flow into the annulus is also increased with the lower back pressure in the annulus.

Since the conductivity of the sand is so high, that is not possible. Another limit is that which is flowing in the formation within a significant area near the borehole, but in an area larger than the borehole. The question then is, what is the area of the subsurface at the formation conductivity which would provide a vertical flow of the same volume per second as occurs in the annulus? If that is a large area, the calculated flow in the annulus with the seals is significant. The equation of that area of the formation that provides the same flow as in the annulus is easily calculated as of a radius (R) such that the annulus vertical flow is equal to the original flow in that larger area. The table in Figure 14.8 shows the radius (R) of that area of equal flow in the formation under the same unit gradient. The unit gradient is picked as the driving head of 10 ft. with a seal spacing of 10 ft. but the calculation is independent of the gradient used for both flows. For a 4″ diameter casing in an 8″ hole, that radius is ~4 or 8-ft. diameter. If one uses the reduced flow of the annulus with seals, the radius (R) is much smaller, but the 0.04 l/s is actually very significant when continued for days or months.

ri (in.)	R(in.)		Qf (cc)	Qs (cc/s)	liters/s
2	45.96	in.	42.8	42.8	0.04
3	53.07	in.	57.1	57.1	0.06
4	59.33	in.	71.3	71.3	0.07

FIGURE 14.8   The radius of an area of equivalent flow in the formation to that flow in the sand-filled annulus as a function of the casing size. The flow rate in the annulus and the formation are shown to be identical for this calculated area of radius R in the formation. For a 4″ casing and 2″ annulus, the flow rate is 0.07 l/s.

While the time-dependent calculations are informative, it is difficult with a simple finite difference calculation to get to the steady-state flow conditions that prevail over a longtime period. An uncertainty in the time-dependent calculations is the storativity that should be used. The following description only deals with a steady-state flow assessment that is more relevant to the longtime period of interest.

14.2.3 WHAT STEADY-STATE FLOW CALCULATIONS SHOW ABOUT SIGNIFICANT BYPASS OF THE SEALS

The open hole has a flow rate like that of an open pipe. The steady-state flow rate in that case is:

$$Q_{oh} = \pi r_o^4 \Delta H/(8 \, \mu L) \text{ for laminar flow in a pipe.}$$

where r_o is the borehole diameter and L is the flow path length from the source at the top to the low head at the bottom, $\Delta H/L$ is the gradient. That flow rate for a 10-ft. head in a 200-ft. hole is 2.0e08 l/day. That is with a gradient of only 0.05. Such a flow is not likely, because the inflow and out flow rates are controlled by the formation permeability. However, there is essentially no resistance to flow in the open hole. An equivalent radius for flow in the formation is extreme. Therefore, as is well known, open holes should not be allowed in contaminated sites.

If the borehole is screened throughout its length, and the casing sealed with a liner, as in Figures 14.1 and 14.2, the only flow path of concern is in the sand annulus. The flow rate in that annulus with no seals is $Qa = Cs \, \Delta H \, \pi(ro^2 - ri^2)/L$ where L is the length separating the inflow and outflow aquifers. The radii ri and ro are the radius of the casing and the radius of the borehole hole. The term Cs is the conductivity of the sand pack (0.176 cm/s). Since this is to be compared to the CSC design, a gradient of 1 is assumed again (a very large gradient). For those conditions, the flow rate through a 10-ft. interval of sand is 111,000 liters/month. The gradient is not known, but this assessment deals rather with the attenuation expected due to the addition of a seal every 10 ft.

The equivalent radius (R) of a cylindrical volume in the formation is discussed above. While R is not an extreme area for the sand flow alone, it is large and not acceptable as a measure of significance. If one uses the geometry of Figure 14.2 for a steady-state flow calculation of the flow past the first seal, the flow is only significantly restrained by the conductivity of the formation in the bypass of the seal. However, one can assume that the impedance to flow around the seal is more than that if one only assigned the formation conductivity to the seal, which is a more pessimistic calculation. In the calculations described here, it is assumed that the flow around the seal is through a path length in the annulus equivalent to three times the thickness of the seal. That is not unrealistic if the sealing grout invades the formation for a short distance. Therefore, if the spacing of the seals of 10 ft. is divided into 1 ft. increments and the seal bypass path is 3 ft., the calculation becomes simply 9 ft. of flow through sand and 3 ft. of flow through an equivalent of the formation conductivity. Since there is no radial loss from the borehole, this is a relatively conservative assumption.

The flow in the annulus, with no seals, defined as Qa, is then: $Qa = C_s \Delta H/L\ A$ for the open sand-filled annulus. A is the cross-sectional area of the annulus, and $\Delta H/L$ is the gradient from the top boundary to below the seal (10 ft.). The flow rate in the sand and around the seal, Q_{CSC}, is the same through both the sand and around the seal. The total head drop through the sand and seal bypass is defined as $\Delta Ho = \Delta H_s + \Delta H_f$ for that in the sand and the bypass respectively. The flow rate through the total path from the top of the sand column and through the bypass is Qcsc. The flow area is defined as $A = \pi(ro^2 - ri^2)$. Therefore, the head loss ΔHi in each increment is equal to $Q_{csc}\ \delta z/(A\ Ci)$ for each increment (δz) as used in the calculation. The term ns is defined as the number of increments in the flow path in sand and nf is the increments in the bypass. This makes it easier to change the lengths calculated. The conductivity in the sand is Cs and in the formation Cf.

Therefore, the total head change, ΔHo, is the sum of the head drops in each material in the path and equal to,

$$Q_{csc}\delta z/(A)\ (n1/Ci + n2/Ci + n3/Ci + ...)$$

For the above geometry with 9 ft. of sand and 1 ft. sealed (simulated by 3 ft. of formation outside the annulus, see Figure 14.2), the relationship is:

$$\Delta Ho = Q_{csc}\delta z\ (9/Cs + 3/C_f)/A$$

Solving for Q_{csc}, the steady-state flow rate in the annulus with the bypass of the seal,

$$Q_{csc} = \Delta Ho\ A/\left(\delta z\ (9/Cs + 3/C_f)\right)$$

It is interesting to note that this expression for a single material reduces to $Q_{csc} = Qa = Cs\ \Delta H\ A/L$ since $N\ \delta z = L$. It is also convenient to replace each δz, with a fraction of the path L. in which case the L's cancel. It is also useful to define the equivalent conductivity for the composite flow path as,

$$C_{CSC} = L/(\delta z(9/Cs + 3/C_f))\ \text{or} = 1/(f1/C1 + f2/C2 + f3/C3 + ...)$$

where the f's are fractions of the total flow path for each flow path of different conductivity. This can be useful for assessments of seals in layers of different conductivities.

What is most useful to the question of significant flow in the annulus is the ratio of the open annulus flow to that flow with the seals in the same path. In that case, with Cs = 0.176 cm/s (the conductivity of size 40 sand) and a formation conductivity of 0.001 cm/s as used above, the ratio of the sand only flow, Qa, shown above, to the flow with the sand and seal, Qcsc, is:

$$Qa/Qcsc = Cs/L/\left(1/(\delta z(9/Cs + 3/C_f))\right) = 54\ \text{for the above values.}$$

Since this is the steady-state flow through the first sealed increment as compared to the flow through an equal unsealed increment, the same reduction calculation should occur for the flow through two 10-ft. increments or three increments since the flow

no. seals	flow per month (liters)
1	2.48E+03
5	4.96E+02
10	2.48E+02
15	1.65E+02
20	1.24E+02
with 4" casing diam., Cf=0.001	

FIGURE 14.9 Total flow in the annulus over a month. This may be considered significant. These results are with a 10-ft. driving head. With $N = 20$, (i.e., the borehole is 200-ft. deep).

below the first seal is the inlet for the next interval. Likewise, the outlet of each interval is also the steady-state flow through the full length of the annulus with no other influences such as lateral flow to the formation. If the series is summed, it is possible to add 12 more increments for the second seal to the above equation which cuts the flow by one half. Adding another increment of sand and grout cuts, the flow through the first 10-ft. segment alone by a factor of 2. In other words, the reduction factor, F, is <u>not</u> 54^N where N is the number of seals, but $F = N \times 54$, where F is the factor by which the sand-filled annulus flow is reduced by N sealed intervals. Or $Q_{csc} = Q_a/(N \times 54)$. However, in the ratio above for the two flow conditions, L will also increase for more sealed sections if $\Delta H/L$ is the gradient for the whole length of sealed segments. The net effect is to reduce the flow rate for the flow in the equivalent interval. For a formation C_f of 0.001, a casing of 4″ and 10 ft. driving head, the table in Figure 14.9 shows the total flow after a month. That may or may not be significant compared to the migration that has already occurred.

The table in Figure 14.10 shows the reduction factor per 10-ft. sand-and-seal interval added to the annulus. This calculation uses a very steep gradient of 1.0 for the first sand and seal interval, since $\Delta H/L$ is 10 ft./10 ft. A smaller head difference and smaller gradient would reduce the flow to very low levels quickly decaying to insignificant. If N is 20, the flow in the sand-filled annulus with seals is reduced by

Steady state flow reduction factor vs. C_f					
C_f (cm/s)	0.176	1.00E-02	1.00E-03	1.00E-04	1.00E-05
depth (ft) N	Qa	Q_{csc}	Q_{csc}	Q_{csc}	Q_{csc}*
10 1	1	5	53	441	12,321
50 5	5	26	264	2,204	
100 10	10	52	528	4,408	
150 15	15	77	792	6,611	
200 20	20	103	1,056	8,815	

FIGURE 14.10 Factors in reduction of sand-filled annulus flow, Qa, with the addition of 1-ft. seals to the annulus every 10 ft. as a function of the formation conductivity, Cf. The single factor shown in green for Cf of 1.0e−05 cm/s is for a "super seal" which has half of the 10-ft. interval filled with a grout aggregate.

a ~1000 fold for $C_f = 0.001$ cm/s. The effect is due to both a reduced gradient as L increases and the increased impedance of many seals.

This calculation does not include radial/horizontal loss to the formation, which as shown in the earlier calculations, reduces the bypass. This steady-state calculation shows a somewhat smaller degradation per 10 ft. sealed interval than the time-dependent calculations that were not extended to steady state. The reason is probably because the flow rate in the several intervals was not near the steady-state flow. Where there was a steady-state flow through the series of seals, the pressure gradients would be identical in each 10-ft. interval. Since that is the case in the time-dependent calculation, this steady-state calculation is "conservative" in assessing the effect of the sealed intervals in the annulus.

The time-dependent calculations show that the annulus with seals limits the cross-connecting flow more at earlier times than at later times. If one were concerned about cross-connection after 9 days, it is possible to grout fill the continuously screened casing and annulus after the measurements are done long before any significant flow has occurred out the bottom of the borehole.

Clearly the higher formation conductivity shows a smaller reduction per sealed interval, but in that case, the equivalent radius of the equal flow area in the formation is also much smaller, reducing the significance of flow in the annulus. For a ten-fold increase in the formation conductivity, the flow reduction factor is much less. If a seal falls in a clay bed or aquitard, the reduction factor is much greater due to the low conductivity in the formation at the seal. The inflow into the annulus is still affected by the formation conductivity which is not addressed here. This assessment assumes a high flow aquifer source. A full 2D Modflow calculation should show less cross-connection than seen in these simple calculations due to the impedance in the formation to flow into or out of the annulus. These calculations are mainly to bound the speculation of how much flow can occur in the annulus when interrupted with frequent seals.

14.2.4   OPTIMIZING THE DESIGN

In the actual application of this design, it is useful to identify intervals of very low conductivity in the sonic core. Those intervals tend to be recovered much more intact than the core appearing as a loose bag of sand and gravel. In the Espanola core, it was observed that some core bags contained a solid extrusion of clay-loaded sediment. That seal interval in the annulus would best be constructed with a "super seal" (Figure 14.11). A super seal is a special construction in the annulus as an extra effective seal. The conductivity of clay rich sediments can be very low to nearly impermeable. If that clay layer does not fall at the usual location of a seal in the removal of the casing, the sealed interval in the annulus can be easily increased in length. A simple procedure would be to locate the top of the sand interval below, but near the clay layer. At that point in the backfill procedure, a larger volume of grout can be added for the seal extension of about 3 ft. per 2 gal. of grout addition to the usual which may be only 2 gal. (more than a foot of the annulus in an 8″ hole with 4″ diam. casing). The grout can be extended by adding large grain gravel (3/8″) on top of the grout. The gravel will sink into the grout and with a 33% porosity can be grout saturated

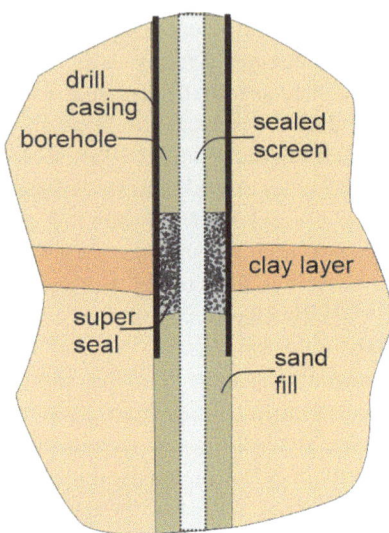

FIGURE 14.11 "Super seal" emplaced at clay layer with extended gravel-filled grout seal to encompass the clay. Drill casing ready for next lift.

an additional 3 ft. for each 2 gal. of grout. In this manner a "super seal" can be constructed to straddle the clay rich bed to provide a nearly complete seal of the annulus since bypass is extremely limited for a formation C_f of less than 1.e−05 cm/s which is more typical of an aquitard/aquiclude. At a C_f of 1.0e−05, the reduction of flow for a single interval with a 5-ft. seal is a factor of ~12,000. This can reduce the annulus cross-connection to insignificant.

Beyond a few seals in the annulus, there is limited cross-connection, because the impedance increases with each seal and the gradient decreases with each seal. There will still be some vertical flow in the annulus, because the seals are not the same as a fully grouted annulus. The fully grouted annulus, down in the dark hole, is not always impermeable. But the question is whether the remaining flow is significant. It is very useful that the low conductivity seals dominate the flow in the annulus, the average flow rate will be nearer that of the formation if a few aquitards are identified and a super seal is constructed. Also, it is true that the conductivity of the sand is less important if the formation conductivity is low. That may allow a more coarse-grained sand pack to settle more quickly if emplaced from the surface. However, a more conductive sand blurs the spatial resolution of FLUTe measurements from inside the casing.

It is useful to recognize that a potential aquitard identified in the core may not be a continuous layer and therefore does not produce an exceptionally large gradient across the less permeable layer. That makes the seal of the annulus less challenging and does not change the advantage of the super seal construction location.

I have not addressed the effect of a high head in the middle of the formation, but the flow, up or down, will be equally dominated by the sealed intervals. The advantage of access to the entire formation with the several FLUTe methods, which also eliminates the flow in the open casing, is obvious. However, it is clear that one should not leave the casing open for any significant time. Some continuously screened cased holes have been left open for many years and some boreholes known to us were left open for more than 10 years in a contaminated area. That is not good.

The steady-state calculation does not include a head loss of the flow in the annulus due to radial flow because if the head in the annulus exceeds the head in the formation, there will be some lateral flow to the formation. That is a concern, but the radial flow can slow the propagation through the annulus from a high head of an aquifer at the top of the well.

14.3   THE CONSTRUCTION OF THE CSC DESIGN

How can such a backfill design be reasonably constructed? It is important that such an interrupted sand pack design can be built using ordinary well construction methods. Past FLUTe test experience suggests that it is certainly within the capability of available well construction methods. A recent installation of six sand packed 3″ diameter 5-ft. long screens in a 3″ casing to 255 ft. inside a 7″ OD sonic casing would have required only some additional procedures to build the interrupted design of Figure 14.1. One means of installing the sealing intervals is shown in Figure 14.12(a–c). In this illustration, Figure 14.12(a), a sonic-driven casing, is used to form the borehole. The cored material inside the drill casing is removed to allow a continuous screened casing to be lowered into the drill casing. An initial sand fill is installed with a length somewhat more than the length of a segment of the drill casing. The drill casing is then withdrawn for one drill casing section (e.g., 10 ft.) as shown in Figure 14.12(b). The sand will settle with the sonic vibration to fill the open borehole beneath the drill casing. A flexible liner is then installed inside the screened casing to seal the interior of the screen. A sealing material (e.g., grout of 5 parts cement to 1 part bentonite with density of 1.35 g/cc) can be emplaced on top of the sand fill in the annulus. An additional sand fill of the same length as the first fill is emplaced inside the drill casing on top of the sealing material, and the drill casing is withdrawn for another drill casing section length. A second grout fill (Figure 14.12(c)) is added to the annulus between the drill casing and the continuous screen. The sealing grout is trimmied into place with a pipe that is filled with the measured grout seal volume (e.g., 1.5 gal. in a 7″ hole) on top of the sand. A "rabbit" can be used to expel the small remaining volume of grout from the trimmie using a water drive, or by simply flushing the trimmie with water. Each time the sonic casing is pulled by the prescribed length (a typical casing section of 10 ft.), the sealant is added.

The trimmie should be fitted with a deflector to avoid jetting a hole in the sand fill with either sand slurry or grout additions. The trimmie can be ~10 ft. above the sand level during the sand fill to better allow separation of the sand size to provide a fine-grained layer on top of each sand fill. For deeper boreholes, the sand addition at the surface may be too slow in settling to the seal depth. The trimmie can be supported on the screened casing during the drill casing removal. The sand depth can be measured each time using the trimmie lowered on top of the sand, then raised just before the grout is added. A sealed end and side vents of the trimmie allows the grout

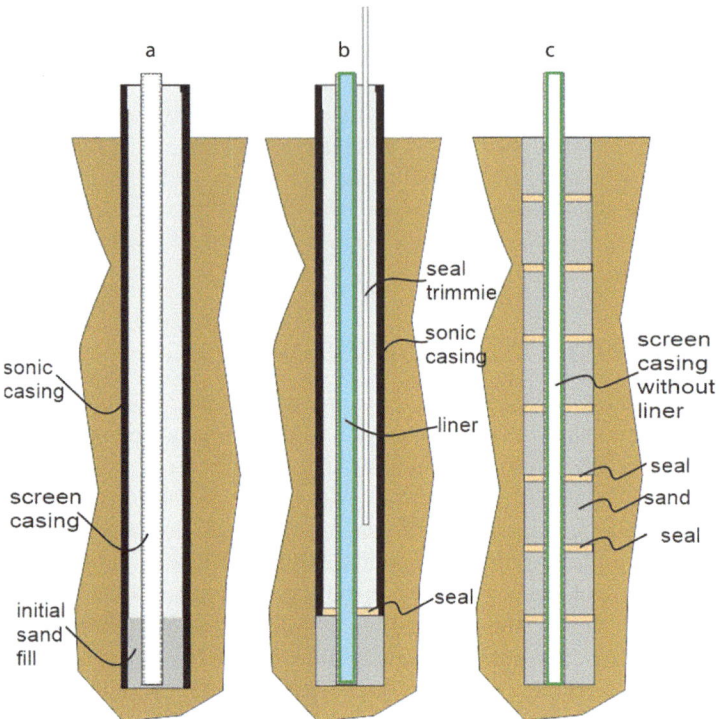

FIGURE 14.12 The backfill sequence possible in sonic casing. (a) Shows the first sand fill after the screened casing installation. (b) A flexible liner (blue) seals the casing. The drill casing is withdrawn and a sealing layer added on top of the sand. (c) Many more sand additions and sealing layers are added to complete the design. The vertical conductivity of the sand pack is greatly reduced by the sealing layers.

addition near the sand and the sand addition after the trimmie is raised. Bentonite pellets are not so well suited for the seals with bridging and with Bentonite loss during development of the screen and poor penetration of the formation.

Each grout fill invades the sand pore space for only a short distance due to the viscosity of the grout and the grout also flows into the slots in the screen but not into the interior of the screen because of the sealing liner. Some grout will also flow into the formation, sealing the formation for a short distance from the borehole wall. Note, the liner is filled with water above the formation water table and prevents any grout flow into the interior of the screen. Subsequent sand additions as shown in Figure 14.12(c) will allow the sand added to settle into the grout fill, further driving the grout into the formation, producing a larger diameter seal. Figure 14.1(c) shows the result of the completed well design with annular seals of the sand fill after the interior flexible liner has been withdrawn for measurement access inside the casing. It is useful that the sand poured through the water will cause the finer grains to settle more slowly forming an especially fine grained layer on top of each sand addition. This will reduce the grout invasion of the lower sand increment and enhance the migration into the formation. It may be useful to add the sand as a slurry well above the last grout addition to allow the finer component to settle later even if the sand is added as a slurry. Clearly, the sand addition must not rise above the bottom of the trimmie.

The water in the annulus should be pumped down frequently in order to prevent the rising water level from collapsing the liner. The liner with the casing support can be filled very high above the water table in the formation. Some water loss from the annulus is likely with the withdrawal of the drill casing.

With the design of Figure 14.12(c), the interior of the continuously screened casing is available for measurement of the several formation characteristics through the screen and sand pack. Those are measurements of horizontal conductivity, contaminant distribution, head distribution, and discrete water quality samples using the sealing liner designs by FLUTe.

The access to the continuous well screen in the casing allows the entire formation sequence to be well developed using surge blocks and air lift pumping. However, as for open holes, the development of the well using only pumping tends to clear only the high flow intervals. Surging is recommended to remove the sediment in the formation developed with the drilling process. The choice of the frequency and length of the sealed intervals is optional.

It is fortuitous that the lateral flow of the sealant into the formation is greater in more coarse sediments where a larger diameter seal against bypass is desirable.

While there has been some objection to the difficulty of measuring the top of a grout pour, that is not a problem with this design. In this design, the grout addition is a fixed volume. The top of the sand above the grout is easy to locate after each sonic drill casing withdrawal. If the sand level is above or below the expected elevation, some adjustment is needed for the next sand lift. In some intervals, the borehole moves inward as the casing is withdrawn, which reduces the sand lateral flow and affects the sand level for each sonic casing withdrawal.

The cost of the many seals being added every 10 ft. to a 200 ft. continuous screen cased well were assessed with the comparison of two quotes provided by drillers. The CSC cost was 36% above that of a conventional single-screened well to 200 ft. A comparison to the cost of three cluster wells should be more. The cost for three nested wells may be similar. However, neither of those installations provides the same access to the formation, and the screened intervals must be selected on the basis of other measurements.

The construction described above is not possible without the use of the sealing liner in the continuous screen. The sealing liner is only removed when the measurement liners are ready for installation to minimize the time the casing is open. Note, this is not a monitoring well as defined by NJ regulations, but an exploratory well used for immediate measurements, and therefore, it may not be a significant concern about cross-connection beyond that of the construction of an ordinary well in an open borehole.

14.4   COMBINED OVERBURDEN AND BEDROCK ACCESS

In many situations, the contaminants have originated in the overburden and migrated into the bedrock. There are numerous direct push methods for mapping contaminants in soft overburden, but if the direct push rods encounter refusal, the hole must be drilled. Figure 14.13 illustrates a design combining the overburden and bedrock

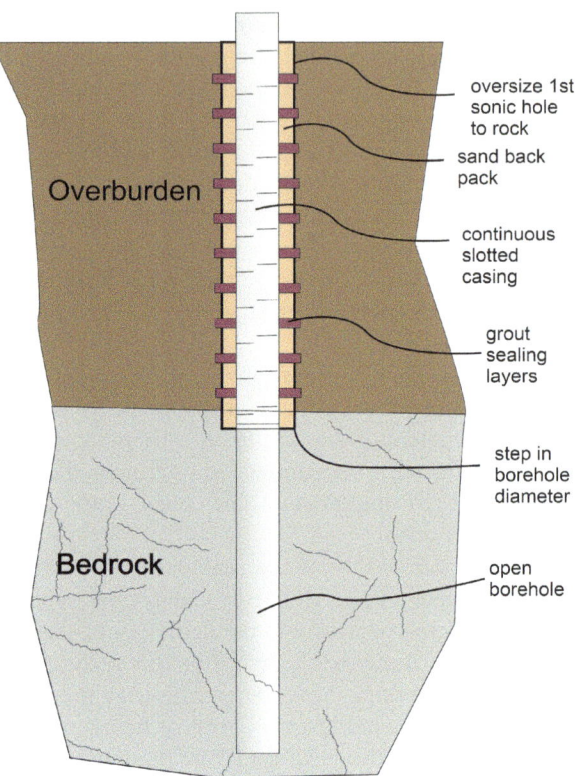

FIGURE 14.13   A CSC through the overburden and an open borehole in bedrock. Liner seals casing during construction of the annular fill from surface to bedrock. Allows access to the entire interval for mapping various parameters with liner methods.

measurements allowing FLUTe methods in both formations. The overburden interval is supported by a continuously screened casing, CSC. The installation sequence is simple:

1. Drill to the bedrock with a larger sonic casing and leave the casing in place.
2. Drill a smaller hole to depth in the bedrock.
3. Withdraw the smaller casing.
4. Lower a continuous screened casing to the bottom of the large casing with an ID of the rock hole, and install a sealing liner in the casing and a short distance into the rock hole.
5. Perform the backfill sequence as described above in the CSC construction (14.3). It may be useful to start with a grout seal at the bottom of the sonic casing.
6. When the CSC construction is complete, remove the sealing liner, develop the entire hole, and replace the sealing liner to later perform the FLUTe sequence throughout the entire hole.

14.4.1 Discussion of the Design Function

Figure 14.14 shows the in situ state of the sealing grout. The initial grout addition invades the sand-filled annulus for a short distance downward due to the grout viscosity. Tests under a 10 psi driving pressure showed invasion of only a few inches of sand with a viscous bentonite-grout. Adding sand on top of the grout pressurizes the grout forcing it into any available pore space in the formation. The more permeable the formation, the further the injection into the formation. The added sand settles into the remaining grout to seal the sand pore space a short distance above the initial grout level. The injection into the formation resists bypass of a seal by any flow in the formation.

The effective vertical conductivity of the seal alone is zero, except for potential bypass of the seals in the formation. Using a more fine-grained sand pack (e.g., 20–40 screen size) than is typically used for water wells, further reduces the flow rate in the sand-filled annulus. The resistance to vertical flow forces any fluid pressure in the annulus to be dissipated by lateral flow from the annulus. This pressure reduction reduces the vertical gradient driving any potential bypass of the seals. The pressure loss from the narrow annulus can greatly limit the vertical migration possible in a sand pack with the seals.

The maximum sand pack length allowed by the State of NJ DEP is 25 ft. An interrupted design of Figure 14.1 with seals every 10 ft. is far less conductive than a continuous 25-ft. sand pack. The CSC design is much less conductive than an MLS with only a series of packer seals in an open borehole as has been a common use.

The continuous screen of Figure 14.1 can be built with extended sealed intervals (super seals) in clay-loaded layers in the formation, as identified in the core, to provide

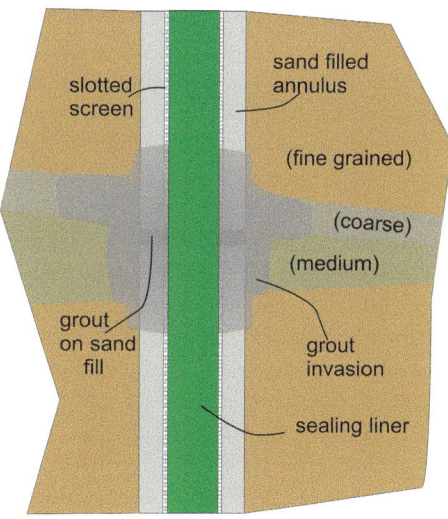

FIGURE 14.14 Grout penetration of permeable materials at the level of grout addition on top of sand lift. Next sand addition causes additional flow of the grout into the pore space of the sand and formation. More coarse layers are sealed further from the borehole.

even more effective sealing of the annulus. Such aquitard layers are not expected to allow contaminant transport but can be confirmed from the open screened interval measurements above and below those layers. This is a modest compromise for the advantage of having most of the formation accessible for measurements as described.

Remember that the interior of the casing is sealed by a liner both during the backfill emplacement and during the subsequent measurements described. Those measurements benefit from the ability to develop the entire open screen and formation of drill cuttings and cross connected fluids from the drilling process.

Variations of the storativity were examined in the time-dependent calculations described above. The two main effects of reducing the storativity were to cause the pressure change to propagate more quickly and therefore to increase the flow past the deeper seals. The second effect was to decrease the flow into the top boundary because of the pressure propagation and the decreased gradient. The increased pressure propagation also increases the lateral flow, but due to the divergence of that flow from the borehole, the potential increase in contamination in the upper regions is small. Ultimately, the steady-state calculations are the more relevant to the longtime effect of the seals in the annulus. The use of a few super seals allows the cross-connection to be less than that possible through the formation, even though the sand-filled annulus conductivity without seals is much greater than the formation conductivity.

14.4.2 Conclusion of the CSC Design

The CSC design includes the installation of numerous seals in the annulus of an otherwise continuous sand backfill of a continuous screened casing during the well construction. This construction allows the full use of the several high-resolution FLUTe measurements of hydrologic characteristics in unstable sediments as is already and frequently done in bedrock wells. The several high-resolution measurements allow a much better effective design of remediation procedures and estimates of risk of propagating contaminant plumes. The resolution is better than the spacing of the seals.

The cost estimate provided for the CSC for a 200 ft. well shows a 36% increase over the construction of a conventional single screened well. Use of this design should allow a significant cost reduction from current practice of site characterizations and remediation injections. This approach offers a much more detailed data return than is available from such measurements as packer testing beyond sonic casing and nested wells. The guidance available for selection of sampling and injection intervals is a significant improvement over the data available from a few single screen wells. The concern about cross-connection in the traditional sand pack of continuously screened wells is reasonable and shown by the above calculations to be an extreme mode of cross-connection even if the casing is sealed with packers.

The CSC method described is not of significant value unless the interior of the screened well is sealed with a continuous flexible liner as with FLUTe's several methods. Even the construction of the CSC system depends on the use of a sealing liner in the casing. It is noteworthy that sonic drilling techniques are especially useful for installing long sand pack columns in slender boreholes because of the vibration of the moving sand available with the sonic method to assure a predictable emplacement of the sand between the seals. The seals are potentially more effective if the grout is able to penetrate the more permeable intervals of seal locations, since the formation conductivity is an important factor in the seals reduction of annular flow. The flow reduction of over a thousand-fold in the annulus is particularly helpful. Even a high formation conductivity, while providing less impedance at each seal, also causes a reduction in the gradient in the sand-filled annulus. It is important to recognize that the flow in the annulus is not zero as is believed for a bentonite or grout seal of a long solid casing, but sufficient with the frequent seals to reduce the flow in the annulus to insignificant as compared to the natural flow in the material removed in drilling of the borehole. In comparison with other measurements in boulder-filled unstable sediments, the measurements in the CSC system should provide much more complete vertical data. The sonic drilling capability in high velocity sediments and glacial tills with boulders is the more useful method for this design. Use of the CSC for remediation injections or extractions is another advantage. If the measurements in the CSC suggest significant migration in the annulus, after obtaining the measurements desired, the casing can be grouted or drilled out of the borehole.

14.5   T PROFILES IN CONTINUOUS SCREENED CASING

14.5.1   Bypass of the Liner in the Sand Pack

While the T profile is normally done in open stable boreholes, it was asked if the T profile can be done in a continuously screened casing. There are many continuously screened 2″ casings in place at old contaminated sites.

First, it is necessary to understand how the FLUTe transmissivity profiling method works. That is described in the paper by Keller et al (2013). The method has been used in open stable boreholes usually in fractured rock for 300–400 different boreholes in the last ten years. Briefly, the everting liner descends at a rate controlled by the transmissivity of the borehole. As the everting liner sequentially seals each flow path from the top-down, the liner descent velocity decreases with the loss of borehole transmissivity. The velocity is recorded to measure the velocity change due to the sealing of each fracture. From the velocity data are deduced the location and flow rate of each fracture/flow path. The flow rate into the fracture is just the velocity change, Δv, multiplied by the borehole cross section, A. In this manner, the transmissivity distribution throughout the borehole is determined on the nominal 6″ scale. For a horizontal fracture, the velocity change is an abrupt step change. For a dipping fracture, the change is somewhat less abrupt to a slope. For a permeable interval, the change may be more gradual. These changes are the response in an open uncased hole.

To address the question of whether the method can be used in a continuously screened well with an annular sand pack, we needed a quantitative means of assessing the several contributing factors affecting the flow in the sand pack. Those influential factors were expected to be:

1. The permeability of the sand pack behind the screen
2. The dimensions of the annular sand pack, in particular the casing diameter and hole diameter

3. The flow rate in the permeable feature, a fracture or flow path, in the formation behind the sand pack
4. The liner velocity (dependent on the transmissivity beneath the liner and the casing diameter) The driving pressure/head in the borehole versus the formation head (Figure 14.15)

It was not practical to perform tests in the field to deduce the significance of all of these factors experimentally. The approach described here is to model the flows involved with an analytical solution.

The flows calculated were:

1. The flow rate out of the casing beneath the everting liner which determines the liner descent rate
2. The vertical flow rate in the sand pack which passes the descending liner
3. The flow rate in the nominal fracture or bed behind the sand pack

Figure 14.15 shows the geometry of the model. The descending everting liner seals the screen adjacent to the hypothetical fracture. The liner continues to descend past the fracture. As the liner descends it is driving the water beneath the liner into the formation through the screen and sand pack. The pressurized water in the sand pack is free to flow upward to the fracture after the liner passes the fracture.

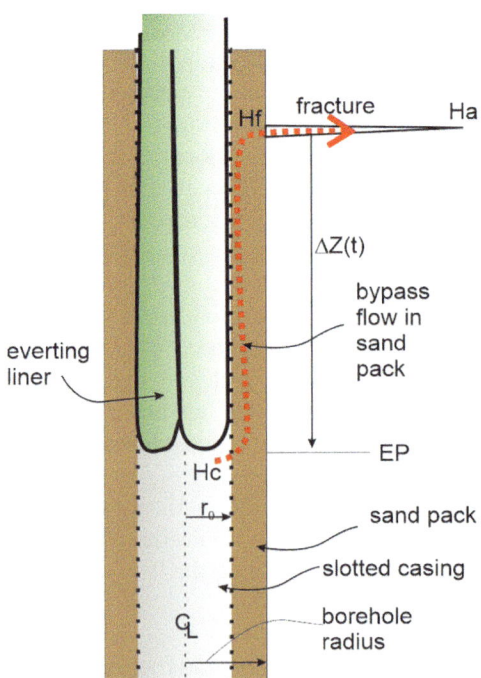

FIGURE 14.15   The geometry of the bypass flow calculation.

Sealing the screen adjacent to the fracture does not seal the fracture or prevent the water flowing in the sand pack from entering the fracture. Therefore, as the liner descends, as a flow meter, measuring the flow out of the casing beneath the liner, the bypass flow from the interior of the casing into the sand pack to the fracture is still part of the liner measured flow. That bypass flow path is shown as the red dashed line from beneath the liner through the sand to the flow path of interest (Figure 14.15).

The normal liner behavior in passing and sealing a fracture in an open hole is to suffer an abrupt liner velocity decrease. The velocity decrease multiplied by the borehole cross section is the flow rate into the fracture. With the bypass of the sealing liner as shown by the red dashed arrow in Figure 14.15, the liner velocity change is not an abrupt step, but rather a slower decay of the liner velocity as water still flows from inside the casing to the fracture through the sand pack.

The flow in the sand pack is defined as the bypass flow. The flow in the fracture after the liner passage of the fracture elevation is called the reduced flow. The liner after passing the fracture elevation propagates for a distance $\Delta Z(t)$ below the fracture depending on the velocity of the liner and the time after passing the fracture. The liner velocity is recorded in time and with depth in the hole, so the liner velocity is known as a function of the depth in the hole.

The flow in the fracture after the liner passes the fracture elevation is the bypass flow from beneath the everting liner depth, called the EP depth, below the surface. That EP to fracture distance, ΔZ, increases with time as V dt where V is the liner velocity, and dt is the time after the liner passes the fracture elevation. The increasing length of the bypass path causes the bypass flow to diminish in time.

The flow to the fracture is considered as two parts: (1) The bypass flow in the sand pack and (2) the flow into the fracture. Those flows are equal since all direct flow through the screen into the fracture is stopped with the liner passage of the fracture elevation. Also prevented by the liner is the flow through the screen between the EP and the fracture.

The flow in the sand pack is assumed to be only the flow from the casing to the fracture. Possible flow into or through the formation will be discussed later. That bypass flow in the sand pack is called Q_{bp}. The flow from the sand pack into the fracture is called Q_f. In this model, $Q_f = Q_{bp}$. If the matrix is of relatively low permeability compared to the fracture, the vertical flow in the annulus is:

$$Q_{bp} = A \ C \ \Delta H_{bp} / \Delta Z(t) \tag{14.1}$$

where A is the cross-sectional area of the sand annulus, C is the sand pack conductivity, and ΔH_{bp} is the head difference between head in the casing, H_c, and the head at the entrance to the fracture, H_f. $\Delta Z(t)$ is the distance the liner has passed the fracture and the length of the sand pack flow path as a function of time. Since $\Delta Z(t)$ is a distance in time, the flow bypass flow rate is also related to a depth in time where $D = D_0 + V$ dt, where D_0 is a reference depth (usually below the GS) and V is the velocity of the descending liner after it passes the fracture. This description only treats the bypass effect upon passing a single fracture or other flow path and does not treat the actual change in V below the fracture. That assumes the magnitude of the full step velocity change is small compared to the actual liner velocity. Were the actual velocity to be considered for the depth calculation, the depth scale would be compressed below the fracture as the liner slows as the bypass decays. That would be the case for a large fracture near the bottom of the hole.

In the fracture, the flow from the sand pack bypass is assumed to be treated by the Thiem equation as used for straddle packer assessments:

$$Q_f = T\Delta H_f\, 2\pi/\ln(R) \qquad (14.2)$$

where T is the effective transmissivity of the hole wall interval containing the fracture, ΔH_f is the head difference between the sand pack head at the entrance to the fracture, H_f, and the distant ambient head in the formation, H_a, or the heads at r_0 and r_a respectively. R is r_a/r_0. The assumption is that the fracture flow behaves as described by the Thiem equation for a permeable medium. The bypass flow is defined as the flow into the fracture. Therefore,

$$Q_{bp} = Q_f \qquad (14.3)$$

In order to assess the effect of the bypass, it is useful to divide the bypass Q_{bp} by the original flow, Q_0, into the fracture flow when the liner has not yet passed the entrance to the fracture. The head in the casing is defined as H_c and assumed to be essentially constant as the liner is descending.

The original flow into the fracture before the liner passed was therefore,

$$Q_0 = T(\Delta H)\ 2\pi/\ln(R), \text{ where } \Delta H \text{ is } H_c - H_a \qquad (14.4)$$

$$Q_{bp}/Q_0 = Q_f/Q_0$$

It is noteworthy that the sum of the pressure drop in the sand pack plus the pressure drop in the fracture is the same total pressure drop

$$\Delta H = H_c - H_a = \Delta H_{bp} + \Delta H_f \qquad (14.5)$$

Upon dividing Equations (14.2) by (14.4), note that $Q_f/Q_0 = \Delta H_f/\Delta H$ = the ratio of the bypass to the original flow as the liner descends.

From the above equations and simple algebra, one can deduce that the ratio of the bypass to the original flow is:

$$\text{Ratio} = AC/(2\pi\ T\ \Delta Z(t)\ln(R) + AC), \text{ where } \Delta Z \text{ is increasing with time, t.}$$

From this, it is clear that the bypass flow is relatively small if T or ΔZ are large.

The bypass is large if A or C are large. In other words, if the sand is coarse or the borehole ID is much larger than the casing OD.

Since the ratio is the fraction of the normal flow into the fracture without a sand pack, the original flow/velocity change measured without a sand pack is reduced by the bypass flow. So, instead of an abrupt change in velocity (flow rate as measured

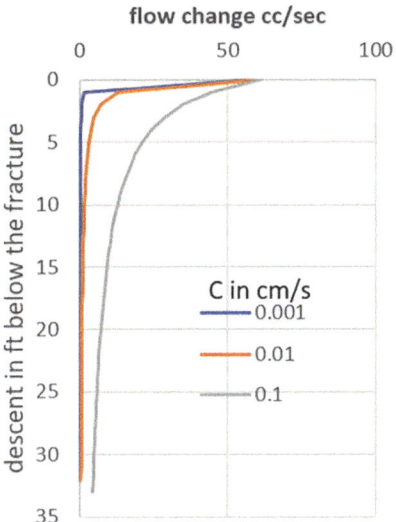

FIGURE 14.16 The liner velocity below the fracture with different values of sand pack conductivity. A low conductivity (<0.1 cm/s) sand pack allows easy identification of the fracture and flow capacity.

by the liner descent rate), the velocity change becomes a slow and gradual change as the ΔZ increases and the bypass flow becomes less.

The liner descent with a sand pack should experience the normal change due to sealing the fracture, but distributed in time and distance because of the bypass flow:

$Q(t)$, which is the liner velocity $V(t)$ times the area of the borehole, A, is the liner displacement rate or flow rate out of the casing and is equal to $V_0 A - Q_0 (1 - \text{Ratio})$, where V_0 is the velocity just before sealing of the casing at the fracture, and Q_0 is the full flow rate in the fracture before the liner passes the fracture. Note that Ratio is time-dependent and decreases as the liner descends further.

Figure 14.16 shows the effect of different values of C for a fracture with a transmissivity, T, of 0.2 cm²/s. C (the sand conductivity) is varied from 0.001 to 1 cm/s or from fine silty sand to very coarse sand to small gravel. The annular space is from 1″ to a 3″ radius.

At the low conductivity which is not common for sand packs unless silt has clogged them, the change in the transmissivity profile measurement velocity is very abrupt with a small tail due to bypass.

For the larger sand pack conductivity, the decay is very long and would reduce the spatial resolution of the measurement of the formation T, unless there were no nearby fractures.

Figure 14.17 shows the effect of different transmissive feature T values on the effect of bypass and the liner measurement. The sand conductivity, Cs, is 0.01 cm/s and the annulus is the same 1–3″ radius.

A flow zone with high T stays connected for a significant depth below the flow feature. However, the flow zone location is easily recognized.

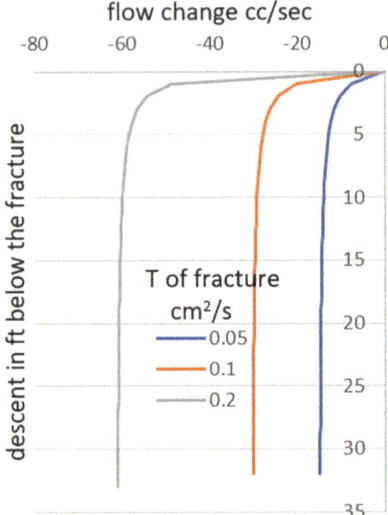

FIGURE 14.17   Effect of transmissivity of flow zone on the decay rate of the step change of uncased hole. Cs = 0.01 cm/s and annulus radius is 1–3″ (2″ in 6″ hole). Transmissivity of fracture does not affect the resolution of flow location or flow capacity much.

An interesting feature of the model is that it does allow one to assess the effect of a vertical conductivity in the formation adjacent to an uncased borehole. The effect of upward flow in the formation to the fracture is to produce a tail in the decay of the liner velocity after sealing a fracture. The presence of a short tail has been noted in some velocity/flow measurements while performing FLUTe T profiles in uncased boreholes.

It would be interesting to review the data from measurements with significant vertical conductivity, or to determine for measurements already done, what is the probable vertical conductivity that would match the decay in velocity seen in the data.

14.6   CONCLUSION

The flow in the sand pack is expected to prolong the usual abrupt change in the liner velocity as normally seen in the uncased borehole transmissivity profile. It should still be clear where the high flow zones exist and one can estimate the total flow change at a prominent flow zone if the extrapolated decay is clear in the recorded liner velocity. Whereas the method resolution is less than for an uncased hole, the resolution should be much better than with packer testing or flow metering that have limited resolution in uncased holes. Actual test data would be very helpful in determining the response to flow paths in the formation. Any difference from a straight line decrease in velocity to the bottom of the hole, as expected for a uniform medium, would be helpful to assessment of higher flow zones. A discrete flow measurement and water quality measurement can be obtained with the FLUTe CHS (Cased Hole Sampler) MLS design especially if it is deflated and moved to a different elevation.

That system can be equipped with large diameter tubing for flow rate measurements and purged simultaneously which avoids the flow in the sand pack problem.

In unstable sediments, the continuously screened well of the CSC design is useful especially since one often does not know where the major flow zones lie. A useful aid to mapping the flow zones behind a continuous screened casing is to use a sand pack of relatively fine material to reduce the vertical bypass for discrete measurements of transmissivity or water quality. It is also useful to not use a borehole larger than needed to emplace the sand pack. This is contrary to the usual water well `design. The fine grained sand pack may be more easily emplaced with the vibration of a sonic drill casing emplacement of the screened well than by simply pouring the sand down an annulus. It is still good to develop the well to reduce the clogging of the formation natural conductivity. The fine grained sand pack and sealing the casing with a continuous liner greatly limits cross-connection. The screen slot size must be considered in the sand pack emplacement. The common 20–40 sand pack has a conductivity of about 0.08 cm/s. A somewhat smaller sand size is suggested by the above calculations.

Finally, with the CSC system described above, the FLUTe T profile would still have a resolution of less than 10 ft., which is the spacing of the CSC seals.

15 Other Applications of Liners

Hereafter is described the theory and calculations of installations in crooked boreholes and pipes. The methods are used to implement installations under buildings, in landfills, and for liner augmentation of horizontal drilling for emplacement of FLUTe systems.

15.1 USE OF LINERS IN ANGLED, HORIZONTAL, AND TORTUOUS BOREHOLES OR PIPES

Flexible liners can be driven with a variety of fluid fills. Some liners are installed in boreholes using heavy mud to overcome artesian conditions. Others are installed in angled holes with water. Some FLUTe liners have been installed under buildings in horizontal pipes laid in trenches prior to construction of the building. The pipes were connected to vertical wells beneath the building. Figure 15.1 shows the geometry of such an installation. The near horizontal portion of those installations under buildings had several 90-degree turns. The model for calculation of the driving pressures as a function of such a pipe trajectory is described in Section 16.1.

15.1.1 THE LAHD HISTORY

LAHD stands for Liner Augmented Horizontal Drilling. The liner augmentation method was one of the earliest liner methods developed by the author. After the original patent was purchased by Eastman Cherrington Environmental, a horizontal drilling company with the intent of melding flexible liner methods with their drilling methods, a shop to manufacture liners was set up in their Houston warehouse. Unfortunately, the Eastman Cherrington drilling method incorporated the most expensive heavy equipment for horizontal drilling designs developed by Cherrington for pipelines under rivers. Other less expensive methods such as the Flow Mole and Ditch Witch methods gained the larger market share and Eastman Cherrington closed their business in 1995 orphaning the flexible liner technology. In 1996, Carl founded FLUTe and continued to build liner systems in Houston for a few years completing contracts already in place like one with the USGS for Yucca Mountain measurements. In 1997, he was invited by a Ditch Witch representative to meet with Ed Malzahn, the inventor of the Ditch Witch trenching machine, at their plant in Perry, OK, to consider Ditch Witch use of flexible liners. Ditch Witch had a number of compact horizontal drilling machines. They suggested at that meeting that rather than developing a Ditch Witch test of Carl's methods in Perry, Carl should ask Roland Davis in Mustang, OK, to try the method. Roland Davis was an adventurous

DOI: 10.1201/9781003268376-15

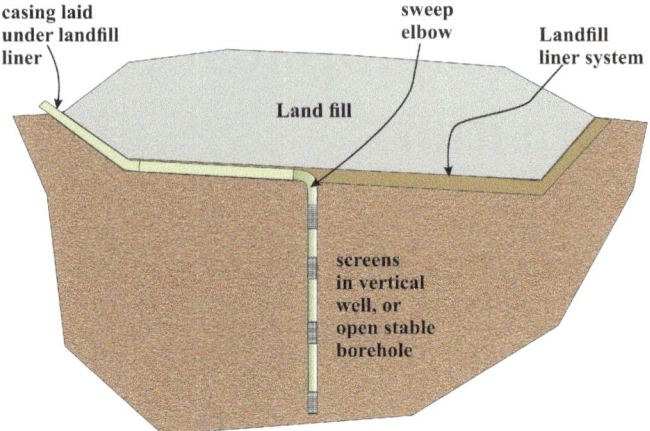

FIGURE 15.1 Flexible liners can be everted into horizontal boreholes and piping containing numerous sweep elbows. This allows the transport of a variety of devices for measurements in tortuous passages as shown.

driller with Ditch Witch drilling machinery who invited Carl to test his methods at his place in Mustang. Carl built the system and took it to Mustang where Roland drilled horizontal holes of 100 ft. long in a field adjacent to his front yard.

The liner system was installed following Roland's reamer through the borehole as described in the following section. The liner was installed from a hose canister as described in Section 3.5.2. The photo in Figure 15.2 shows Roland next to the FLUTe liner as it propagated across his field after exiting the horizontal borehole. The LAHD technique for the installation of a casing was also tested at Roland's

FIGURE 15.2 Roland Davis at the first test of LAHD in Mustang, OK, with the liner everting across the surface beyond the borehole exit after the reamer is removed from the borehole.

place. Another design tested there was the method for inverting the carrier liner from the outside of casing in a horizontal hole, as is necessary for a casing with screens left in place requiring access to the formation.

15.1.2 THE LAHD METHOD

The concept is to drill a horizontal hole with one of the several horizontal drilling methods. After the pilot bit exits the formation far from the entrance of the drill, the pilot bit (typically 4″ diameter) is replaced by a larger reamer bit, which is drawn back through the pilot hole, enlarging the borehole, and often towing a casing into the borehole. However, the reamer leaves the mud and cuttings in the borehole that can often impede the casing travel and if the drag is sufficient, can cause the casing to tear apart, especially if the casing contains screen sections.

The LAHD method adds a pig (tapered cylindrical plug) of the reamer diameter to the reamer as shown in Figure 15.3. As the reamer is drawn through the borehole, a flexible liner is everted against the pig following the reamer. The pressurized liner forces the mud and cuttings from the reamer out the entrance of the borehole. The liner also supports the borehole against collapse in unstable formations, which is the usual function of the drilling mud. As the reamer is withdrawn from the borehole, the liner continues to evert out of the borehole as seen in Figure 15.2. Figure 15.3 shows the liner installation sequence with the liner everting from a reel canister that was later used in an installation under a landfill in Indianapolis, Indiana, as shown in Figure 15.4.

The same liner in Figure 15.4(a) installation photo is shown in Figure 15.4 as it exits the borehole after 400 ft. and everts out of the mud pit beyond the borehole clean of any mud. The liner in Figure 15.4 was built with eight sampling intervals that were used for extracting pore water samples from the formation beneath the landfill.

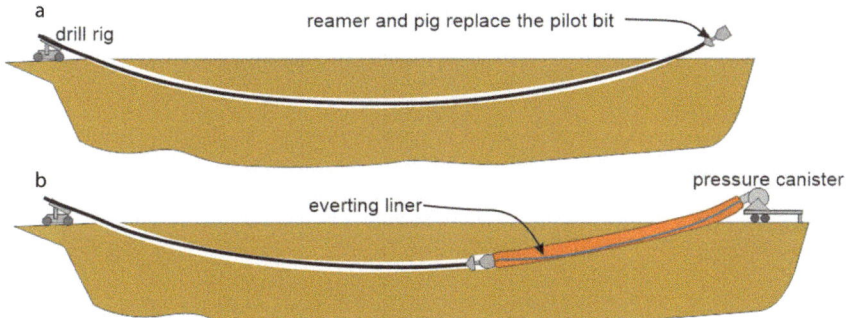

FIGURE 15.3 (a) The horizontal borehole is drilled with a pilot bit exiting at a distance. The pilot bit is replaced by a reamer and pig. (b) As the reamer cuts a larger borehole, the liner is everted against the reamer, forcing the drill mud and cuttings out of the borehole entrance. The liner supports the borehole wall preventing collapse. Various devices can be emplaced with the everting liner.

FIGURE 15.4 The top photo shows the liner installed for 400 ft. from a pressure canister under a landfill in Indiana carrying eight sampling intervals. The bottom photo shows the everting liner exiting the borehole after the reamer is removed. It is clean exiting out of the mud pit because it is everting.

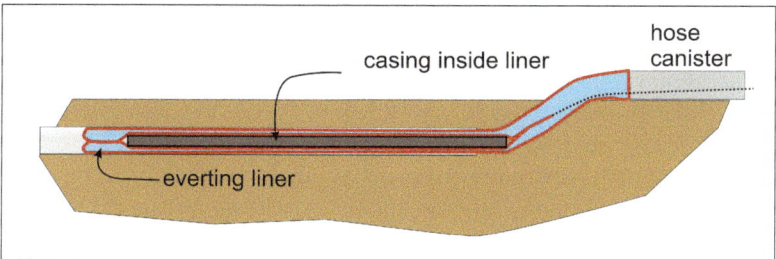

FIGURE 15.5 Liner everting through borehole carrying a casing in the inverted liner. As the eversion continues, the casing extends from the liner usually at the surface.

However, the first LAHD liner of Figure 15.2 was installed from a hose canister as described in Section 3.5.2. Later a casing was carried into the borehole inside the liner as shown in the drawing of Figure 15.5. The hose canister needed to be twice the length of the casing. The liner carries the casing while supporting the borehole against collapse on the casing. A subsequent third design of the liner, Figure 15.6, allowed the liner to be withdrawn from off the casing and borehole, leaving only the casing in place, as is useful for extraction or injection in the horizontal borehole. The liner in Figure 15.6 was inverted simultaneously off the casing and from off the borehole wall using a cylindrical attachment (a cylindrical tether) to the fully everted end of the liner. The liner was everted until the attachment point was visible. The liner withdrawal from off the casing or pipe allows the formation to collapse on the screened casing sealing the annular space outside of the casing. The double inversion of Figure 15.6 was done with a reel canister at the wellhead that provided the pressure to invert the liner and to house the cylindrical tether and liner as they were removed. The cylindrical tether was an uncoated permeable fabric cylinder, which

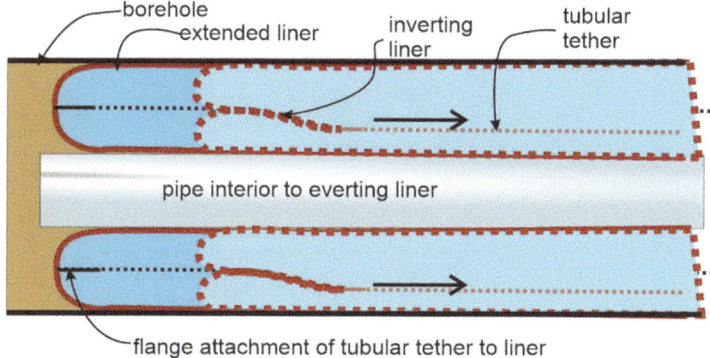

FIGURE 15.6 Geometry of tubular tether attachment to liner that allows a double inversion of the liner into itself by peeling the liner simultaneously off the hole wall and the casing or pipe avoiding the drag of attempting to pull the liner directly off the casing. The normal closed end of the liner is at the end of the pipe.

was drawn back into the reel canister, peeling the liner off the pipe and peeling the liner off the hole wall at the same time to avoid the prohibitive drag otherwise resisting the liner removal under pressure. All three methods, as tested in Mustang, were patented. Neither of the methods in Figure 15.5 or Figure 15.6 have been used except in various other liner designs.

The horizontal drill rig used in Indiana (Figure 15.4) several years after the first test in Mustang was a Roland Davis rig and his crew from Mustang, OK.

15.1.3 Advantages of the LAHD Method

The supporting liner provides two or three functions. One is to force the mud and cuttings out of the entrance of the pilot hole. The second is to support the borehole against collapse on the casing which was usually pulled with the reamer into the horizontal hole through the mud and cuttings. The collapse of the formation on the casing was a limiting factor in the length of casing that can be pulled directly into the hole, especially if the casing has weak slotted screen intervals in the casing for access to the formation. The casing installation required a very high pulling strength of the drill rig as was achieved by Cherrington for under river crossings (said to be over a million lb. pulling capability). The casing used for fluid flow or routing of electrical cables does not require screen intervals but is also limited by hole collapse. The LAHD technique could even be used to emplace typical short sections of clay field tile. No tensile strength of the casing is required. A third function is to emplace sampling tubing and spacers, as was done for the Indiana site at eight locations (Figure 15.4).

16 FLUTe Calculational Models

FLUTe has developed a variety of calculational models as an important part of the use of various liner methods. Most of the theory and mathematical models of the FLUTe measurements are earlier in this text. A mathematical model developed for the installation procedures is described in this chapter. Other models described hereafter are used in reducing the data and in use of the data. Since I, the principal scientist, spent many years developing a variety of models for flow in the earth and programing in Fortran, it was logical to build models for the methods invented for use of flexible liners. Most of the models used by FLUTe now use spreadsheets to do the calculations even for finite difference time dependent calculations. This chapter is a brief description of the assumptions in those models and applicability of some models for complex problems such as installations in crooked pipes. The models have been an important basis for judgment of the relevance of the measurements and the results have been compared to many test results and actual field data sets. Some models are simple mechanical models and some are models of flow as part of the measurements. Mechanical models for the design of the machines developed for use with liner methods are not described here but were used to select the components of those machine designs. Nearly all FLUTe models can be refined with more detailed calculations, but such has not been necessary. Tests against experimental data are important to assess the validity of the models. Comparisons with other kinds of measurements have also been done and are not described here, but they have been done and published by others (e.g., packer tests versus T profiles) with good results.

16.1 THE CROOKED PIPE MODEL

16.1.1 HISTORY AND PURPOSE

The first everting liner conceived by the author was for the purpose of installing absorbers in a horizontal borehole in the vadose zone. Early tests showed the hole diameter to be an important factor in the installations as described in Chapter 3. Drag of the inverted liner on the everted liner was also recognized in that longer liners installed horizontally required a higher driving pressure the further the liner was everted. In order to assess the effect of the liner diameter on the ability of the liner to tow instruments into boreholes, tests were done to measure the towing force at different driving pressures. The results were plotted and a model was developed as described in Section 3.2.4.4. One of the impediments to deploying everting liners was the drag. Liners made of silicon rubber for high-temperature environments had very high friction coefficients unless dusted with talc on the inside surface of the liner.

DOI: 10.1201/9781003268376-16

Where the drag was a major factor was when the liner turned a corner in the passage of a pipe or borehole (e.g., a curved horizontally drilled hole). Like pulling a garden hose around a tree, drag in a turn depends on the angle of the wrap on the turn. A 90 degree turn of the hose around a tree provides much more resistance than a 10 degree turn. Vertical upward deployments of heavy liners as in the Smithsonian Museum of Natural History from the basement to the roof through crooked pipes needed detailed assessments to determine if the required driving pressures exceeded the air canister capacity.

16.1.2 The Drag Model in a Crooked Pipe

As with many liner problems, FLUTe derived a model for the drag on a liner in a turn. Of course, the drag depended on the pressure against the curved surface, the friction coefficient, and the tension on the liner. It was assumed that it depended on the radius of the turn, but it actually didn't. As is often the case, the resulting equation derived was already well known in the engineering community as the pulley equation. This was not the first time that FLUTe had derived a relationship already known in the engineering or hydrologic community, but it was a reassuring confirmation of the model component. The simple form of the equation is:

$$T = T_o e^{F\phi}$$

where F is the friction coefficient and ϕ is the angle of wrap on a cylinder in steradians. The To is the tension on one side of the cylinder (e.g., a hanging weight) and T is the tension required to overcome To or lift the weight with the friction on the cylinder (see Figure 16.1). The term $e^{F\phi}$ is the multiplier of the tension To due to friction. It is interesting that the radius of the cylinder is not in the equation, but the pressure of the liner against the cylinder does increase for smaller diameters, and hence, the friction coefficient may increase with the radial pressure. It is useful to note that with a tension measurement at T and a weight at To, one can measure the effective F, the friction coefficient, for a variety of materials sliding on other materials. FLUTe uses that technique to determine F for different materials such as liners sliding on each other.

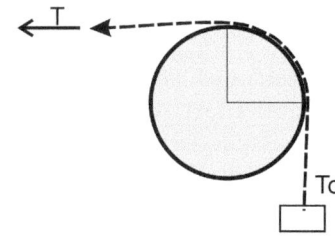

FIGURE 16.1   Drawing of 90 degree bend of rope or liner over a stationary cylinder. To is the tension and T is the tension needed to move the rope on the cylinder.

16.1.3   PARAMETERS FOR A CROOKED PIPE CALCULATION OF LINER TRAVEL

The actual mathematical model developed is not shown here, but the model has these features: The model breaks the passage, or pipe, into sections. Each section has a diameter and the liner is assumed to dilate to the full diameter of the section. Each section has a slope from vertically upward to downward. At the end of each section is an elbow turning a specified number of degrees and a turning radius specified. The gravitational load is included in proportion to the weight of the liner or the nominal tether which can also be a heavy cable. The liner has a friction coefficient on itself. The tether has its own friction coefficient on the liner and both components have a weight per unit length. The tether may actually be a rubber-covered cable with a high friction coefficient. The liner has a P_{min} assigned and other characteristics as shown in Figure 16.2. Pressure is in psi and tension in lb force. The end result is a calculation of the minimum driving pressure necessary to cause the liner to evert given the position of the liner in the passage. In other words, as the EP of the liner is extended, the length of liner, the turns of the liner, the turns of the tether, and the location of each are used in the calculation as each section is exited. The table in Figure 16.2 shows an example of the input data for the several sections that are calculated. The plot in Figure 16.3 shows the minimum pressure needed to evert the liner after each turn and the sum of the angular turns traversed as a function of the distance traveled in the passage. The plot in Figure 16.4 shows the vertical variations of the passage

Tether			Liner				minimum	Initial tension	
friction coe	wt/length		length	Friction coeff	wt/length	bulk drag	ever. Press	(lb)	
0.3	0.3		350	0.3	0.2	0.05	0.5	5	
	lb/ft		ft			lb/ft	lb/ft		

Passage parameters

section no.	section length (ft)	section slope	turn (degrees)	section diameter (in.)	sweep radius (in.)	vertical travel (ft)	horiz. Travel (ft.)	minimum driving pressure (psi)	total turns (deg.)	length along passage (ft)
inlet						0	0	0.5		0.1
1	1	45	45	4	2	0.707106	0.707107	0.5	0	1
						0.707106	0.707107	1.5	45	1.1308996
2	100	0	30	4	2	0.707107	100.7071	2.5	45	101.1309
						0.707107	100.7071	2.8	75	101.21817
3	50	45	0.01	4	2	36.06242	136.0625	4.3	75.01	151.21817
						36.06242	136.0625	4.3	75.02	151.2182
4	50	0	90	4	2	36.06242	186.0625	5.1	75.02	201.2182
						36.06242	186.0625	7.9	165.02	201.47999
5	100	-60	80	4	2	-50.5401	236.0625	7.0	165.02	301.47999
						-50.5401	236.0625	10.3	245.02	301.7127
6	25	60	90	4	2	-28.8894	248.5626	10.3	245.02	326.7127
						-28.8894	248.5626	16.3	335.02	326.9745
7	25	0	0	4	2	-28.8894	273.5626	17.2	335.02	351.9745
						-28.8894	273.5626	17.2	335.02	351.9745

FIGURE 16.2   The input parameters of the calculation and the results. Each section of the pipe has its own slope and turn angle at the end of the section. Turns are not identified regarding direction. That is inferred from the slope. That does not affect the drag calculation.

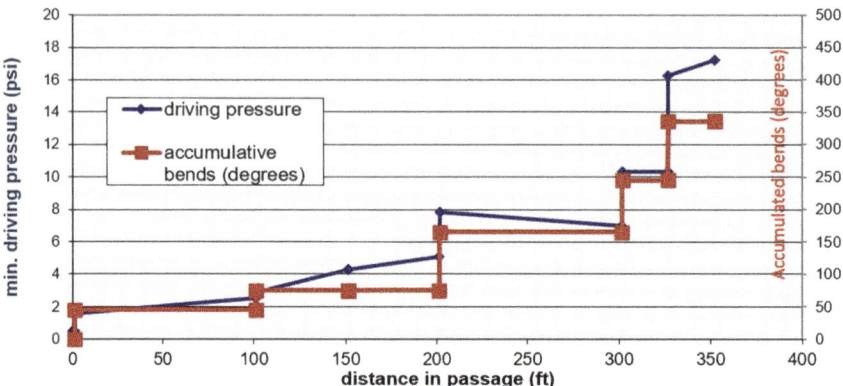

FIGURE 16.3 Minimum driving pressure to propagate the liner versus distance along the passage (blue curve). The red cure is the accumulated angle of bends passed in degrees (right hand scale). Large angle turns generate more drag and more drag generates a higher driving pressure after the turn. If the liner is traveling downhill, the weight reduces the driving pressure needed as from 200 to 300 ft.

with length or with horizontal distance traveled. The horizontal distance is less than the distance in the passage since the trajectory is not linear. It is useful to note that the angular turns are not specified as to what direction is the turn. Only the slope of the next section is indicative of the turns as up or down, but no distinction is made as to right or left, since the driving pressure does not depend on the direction of the

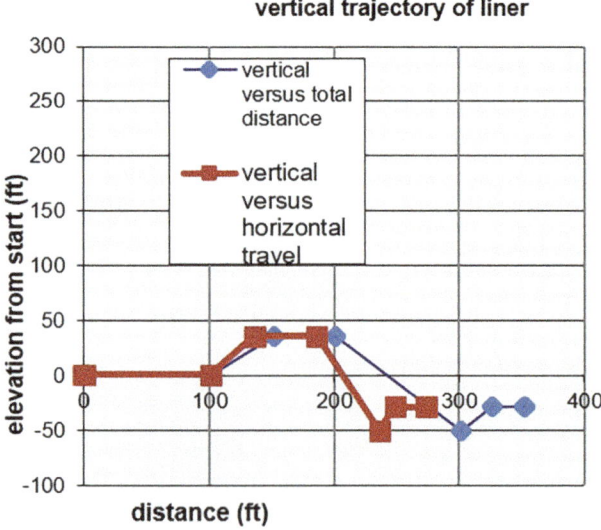

FIGURE 16.4 Trajectory of liner in the pipe/passage as a function of both the horizontal distance traveled (red) and the distance traveled in the pipe (blue). The blue curve is nearly the slope of the pipe section. Note the 100 ft. section drops below the starting point.

turn. The slope is used to calculate the friction on the passage due to the weight of the components. The slope also is used to determine if the drag tension is reduced by a downward slope or increased due to an upward slope. In other words, the weight effects are included in both friction and tension on the liner.

This calculational model was used for a variety of actual and potential applications. Those were as follows:

1. Air driven liners were used to carry a cured-in-place resin-soaked fiberglass liners into cast iron piping of the Smithsonian Museum of Natural History 100 ft. from the basement to the roof. The piping had several elbows in the passage. The resin cure was initiated with electrical heating woven into the resin-soaked liner.
2. Many liners were installed under a Lowes hardware store in Paterson NJ to carry sampling systems in horizontal piping with various sweep elbows into vertical wells beneath the building. Some vertical wells were only 2″ diameter. The horizontal piping was 3 and 4″ diameter.
3. Liners were installed using the LAHD method (Section 13.2) under a landfill in Indianapolis, IN for 400 ft. with eight sampling ports.
4. A demo of a liner installation upward from an underground tunnel was performed in the INCO mine in Sudbury, Ontario. The proposal was to use the liner to tow logging tools into the boreholes in the back of the tunnel without concern of hole slough on the sonde.
5. A liner was used to follow the retreat of a guided mole (using a built-in hammer for propagation) at a site in Oak Ridge National Laboratory (ORNL) in Tennessee.
6. Designs have been assessed for a variety of applications from firefighting to sewer installations. One novel application that FLUTe was asked to consider was to create access to the young cave explorers trapped in a flooded cave in Thailand. The cave was too long for a practical application, but the model was used to assess the feasibility of an everting liner to provide air using a liner after the rising water level sealed the cave. Those stranded were in the meantime rescued by divers.

Some liners have been constructed with elbows fabricated in the liner to turn the liner to travel along an underground coal mine in Pennsylvania. A recent patent application includes a means of remotely turning a liner.

One example of an air driven liner traversing a crooked pipe is the video: (https://vimeo.com/flut/crookedpipe) showing a 4″ liner traveling through 450 degrees of turns. That test was to validate the model. It worked very well. The liner was deployed from an air canister.

In the video, the 4″ liner is propagating though standard 4″ 90 degree elbows of a very short turning radius, ~2″. The central leader seen with a small amount of tension can deflect the EP to start the eversion in the new direction. Smaller turns like 45 degrees or sweep elbows are easily passed with the everting liner without any help. If the flow in the piping system can be controlled, even the turn of a liner in a 90 degree Tee fitting can be achieved. We have not tried using a vacuum to turn a sharp corner.

If the 90 degree turns are sweep elbows, the propagation through the turn is especially easy. If a sharp 90 degree turn, as in some piping, the liner may need to be led through the turn by deflecting it with a leader built into the liner and extending out the far end of the passage. Clearly, that is not possible for dead-end holes as drilled by some horizontal drilling, but those do not have bends, except perhaps at the entry. The LAHD installation under the landfill in Indiana used a leader as described in Section 6.1. The leader ran through a sheave on the back of the pig of the LAHD system to help to overcome the drag of a 400 ft. horizontal installation. The leader was not needed to guide the liner. As shown in Section 13.2, it worked well.

In general, it is surprising how well an everting liner can traverse a crooked passage.

16.1.4 ADVANTAGES OF THE CROOKED PIPE MODEL

When considering the installation of a liner into a passage whether a pipe or borehole such as a vertical upward borehole from a tunnel, the question arises of whether the eversion method considered can provide a sufficient driving pressure for the completion of the installation. If the trajectory of the passage, the liner weight, and weight of attachments are known, the crooked pipe model allows that question to be addressed. For example, if the liner is towing a logging tool and the cable attached has a high friction coefficient or is heavy, the driving pressure that is required may be a burst hazard for the canister from which the liner system is being deployed. Liners carrying cured-in-place liners can be very heavy per unit length or curves in the passage may be too many or of too large a turn to allow the system to propagate as desired. Vertical downward installations may not have a sufficient excess head available to propagate the full path length needed. If the cable being towed is very heavy, it may buckle behind the liner for a downward passage. In that case, the calculated driving pressure would be negative suggesting the potential problem if the cable is not kept sufficiently taut to support the cable. The tension on the liner or tether/cable is included in the calculation.

For these questions of feasibility, the model is very helpful in the determination of feasibility of everting liners. The more crooked the passage, the greater the need for the model. Another important factor in the model use is the minimum eversion pressure required, which limits the towing force available to move the liner through the passage. For small pipes, the pipe diameter can greatly limit the towing force available. All these factors are included in the model calculation of minimum driving pressure.

The main limitation of the model is only the knowledge of the parameters required in the calculation.

16.2 TRANSIENT CORRECTION MODEL OF THE T PROFILE METHOD

The transient correction is a model developed to make a first order correction to the transmissivity profile. The transient seen at the beginning of a T profile is described in Section 10.3.6. The model to make the correction is a relatively simple hydrologic calculation using Darcy's equation for water flow in a porous medium. Whereas fracture flows are the norm in T profile measurements, that flow is often addressed with the simple porous flow model. The transient occurs because the T profile method or model assumes that any drop in the velocity of the liner is due to the sealing of the flow paths into the borehole wall and therefore reductions of the remaining transmissivity of the borehole beneath the everting liner. The model used is the Thiem equation that assumes steady state flow into the hole wall over the entire borehole as the liner is everting until sealed by the everting liner. However, the initial flow when the liner eversion begins, with release of the liner tension and high excess head in the liner, the flow into the entire borehole wall is not steady state. The flow into available paths starts with an extremely steep gradient into all flow paths and then the gradient decays as the liner propagates under a constant driving pressure to a nominal steady state. The flow into each flow path intersecting the borehole may be different. As that flow approaches steady state, the liner continues to evert down the borehole with a decaying velocity proportional to the sum of the decaying flows into all the flow paths. This approach to a steady state occurs for a wide variety of flow measurements, including that of straddle packer measurements. Unless the flow model includes the transient state throughout the formation, the average conductivity of the formation over any interval is not accurate. This is true in the fine detail because as the flow rate changes, the effective conductivity also changes because the flow rate usually depends on the velocity and state of flow in the flow path whether laminar or turbulent.

Because of the initial variability of the flow rate and descent rate of the liner and the fact that steady state is approached at different rates in different fractures and true steady state never is achieved, one might be discouraged from using the T profile method. However, the transient decays in time as the borehole approaches a steady state and after the liner has descended a reasonable distance, the flow into the borehole wall can approach a nominal steady state for permeable intervals and the flow rates into all flow zones approach steady state, even if never achieved. It is also true that one need not obtain the precise conductivity for any feature but an approximate conductivity and to determine where the highest flow zones exist and where aquitards may exist.

The initial transient is still a major perturbation of the T profile method and masks the resolution of the conductive zones while the transient persists. A possible solution to the problem is to add water to the borehole until the steady state is reasonably achieved and the head in the borehole is measured for the steady state and used to drive the liner after the steady state is developed. However, that has been dismissed for several reasons, the main one being that the regulators do not allow the addition of large quantities of water to the borehole and that the flow rates may be very high for some boreholes to achieve that nominal steady state with a useful driving head in the liner.

The solution used by FLUTe is to develop an estimate of the actual transient flow in time and to subtract it from the flow rate measured by the descending liner at the beginning of the T profile. Because there are many different transient flow rates in the borehole, it may be considered impossible to develop the correct equivalent flow without modeling the entire borehole flow. A consoling fact is that if one calculates a transient flow for many different conductive features, the sum of those transient flows in time looks like a typical transient flow for some equivalent conductive feature. That means that we needed to estimate the equivalent transient flow rate for the sum of all the transients.

It is fortunate that a T profile initiated at a water table high in a casing decays in velocity without any flow through the wall of the casing. Therefore, the transient velocity of the liner in a casing under a constant driving head is the total of all transients. The only need remaining is to reproduce that transient and subtract it from the liner measured flow rate in and beyond the casing. That is relatively easy.

A simple Darcy flow into a plane in 1D radial flow with a suitable storage coefficient and conductivity was modeled in a finite difference calculation in a spreadsheet as one of the sheets in the spreadsheets used to reduce the entire T profile data.

Fortunately, the T profile velocity measured and the calculation from the profile deeper in the borehole below the transient affected interval is known from the T profile calculation. The driving excess head is known as well as the borehole dimensions and estimates are made of other parameters as needed in the calculation. The time step used in the finite difference calculation and zone size can affect the result and are picked to prevent the usual instability of a time step too large. The refinement of a better mathematical model in implicit form was not used. That can be an improvement if considered necessary.

Using the conductivity or transmissivity of the borehole below the transient and other parameters available, or estimated, the transient is determined in time and subtracted from the liner transient at the same time at each time step. The effect is to reduce the measured velocity of descent of the liner in the interval where the transient is apparent. That is plotted with the actual liner velocity. If the transient occurs primarily in the casing, the ideal transient would reduce the liner velocity in the casing to a constant velocity. That is approximated, but not always the case. Adjustments to the driving pressure, conductivity, and storativity can be made to approach a better match to the constant velocity expected. Another constraint on the correction is that the liner velocity in the casing or any relatively impermeable interval will not increase. An overcorrection of the transient gives an obvious increasing velocity in some intervals especially in the casing. The storativity increase tends to increase the time of the transient effect. The main constraint on that parameter is the match in the casing or the observation that the transient correction causes increasing velocity in time or the liner velocity no longer appears to be decaying in the typical transient manner.

In this manner, the first order transient correction can be made. The advantage is that with the effective transient correction, active fracture flows associated with a drop in the liner velocity can be much more easily recognized during the transient. The transient corrected velocity is then fit with the monotonic curve to ignore borehole enlargement effects and the T profile model is then applied to the monotonic fit data.

Whereas this is not a perfect solution for the transient, it works well to identify the main flow zones even high in the hole. However, it is common that the very first part of the liner velocity may still show a sudden drop in velocity. It is not easy to explain that drop although the conductivity in the shallow bedrock can be high, and the effective storativity may be much higher in an unconfined aquifer near the water table. Therefore, the T profile is not a reliable measurement of the flow for that very short distance when the profile is initiated. Unless the transient is initiated in the casing, it is difficult to judge the first transient extreme. The data report includes a caution for the less reliable interval of the measurement at the beginning of the transient.

Finally, it is a pleasant surprise how well the transient correction seems to work in allowing the effect to be removed from the raw data.

16.3 EXTRAPOLATION TO THE EQUILIBRIUM ASYMPTOTE FOR THE RHP

The reverse head profile (RHP) model described in Section 10.4 refers to the model used to estimate the asymptote of equilibrium pressure or head each time the liner is inverted to a new stopping point in the procedure. Like most of FLUTe's methods, new mathematics is not invented, nor are new hydrologic principles developed. Rather, FLUTe methods are simple applications of existing concepts combined with observations and other experiences.

What is described here is the method used to estimate the actual asymptote limit expected for the exponential decay of the rising head beneath the liner upon halting the inversion during the RHP measurement of formation head. The calculation of the formation head in each measured interval is easy if the equilibrium head is known for each stopping point (see Section 10.4). However, the rate of rise to equilibrium in each newly uncovered interval varies a great deal in our experience. It is not obvious why that occurs. It may be that high flow intervals equilibrate more quickly or that longer borehole intervals adjust to the new equilibrium more slowly. The typical rise to equilibrium is shown in Figure 16.5 for five stopping points each for about 1 h. Note the longer open borehole intervals equilibrate more slowly even though they should be of higher total transmissivity.

The model is mathematically simple once the algebra is done. The first assumption is that the decay to equilibrium is exponential with a time constant that is constant

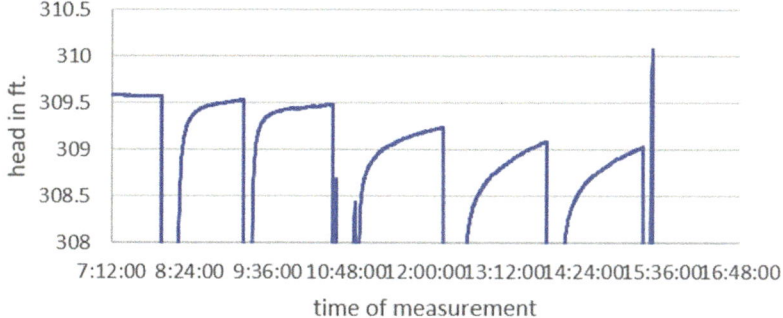

FIGURE 16.5 Actual head history for rise toward equilibrium at five stopping points. Although the time at each point is about 1 h, the stops at higher elevations were slower to equilibrate. This necessitates the calculation of the asymptote at each stop.

in time. Differencing the data in the curvature to equilibrium and solving for the time constant allows one to estimate the actual asymptotic limit without the need to measure a value that theoretically never occurs.

16.3.1 HOW TO CALCULATE AN ASYMPTOTIC LIMIT FOR AN EXPONENTIAL APPROACH TO EQUILIBRIUM

If the approach to the asymptotic limit of HF is given by $H = HF - Ho\ e^{-At}$, where H is the head or level of water in a tube rising at a rate proportional to HF $-$ H, the asymptotic limit of the rise HF is of interest.

How can HF, Ho, and A be determined if only values of H are known in time such as the measured water level rise in a Water FLUTe pump tube or the rise to equilibrium in a RHP step?

I have devised the following since I could not solve the equation directly:

Let the derivative of H be $dH/dt = 0 - A\ Ho\ e^{-AT}$
using three equally spaced values of H at H(t1), H(t2), and H(t3 called H1, H2, and H3). The times t1–t3 should be equally spaced in the time of the measurement seconds or minutes.
The midpoint of the interval of $t1 - t2$ is $(t1 + t2)/2 = t1m$ and for $t2 - t3$, it is $(t2 + t3)/2 = t2m$.
The derivative, or slope, of H at t1m is $(H2 - H1)/(t2 - t1) = dH1$
The derivative of H at t2m is $(H3 - H2)/(t3 - t2) = dH2$
dH/dt at t1m is $- A\ Ho\ e^{-A\ t1m} = dH1$
dH/dt at t2m is $- A\ Ho\ e^{-A\ t2m} = dH2$
Dividing dH1 by dH2 gets rid of Ho and one A coeff.
Then $e^{-A\ t1m}/e^{-A\ t2m} = e^{-A(t1m - t2m)} = dH1/dH2$.

Taking the natural log of both sides:

$$-A(t1m - t2m) = \ln(dH1/dH2), \text{ or } A = \ln(dH1/dH2)/(t2m - t1m)$$

Given the value of A, and $H = HF - Ho\ e^{-At}$, then subtract H (t2) from H(t1) to avoid HF. Solve for $Ho = (H2 - H1)/(-e^{-A\ t2} + e^{-A\ t1})$

$$\text{Then } HF = H1 + Ho\ e^{-A\ t1} \text{ or } H3 + Ho\ e^{-A\ t3}$$

This is the value of the asymptote. This checks out against the spreadsheet and when generating such a function and picking points on the curve.

Using actual head rise data, it is best to pick points nearer the asymptote rather than early in the rise since the flows may not have stabilized earlier. However, if the points are picked too near the asymptote, the noise in the data can lead to an erroneous coefficient, A.

This calculation can be used with transducer data, but it is best to smooth the transducer data due to noise near the asymptote. This can be entered into the transducer

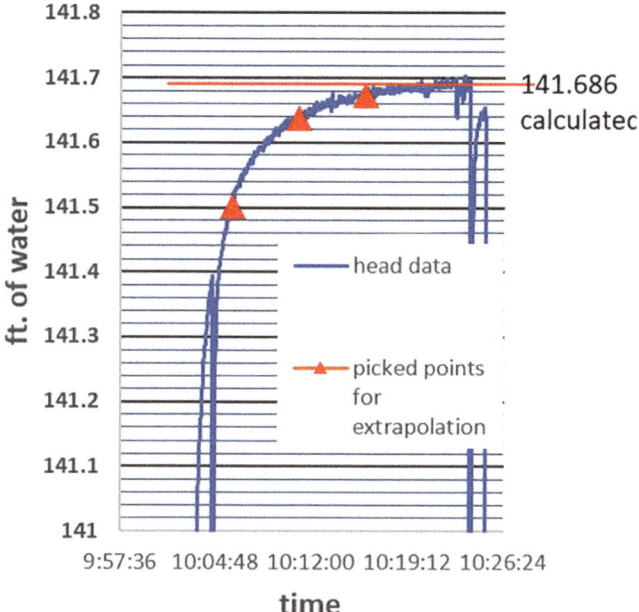

FIGURE 16.6 Actual measured rise to equilibrium beneath a liner. The data points picked to calculate the asymptote are shown as red triangles. Note the transducer noise as it reaches its resolution limit near the asymptote.

data sheet to calculate the asymptote, HF, throughout the approach to equilibrium. Ho is only relevant to where the decay started, the theoretical $t = 0.0$ place, which is not important or relevant.

An actual pick of points on an equilibration data curve is shown in Figure 16.6. Note how noisy is the data as the transducer measures little change at the limit of its resolution. The asymptote of 141.7 ft. of head seems about right, but slightly higher is more likely. The third data point picked looks to be a bit low, which would reduce the asymptote slightly. This model was tested using this method for a plot of a mathematical equation that gives the actual asymptote.

16.4 FRACTURE APERTURE CALCULATION MODEL USING THE T PROFILE DATA

The aperture calculation is suggested in Section 10.3.5. Effective aperture dimensions are reported as part of the T profile results report provided by FLUTe. The assumptions of the model are the common ones. The refinements that can be made are many, such as assuming a degree of wall roughness, or variability of the actual fracture dimensions with distance from the source of flow, and many other refinements. However, being content with a concept of effective aperture based on a flow measurement seems most advantageous. If the flows measured are turbulent or laminar, some correction is useful for application to more distant laminar flows, but for the data reported for the measurements described in Section 10.3, the more useful feature of the effective aperture size is the volume estimate with range from the borehole. For an estimated cross section with distance from the borehole, one can estimate from what range a sample is obtained after purging a known volume prior to the sample collection. This is most useful for monitoring wells below landfills or other sample collection wells where the groundwater contaminants more distant from the actual well location are important.

It is well known from many measurements that the flow through granular porous media of deliberately uniform size is still along very heterogeneous flow paths. This is especially true for gravity driven flow like infiltration. However, for more uneven flow paths like old braided stream beds, the contaminated flows can be easily missed with a few wells. That is true but to a more uncertain extent in fracture rock where preferred paths depend on fracture aperture and the 3D gradient shape.

A significant feature of the aperture calculation is in the comparison of the effective range of the sampled volume difference in a fractured medium with a sand-packed well, or one with only a few packers, in an otherwise open borehole given the same purged volume. In sand-packed wells, an inherent assumption is that the water in the sand pack is representative of the water at a distance. The same assumption is for a sample collected between two packers. That may be a very bad assumption.

The model for the fracture aperture is described in detail in Section 10.3.5. The important point is that the T profile data may actually define flow rates in single fractures. The number of fractures measured in a flow measurement between widely spaced packers is unknown, and the flow rate in each of the fractures in the straddled interval is also unknown. A typical assumption is that there are N fractures of equal aperture and the apertures may actually be estimated from an optical televiewer, which is indeed a very poor measurement of actual width more distant from the drilled surface of the borehole. Most apertures will be enlarged at the hole wall by the drilling process.

16.4.1 The Model

The aperture of fractures is most useful in models which calculate propagation of groundwater under assumed gradients. While it is well known that fracture apertures vary significantly along a natural fracture, it is useful to estimate an effective fracture from actual flow data. The measured flow includes the effect of how rough is the fracture wall and the variations of aperture. One can estimate the effective aperture using a realistic model which includes viscous effects and an idealized geometry of the flow. For a fracture intersecting a borehole, and when the borehole is pressurized, it is reasonable to assume that the flow is radially outward from the borehole for some distance. Such a flow model is provided in the Bird, Stewart, and Lightfoot text titled, "Transport Phenomena". The flow is assumed to be one dimensional radial outward flow between two parallel plates. The model provides a flow, Q, as:

$$Q = 4\pi (P_1 - P_2) b^3 \rho/(3\mu \ \ln(r2/r1))$$

Where ρ is the density and μ is the viscosity of the fluid, in this case that of water. In the CGS (centimeter gram second) system, μ is 0.01 Poise and ρ is 1 g/cc. The term b is the half width of the space between the parallel plates. P1 is the pressure in the central hole, the borehole of radius r1, and P2 is the pressure at a radial distance r2.

From this equation, one can calculate the half aperture b if the values of P, μ, and ρ are known for a measured flow rate Q.

Given the known values of μ and ρ, the value of 2b, the fracture aperture, is:

$$2b = 2(Q(.03 \; \ln R/(4\pi(P1-P2))^{1/3}$$

Since the FLUTe T profile method uses the Thiem equation to deduce a transmissivity from a measurement of flow into the hole wall, the T values can be used in reverse to deduce the flow into the hole wall.

Then from the Thiem equation, $Q = T\Delta P \; 2\pi/\ln(R)$. Then T as deduced from Q can be substituted into the equation for 2b above for a greatly simplified result for 2b:

$$2b = 2(0.015 \; T/(980N)^{1/3}$$

This is in the CGS system and C and T are in cm and seconds.

The 980 is the acceleration of gravity in cm/s^2 because hydrologists use head instead of pressure. And T is defined as $C \times H$ with C in cm/s and H in cm. and T is in cm^2/s. C includes the value of acceleration due to gravity. The 2b equation above assumes P_1 and P_2 are pressure not head. N is the number of fractures assumed.

Individual fracture apertures are seldom used in hydrologic calculations because there are many fractures of different sizes in a typical representative volume of those calculations. However, the FLUTe T profile results of Section 10.3 provide transmissivity calculations for very short intervals in the borehole based upon flow rates into the hole wall measured for short intervals. In many cases, there is no more than one significant fracture in the interval of interest as seen in the optical televiewer image.

The interval of interest that is more useful to this model is the length subtended by the spacer of a FLUTe MLS system. If a sample is drawn from a single screen, the sample may be drawn from anywhere above or below the screen. In a Westbay system, the sample is drawn from a portion of the straddled interval. For a sand-packed interval, the answer depends on the volume of the sand pack and porosity, but in many cases, the sample comes mainly from the sand pack unless a large volume is purged. For a FLUTe system, the range from which the water is drawn is more relevant to the fracture aperture since there is no sand pack or open hole from which the water would be drawn.

For the FLUTe systems the range of extraction is dependent on the fracture volume and the volume purged before sampling. If the aperture is known and the best guess is a simple fracture approximated by two parallel plates and radial flow, the 2b value can be used to estimate the range from which the sample is drawn. The volume, V, of the water purged from the fracture may be approximated by:

$$V = \pi r^2 \times 2b,$$

10 ft interval	mid depth (ft)	T of interval (cm²/s)	2b aperture in cm	range of extraction (ft)
1	149.0	0.0655	0.020	13.967497
2	147.3	0.0076	0.010	20.009708
3	141.85	0.0740	0.021	13.685075
4	131	0.5086	0.040	9.9248189
5	118.5	0.0503	0.018	14.59233
6	104	0.1243	0.025	12.551673
7	88	1.2960	0.054	8.4923277
8	75	0.3372	0.035	10.628558
9	65	0.3357	0.035	10.636101
10	58.325	0.0627	0.020	14.067798
11	51.325	1.1798	0.052	8.6263022
12	38	0.4214	0.037	10.240856
10 ft intervals for spacer 11.5 liter (3 gal.) purge				

FIGURE 16.7 Range of extraction from a 10 ft. interval in the NAWC 94B borehole described in Section 10.2.2 using the continuous T profile to calculate the aperture for one fracture in each 10 ft. interval. This is useful to the question of how far from the borehole is the reach of a 3 gal. purge for a FLUTe MLS. For more fractures, the reach is less.

where r is the outer radius from which the purge water was drawn. The calculation was done for the spacer intervals and the volumes purged for the SWF system used in the borehole 94B at the NAWC site described earlier in this text (Section 10.2.2). The range of extraction depends on how many fractures share the flow. It is more likely that the larger fractures are providing the flow. Larger fractures will have a shorter range, or reach, of sample extraction. If there is more than one fracture, the reach is less. For the assumption of a single dominant fracture at each spacer, and the spacer length of 10 ft., the range of extraction of the purge volume is as shown in the table in Figure 16.7. The spacer length is important because it defines what portion of the continuous transmissivity measurement is relevant. For very low transmissivity and a thin fracture the range of extraction is large.

It is interesting to note that a 6″ casing in a 10″ hole with a 5 ft. screen and 35% porosity has about 17 liters of water in the pore space. This is more than the prescribed purge of 4 gal. for the WF and SWF systems. In that case, the range of extraction with 4 gal. would not remove the water in the sand pack. That makes it necessary that the water in the sand pack is in equilibrium with the formation, but would require a very large purge to extract from a significant distance from the borehole.

Shorter spacer intervals intersecting fewer fractures provide the ability to monitor further from the borehole. The time to extract a prescribed purge volume is inversely proportional to the transmissivity.

An important consequence of this estimate is that monitoring wells down gradient from a landfill must be relatively closely spaced to reliably intersect a plume of landfill leakage. This is even more true in sediments since the reach of a purge volume is much less in a high porosity medium.

16.5 DATA REDUCTIONS OF T PROFILE

16.5.1 WHO DOES THE DATA REDUCTION

When a T profile is performed by FLUTe, the data recorded is the velocity of the liner decent as a function of time and the driving conditions. The driving condition of primary importance is the head in the liner or more importantly the head beneath the liner. The head in the liner is recorded with the tension on the liner. From that data a calculation can be performed for the probable head beneath the liner. However, a more useful measurement is of the head beneath the liner in the same time frame as the velocity of the liner. As described in Section 10.3.2, the velocity profile identifies the location of the liner and the flow change as a flow path is sealed. The measured head beneath the liner provides the information needed to calculate the flow rate per unit driving pressure for the flow into the hole wall. A large spreadsheet was generated which uses all the conditions recorded for the installation to deduce the transmissivity of the borehole with depth. FLUTe does the data reduction.

16.5.2 WHEN IS THE DATA REDUCED TO A T PROFILE

After the data is collected at the borehole, the data is sent to FLUTe for conversion to the reported parameters that can be deduced from the data. Those are reported in a file labeled "results" sent to the customer. At that time, if the RHP is to be done, the stopping points are selected from the T profile for the RHP and provided to the field staff doing the RHP.

16.6 DATA REDUCTION OF RHP

The RHP data is collected in the field measurement procedure. When the liner inversion is paused at each stopping point and the head history beneath the liner is recorded until near equilibrium, the downhole head is recorded continuously. That head data is used with the transmissivity profile to calculate the vertical head distribution in the formation as described in Section 10.4. The FLUTe staff uses the data in a data reduction procedure programmed in a spreadsheet to determine the formation head distribution for each interval straddled by the stopping points. That result is then provided to the customer. The RHP and T profile also allow the calculation of the flow in each interval into or out of the borehole. From that flow data, a synthetic flowmeter log is calculated for the borehole. Comparison of the synthetic log with any actual flow log done in the geophysics measurement suite is a useful measure of the reliability of the T and head profiles. The RHP has the additional advantage of use in the T profile to refine the T profile and the RHP calculation as described in Section 10.4.

16.7 DATA REDUCTION FOR THE ACT

The ACT measurement is described in Section 9.5.1. The ACT transducer data is of the pressure history in the air column above the water level in the slender tube. With the measured initial head and the ACT data, the water table history in the formation at each port is calculated as described in Section 9.5.1. That data reduction is performed in a FLUTe derived spreadsheet which also calculates a temperature correction as appropriate. The resulting spreadsheet is then sent to the customer for each port measured. That spreadsheet can then be used for subsequent data sets recorded with the ACT transducers at each port.

16.8   FACT DIFFUSION MODELS

FLUTe has developed a simple diffusion calculation model that is used with all available data to estimate the rate of diffusion of contaminants into the borehole wall prior to a FACT installation. The FACT method is described in Section 10.2.2. The model is compared to analytical solutions for the boundary conditions and diffusion coefficients for the model conditions to assure the quality of the model developed. The same model is also used to assess the rate of diffusion of contaminants into the FACT carbon. Those results are compared to the actual adsorbed mass and are also used to assess the effect of various parameters on the FACT analytical results. Some of those results are reported in the paper by Keller et al (2017). That paper addresses the significance of various possible perturbations of the FACT measurement method.

17 Installation Procedures of Many Kinds

Most of the FLUTe methods require unique installation procedures. Specific procedures have been developed for the following methods. Some procedures are performed by the customer, but most are not. FLUTe has been evolving to systems designed to be installed by the customer. However, some measurement methods use procedures and equipment beyond the typical customers' interest. For those applications, FLUTe has trained other than FLUTe personnel to perform, but some methods are usually still performed by FLUTe staff or FLUTe-licensed distributors. Brief summaries are available from FLUTe to meet the requirements for standard operating procedures (SOPs), required by many regulators and customer safety personnel. SOPs are available for the commonly used methods:

1. Blank liner
2. NAPL FLUTe
3. FACT
4. T profile
5. RHP
6. WF
7. CHS and pdCHS
8. ACT

For copies of these current SOPs, contact FLUTe at info@flut.com or call the FLUTe field office.

DOI: 10.1201/9781003268376-17

18 The Manufacturing Machines and Facilities Developed for Liner Fabrication

FLUTe liners and the many associated manufacturing devices were not available on the commercial market. As a new technology for underground measurements, the first need was to build and test the concept. The building of FLUTe systems led to a series of special methods and the equipment designs needed to manufacture the systems.

18.1 SPECIALLY DESIGNED RF WELDING MACHINES

Since tubular liners of urethane-coated fabrics are not readily available in the diameters and lengths of use by FLUTe, it was necessary to design and build devices for the special fabrication needs. The first tubular liners were manufactured by hand from flat role stock with hot air guns. A few thousand feet of liners were welded into a tubular form in that manner, but the labor costs were prohibitive. However, the quality of the work of the chief welder was impressive and she could weld many complex shapes as needed occasionally. The next step was to employ a subcontractor who welded urethane-coated fabrics into custom designs. Finally, because of quality issues and the need for special welded shapes, FLUTe finally designed its own custom RF welding machine and trained its own operators. The quality improved greatly, eliminating the defects of the subcontractor. That machine was installed in the FLUTe plant in a special environmentally controlled space and with custom-designed support equipment. A second RF welding machine was installed in 2021.

18.2 DYE STRIPING MACHINE

The NAPL FLUTe evolution to the dye striped cover material required the dye striping of the cover material. FLUTe designed and built the machine that would perform the dye striping operation in a very uniform pattern at a controlled rate.

18.3 COMPRESSION WRAPPING MACHINE

The need for compression that wraps the NAPL FLUTe system for lowering the assembly in a direct push rod also led to a FLUTe design and construction of a machine that performs that operation with the feed rate and wrapping material very uniformly compressing the tubing, liner and cover assembly.

18.4   AIR-DRIVEN CANISTERS

Air-driven canisters for liner installation were not commercially available of the right dimension, materials, pressure capacity, and function. FLUTe designed the aluminum pressure vessels which are preferred, using a local machine shop and materials of special manufacture such as spun aluminum. Smaller and much larger canisters were manufactured of steel. An example of a large mobile canister is shown in Section 15.1.

18.5   EP MARKING METHODS

The everting liners are marked for the depth in the borehole. That procedure was automated with a single dedicated operator using a spreadsheet guiding markings for variable spacings.

18.6   PORT WELDING MACHINES AND OTHER ATTACHMENTS

The many weldments of liners and welded attachments to liners require a very careful control of the classic temperature, pressure, and time of contact welding procedures. FLUTe has built the systems that control those parameters for many different components attached to the exterior and interior of systems with the different materials welded. Extensive testing and heat flow modeling has been used to develop the most reliable weldments and the fixtures necessary for several kinds of welds.

18.7   LONG TRAYS FOR EVERSIONS

The FLUTe facility includes 2000 ft. of trays for everting liners. While not a special machine, the design of the trays and construction using FLUTe multitalented staff and equipment has been a major advantage in the practical production of liner systems. The very long indoor trays allow fabrication in all weather conditions. The leak checking procedure also requires a long space that is well protected from weather or significant temperature changes. Liners are air pressurized for maximum sensitivity to leakage that can be significant over years.

As might be expected, explicit fabrication procedures have also been written for the construction of each of the liner systems manufactured. Those procedures serve as training materials for new employees and cross-training of many of the fabrication staff in multiple procedures. Also, in hand are explicit test procedures for quality control from those for pressure testing of welded liners and testing of tubing system integrity prior to shipment. All critical attachments and fittings are done with torque control devices. The FLUTe warranty makes manufacturing defects, and the expensive remedy, to be avoided.

Chemical tests are also performed on each new material lot to assure the FLUTe requirements are met.

19 Conclusion

The mechanism of everting/inverting liners has provided unique capabilities for an extremely wide variety of applications. Exploiting the attributes of flexible liners has been the basis of FLUTe technologies. The ability to seal the entire borehole has been an improvement allowing the measurement of hydrologic characteristics of the natural state of underground flows. The traveling everting liner and sealing capabilities have also been useful advantages of FLUTe methods. In order to exploit the full advantages of flexible liners, FLUTe has developed the necessary fabrication, installation, and removal methods. In order to prove the utility of the methods, both in function and in cost, it has taken two decades of evolution to gain the current capability and achieve the acceptance of the FLUTe methods over traditional practice. Now that the level of development has evolved to a common acceptance in some areas of application, it is important that the lessons learned in the development of the art and science be recorded for the preservation of the technology and its use by other practitioners. That was the purpose of this book. Hopefully, it will be helpful in the further refinements of the methods in hand and the exploration of new applications.

DOI: 10.1201/9781003268376-19

References

FLUTe liner systems partial patent list:
5, 176, 207; 5, 377, 754; 5, 803, 666; 5, 816, 345; 5, 853, 049; 6, 0269, 00; 109, 828; 6, 298, 920; 6, 910, 374; 7, 281, 422; 6, 244, 846; 7, 753, 120; 7, 841, 405; 7, 896, 578; 8, 069, 715; 8, 176, 977; 8, 424, 377; 9, 008, 971; 9, 534, 477; 9, 797, 227; 10, 030, 486; 10, 060, 252; 10, 139, 262; 10, 337, 314; 10, 472, 931; 10, 954, 759

Plus, numerous foreign patents in many countries. The remaining such patents are in Europe and Canada. Heavy foreign patent fees discourage the continuation of some of those patents in other countries.

Broholm, M.M., G.S. Janniche, K. Mosthaf, A.S. Fjordbøge, P.J. Binning, A.G. Christensen, B. Grosen, T. Jørgensen, C. Keller, G. Wealthall, and H. Kerrn-Jespersen, 2016. "Characterization of chlorinated solvent contamination in limestone using innovative FLUTe® technologies in combination with other methods in a line of evidence approach," *Journal of Contaminant Hydrology* 189: 68–85.

Beyer, M., 2012. "DNAPL characterization in clayey till and chalk by FACT (FLUTe Activated Carbon Technique)", 30 ECTS Master's Thesis, October 2011 till March 2012, Danish Technical University.

Butler Jr., J.J., C.D. McElwee, and G.C. Bohling, 1999. "Pumping tests in networks of multilevel sampling wells: Motivation and methodology," *Water Resources Research*, 35, no. 11: 3553–3560.

Chapman, S., B. Parker, J. Cherry, J. Munn, A. Malenica, R. Ingleton, Y. Jiang, G. Padusenko, J. Piersol, 2014. "Hybrid multilevel system for monitoring groundwater flow and agricultural impacts in fractured sedimentary bedrock," *Groundwater Monitoring & Remediation*, 35, no. 1.

Cherry, J.A., B.L. Parker, and C.E. Keller, 2005. "A new depth-discrete multilevel monitoring approach for fractured rock." *Ground Water Monitoring & Remediation* 27, no. 2: 57–70.

Driscoll, F. D., editor, 1995. *Groundwater and Wells* 1995 by Johnson Screen, St. Paul., MN.

EPA, 2017. "Technical Fact Sheet *N*-Nitroso-Dimethylamine (NDMA)." Nov. 2017.

Goode, D.J., T.E. Imbrigiotta, P.J. Lacombe, 2014. "High-resolution delineation of chlorinated volatile organic compounds in a dipping fractured mudstone: Depth-and strata-dependent spatial variability from rock-core sampling." *Journal of Contaminant Hydrology* 171: 1–11.

Holloway, O.G., and J.P. Waddell, 2008. "Design and operation of a borehole straddle packer for ground-water sampling and hydraulic testing of discrete intervals at U.S. Air Force Plant 6, Marietta, Georgia." USGS Open-File Report 2008-1349.

Keller, C.E., and B. Travis, 1993. "Evaluation of the Potential Utility of Fluid Absorber Mapping of Contaminants." NGWA Outdoor Action Conference, Las Vegas, NV.

Keller, C.E., 1996. "A Reliable Landfill Monitoring System, Leakage Barrier, Plus Remediation Method in One Design." Geological Society of America Austin TX Conference March 1996.

Keller, C.E., 2013. "Some Solutions to Necessary but Risky Open Borehole Development." NGWA Conference on Groundwater in Fractured Rock and Sediments, Burlington, VT. Sept. 23–24, 2013.

Keller, C.E., J.A. Cherry, B.L. Parker, 2014. "New method for continuous transmissivity profiling in fractured rock." *Ground Water* 52, no. 3: 352–367.

Keller, C.E., 2016. "A new rapid method for measuring the vertical head profile." *Groundwater*, DOI: 10.1111/gwat.12455.

Keller, C.E., B.L. Parker, S. Chapman, and S. Pitkin, 2017. "Overview of the FACT Method for a Continuous Profile of Dissolved Phase Contaminant Distribution in Fractured Rock," NGWA Conference on Fractured Rock and Groundwater, Burlington, VT, Oct. 2–3, 2017.

Lackey, S.O., W. Myers, T.C. Christopherson, and J.J. Gottula, 2009. "In-Situ Study of Grout Materials 2001-2006 and 2007 Dye Tests," Nebraska Grout Task Force. Lincoln, NE: University of Nebraska, October 2009 and subsequent unpublished research.

Munn, J.D., T.I. Coleman, B.L. Parker, M. J. Mondanos, and A. Chalari, 2017. "Novel cable coupling technique for improved shallow distributed acoustic sensor VSPs." *Journal of Applied Geophysics* 138: 72–79.

Parker, L.V., and T.A. Ranney, 1997. "Sampling trace-level organic solutes with polymeric tubing: Part I. Static studies." *Ground Water Monitoring and Remediation* 17, no. 4: 115–124.

Parker, L.V., and T.A. Ranney, 1998. "Sampling trace-level organic solutes with polymeric tubing part 2. Dynamic studies." *Ground Water Monitoring and Remediation* 18, no. 1: 148–155.

Persaud, E., J. Levison, P. Pehme, K. Novakowski, B.L. Parker, 2018. "Cross-hole fracture connectivity assessed using hydraulic responses during liner installations in crystalline bedrock boreholes." *Journal of Hydrology* 556: 233–246. DOI: 10.1016/j.jhydrol.2017.11.008.

Quinn, P.M., J.A. Cherry, and B.L. Parker, 2011. "Quantification of non-Darcian flow observed during packer testing in fractured sedimentary rock." *Water Resources* 47. DOI: 10.1029/2010WR009681.

Quinn, P., J.A. Cherry, and B.L. Parker, 2015. "Combined use of straddle packer testing and FLUTe profiling for hydraulic testing in fractured rock boreholes." *Journal of Hydrology* 524: 439–454.

Quinn, P., B.L. Parker, and J.A. Cherry, 2016. "Blended head analyses to reduce uncertainty in packer testing in fractured-rock boreholes." *Hydrogeology Journal* 24, no. 1: 59–77.

Schaefer, C.E., R.M. Towne, V. Lazouskaya, M.E. Bishop, and H. Dong, 2012. "Diffusive flux and pore anisotropy in sedimentary rocks," *Journal of Contaminant Hydrology* 131: 1–8.

Shapiro, A.M., 2002. "Cautions and suggestions for geochemical sampling in fractured rock." *Ground Water Monitoring and Remediation* 22, no. 3: 151–164.

Sterling, S.N., B.L. Parker, J.A. Cherry, J.H. Williams, J.W. Lane Jr., and P.P. Haeni, 2005. "Vertical cross contamination of trichloroethylene in a borehole in fractured sandstone." *Ground Water* 43, no. 4: 557–573.

Tiedeman, C., W. Barrash, C. Thrash, and C. Johnson, 2015. *A Hydraulic Tomography Experiment in Fractured Sedimentary Rocks.* Newark Basin, NJ: American Geophysical Union Fall Meeting, [San Francisco, CA].

Index

110d, 52
400d, 52
840d, 52

A

Absorbers, 135
Acoustic tele-viewer, 91
Activated carbon, 104
Active fractures, 129
Adsorbers, 105, 225
Advantages, 61
Aeration of samples, 179
Air blower, 12
Air drilling, 154
Air-driven canisters, 300
Air lift pumping, 18
Air vent tube, 15
Aperture calculation, 291
Aquicludes, 172
Aquitards, 172
Arsenic, 43
Artesian conditions, 241
Artesian heads, 241

B

Barite, 37
Baroid, 37
Barometric changes, 83
Barometric corrections, 83
Bentonite mixture, 37
Blended head, 163
Borehole
    angled, 234
    caliper, 151
    cavern, 229
    collapse, 29, 61
    depth, 204, 207
    development, 155
    diameter, 17
    enlargement, 138
    hazards, 63
    horizontal, 234
    open, 17, 36
    seal, 23
Breakouts, 12, 17, 137
BTEX, 43
Bubbler, 154
Buckling, 10, 13

Bundle weight, 187
Bypass flow, 267

C

Calculation of asymptote, 289
Capillary tension, 225
Carbon content, 117, 132
Carbon weight, 105
Central tube, 193
CHS, 191
    benefits, 191
    design, 193
    history, 191
CMT, 191, 213
Comparison of MLS, 213
Compression wrapping, 92
Conductive layers, 137
Conductivity, 137
Contaminant diffusion, 110
Contaminant mass, 113
Contaminant transport, 23
Continuously screened casing, 249
Core, 113, 123
Core drilling, 126, 154
Cover over core, 97
Crooked piping, 56
Cross flow in open boreholes, 208
Cross connection of boreholes, 208
Cross hole mapping
    background, 208
    method, 73, 208
    summary, 212
Crushed core, 113, 123
Cylindrical tether, 277

D

daFACT, 128
Deep water table, 187, 205
Denier, 48
Depth limits, 187, 205
Descent rate, 136
Devices
    air vents, 15
    bulbous wellhead, 37
    canister, 30
    end seals, 193
    green machine, 65
    kellum grip, 65

linear capstan, 69
  reel, 187
  scaffolding, 37
  snatch blocks, 65
  tether, 8
  towing, 30
  wellhead roller, 65
  winch, 65
DI water, 105
Diffusion
  into FACT, 110
  into hole wall, 113
  through liner, 179, 185
Diffusion bags, 128
Diffusion barrier, 179
Direct push installations, 92
  with air, 93
  with water, 92, 93
  direct push problems, 92
  direct push rods, 91
Direction of flow, 129
Discard volume, 179, 199, 223
DNAPL, 91
Drag, 15
Drilling, 155
Drilling methods, 115
Drop in place liner, 191

E

Electric water level meters, 179
Emplacement, 10, 179, 193, 200
Encoders, 152
Entrapment, 35
Equilibrium asymptote, 168, 289
Equilibrium heads, 289
Eversion aids, 87
Eversion through crooked piping, 56, 281
Eversion underwater, 55
Everson on water, 55

F

FACT comparison to core, 123
FACT comparison to water samples, 123
FACT comparison with transmissive intervals,
  123
FACT perturbations, 110
  DNAPL migration in open hole, 118
  drilling effects, 115
  during installation, 116
  fracture flow direction, 129
  handling effects, 105
  liner toluene, 129
  open borehole water, 116
  trapped water, 119

Fiber optic cables, 91
Field construction, 213
Formation head, 8, 20, 163
Fracture aperture, 291
Friction coefficient, 9

G

Gamma radiation, 48
Gas bottle, 180
GCMS, 105
Geophysical logging, 48
Grab sample, 115
Grout, 214
Grout fill, 42
Grout loss, 57
Grouting of casing, 57

H

Head difference, 10, 22
Head profile calculation, 163
Heavy mud, 37
Horizontal drilling, 275, 277
Hose canister, 34
Hydrologic constraints, 137, 148

I

Induction-coupled resistance, 48
Initial transient, 148
Injection, 214
Interior head, 8
Iteration, 151

K

Karst, 229

L

Landfill, 275
Lateral permeability, 136
Lay-flat hose liners, 54
Leakage past liner, 160
Leakage past packer, 160, 267
Liner
  110 denier, 52
  400 denier, 52
  840 denier, 52
  1,4-dioxane, 43
  air filled, 30, 223
  air vent, 15
  balloon, 15
  burst pressure, 63
  carrier, 91, 104

chemistry, 43
cover, 91, 104
denier, 47
depth limit, 184, 189, 202
diameter, 8, 63
elasticity, 19
eversion, 8
excess head, 10
hazards, 63
installation, 10, 30
inversion, 20
materials, 47
mud fill, 37
PFAS, 44
port, 174
removal, 20
spacer, 174, 185
strength, 47, 63
stretch, 19
tension, 8, 13
tether, 8, 193
toluene, 43
transparency, 48
water fill, 10
LNAPL, 91
Logging sondes, 30, 48
Long vent tube, 22
Low flow sampling, 213

M

Magic gland, 35
Manifold, 181, 199
Mapping contaminant, 99, 105, 179
Mass spectrometer, 105
Methanol, 105
Method 8260, 105
Methods
ACT, 71
blanks, 89
CHS, 191
CSC, 247
DEIL, 214
DUET, 238
dye striping, 97
FACT, 104
installation, 10, 187, 193
LAHD, 234
leak checking, 24
NAPL FLUTe, 91
pdCHS, 197
removal, 20
RHP, 161
shallow Water FLUTe, 188
submerged standpipe, 244
T profile, 136

TACL, 229
VWLM, 190
Water FLUTe, 172
WILD, 242
Minimum tension, 13
Models
ACT, 71
ACT data reductions, 83
crooked pipe, 281
eversion process, 8
linear capstan, 67
mud pressure, 37
stretch calculation, 19
T profile reductions, 295
T profiles, 135
towing force, 8
Monotonic fit, 138
Mud column, 37
Multi-level, 172

N

NAPL sandbags, 98
NAPL stain history, 91
NAPL stains, 99
direct push, 92
open borehole, 96
NDMA, 45
Nearby pumping, 63
Nearby wells, 208
Nested wells, 213
Neutron moisture log, 30
Novel applications, 55
Nylon liners, 48

O

Over pressure, 63
Oversize casing, 10

P

Packers, 159
PCE, 91
Peristaltic pumping, 188
PFAS, 44
Photo ionization detector, 97
Plastic film liners, 53
Poly film, 53
Polyester liners, 48
Port to pump tubes, 174
Positive displacement pumping, 179, 197
Potassium permanganate, 63
Pressure transducers, 71, 174
Progressive packer, 234
Pump and drag removal, 21

Pump and treat, 214
Pump tube, 174, 197
Pumping tests, 208
Purge pressure, 179
Purge stroke, 179
Purging, 179, 196, 199

Q

Quick connect, 179

R

Radial flow, 140, 159
Radwaste, 30
Recharge, 179
Recirculation of drilling fluid, 110
Regulator, 179, 196
Remediation, 214
RF welding, 47
Rough hole, 151

S

Sample collection, 179, 199
Sample pressure, 179
Sample pumping, 179, 199
Sample tube, 174, 193
Sampling, 179
Sampling intervals, 174
Sand pack, 274
Screened intervals, 174
Sealing liner, 23
Silicon rubber liners, 48
Simultaneous purging, 179, 199
Site conceptual model, 208
Solar panel, 223
Sonic casing, 97
Spatial resolution, 141
Steady state, 165
Stopping points, 167
Storativity, 148
Straddle packers
    bypass, 157, 160
    comparison with RHP, 172
    comparison with T profile, 156
    insufficient purge, 128
Striping methods, 299
Strong oxidizer, 63
Super seal, 259

Surface casing, 10
Surge block, 154

T

Tag tube, 176
TCE diffusion rate, 113
Temperature effects, 76
Tensile strength, 22, 63
Tension control, 152
Thiem equation, 140
Three way valve, 199
Toluene, 43
Tortuous passages, 56, 281
Towing force, 8
Transient, 148
Transient correction, 148
Transparent liners, 48
Turbulent flow, 156

U

Unstable sediments, 247
Urethane coating, 48
UV camera, 48

V

Vacuum pump, 15
Vadose sampling, 223
Vadose zone, 223
Vault, 178
Venturi vacuum pump, 15

W

Water FLUTe
    geometry, 174
    installation hardware, 175, 187
    sampling method, 179
    simultaneous purging, 179
    spacer design, 174
Water table, 20
Water table depth history, 71, 174
Wellhead, 178
Westbay, 213
Wet film adhesion, 9
Wire wrapped screens, 63

X

Xylene, 99